STUDENT LECTURE NOTEBOOK

BIOLOGY OF HUMANS

CONCEPTS, APPLICATIONS, AND ISSUES

JUDITH GOODENOUGH
University of Massachusetts, Amherst

BETTY McGUIRE
Smith College

ROBERT A. WALLACE

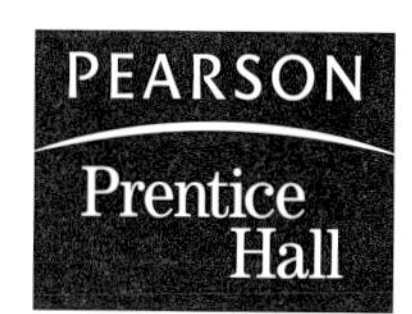

Upper Saddle River, NJ 07458

Project Manager: Crissy Dudonis
Executive Editor: Gary Carlson
Editor-in-Chief: Dan Kaveney
Executive Managing Editor: Kathleen Schiaparelli
Assistant Managing Editor: Becca Richter
Production Editor: Diane Hernandez
Supplement Cover Manager: Paul Gourhan
Supplement Cover Designer: Joanne Alexandris
Manufacturing Buyer: Alexis Heydt-Long
Cover Image Credits: Volleyball player: Michael Kevin Daly/Corbis/Bettmann;rollercoaster: Robert Landau/Cobis/Bettmann; father and son: Jim Erickson/Corbis/Bettmann; gymnast: Brian Snyder/ Corbis/Bettmann

Pearson Prentice Hall
Pearson Education, Inc.
Upper Saddle River, NJ 07458

Printed in the United States of America

10 9 8 7 6 5 4 3 2 1

ISBN 0-13-188738-6

Pearson Education Ltd., *London*
Pearson Education Australia Pty. Ltd., *Sydney*
Pearson Education Singapore, Pte. Ltd.
Pearson Education North Asia Ltd., *Hong Kong*
Pearson Education Canada, Inc., *Toronto*
Pearson Educación de Mexico, S.A. de C.V.
Pearson Education—Japan, *Tokyo*
Pearson Education Malaysia, Pte. Ltd.

TABLE OF CONTENTS

CONTENTS

TO THE STUDENT

The Student Lecture Notebook is designed to assist you during your biology course. It includes all of the art from the textbook with available space for your note taking needs. Since you wonít have to redraw the art in class, you can focus your attention on the lecture while you mark up the pages.

Additional help, including self-grading quizzes, animations, and Internet links are available on the Companion Website at http://www.prenhall.com/goodenough

NOTES

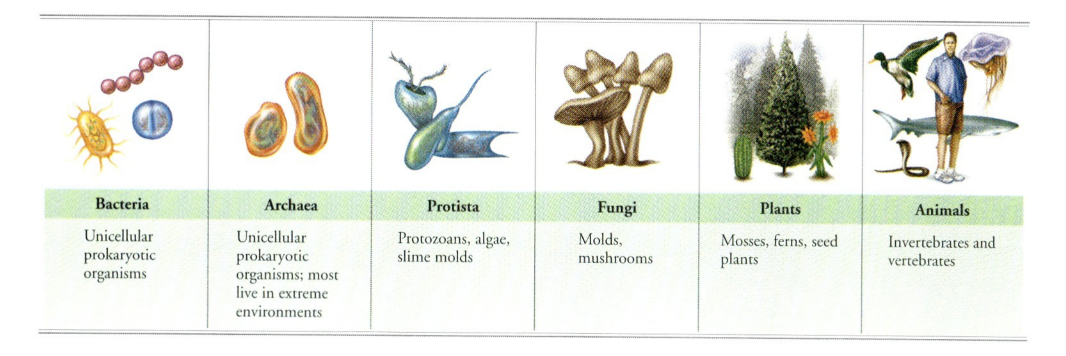

Fig. 1.3

Kingdom	Animalia (animals)	Multicellular; heterotrophic (ingest food); lack cell walls; have nervous and muscle tissue; most reproduce sexually.
Phylum	Chordata (chordates)	Possess a notochord (a dorsal, hollow nerve cord); pharyngeal gill slits; tail that extends beyond the anus.
Subphylum	Vertebrata (vertebrates)	Possess bone and a backbone; at least two semicircular canals in inner ear; paired fins or limbs.
Class	Mammalia (mammals)	Have mammary glands; hair; maintain constant body temperature; three middle ear bones; mature red blood cells lack nucleus.
Order	Primates	Have rotating shoulder joints; grasping feet and hands; fingernails (instead of claws); forward-looking eyes; well-developed cerebrum (the part of the brain most directly related to intelligence); reliance on learned behavior; small litter size.
Family	Hominidae (human family)	Walk upright on two legs; have well-developed brain; language; culture.
Genus	*Homo*	Walk upright on two legs; continued development of brain; language; culture.
Species	*Homo sapiens*	Walk upright on two legs; continued development of brain; language; culture.

Fig. 1.4

NOTES

Molecule	The chemical components of cells	
Cell	The smallest unit of life	
Tissue	A group of similar cells that perform the same function	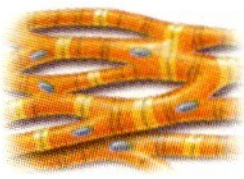
Organ	A structure with two or more tissues working together to perform a function	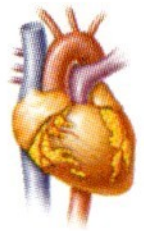
Organ systems	At least two organs working together to perform a function	
Individual	A single organism	
Population	All individuals of the same species in an area	
Community	All the species in an ecosystem that can interact	
Ecosystem	A community and its physical environment	
Biosphere	The part of the earth that supports life	

Fig. 1.6

NOTES

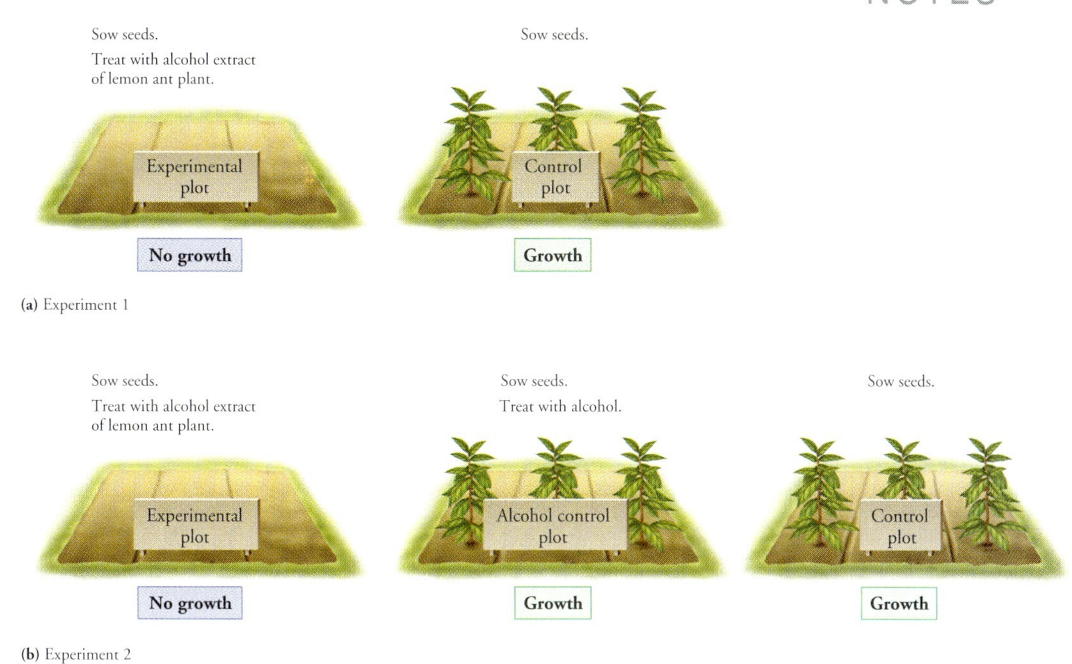

Fig. 1.8

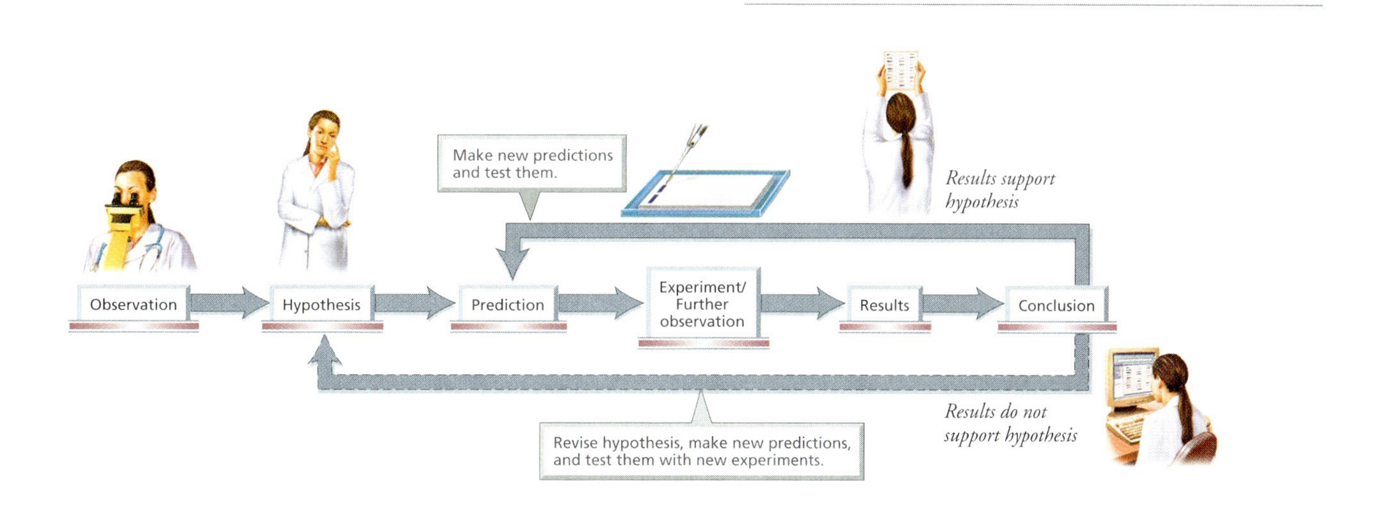

Fig. 1.9

NOTES

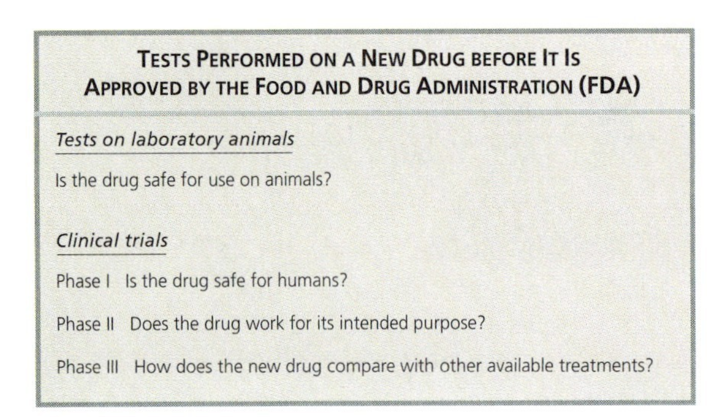

TESTS PERFORMED ON A NEW DRUG BEFORE IT IS APPROVED BY THE FOOD AND DRUG ADMINISTRATION (FDA)

Tests on laboratory animals

Is the drug safe for use on animals?

Clinical trials

Phase I Is the drug safe for humans?

Phase II Does the drug work for its intended purpose?

Phase III How does the new drug compare with other available treatments?

Table 1.1

SOURCE PLANT	COMPOUND	MEDICINAL USE
Willow bark and meadowsweet	Salicylic acid (active ingredient in aspirin)	Pain relief, fever reduction
Foxglove	Digitalin	Treatment of heart conditions (increase intensity of heart contraction, slows heart rate)
Quinine tree	Quinine	Malaria preventive
Opium poppy	Morphine	Pain relief
Eucalyptus tree	Menthol	Decongestant
Rauvolfia root	Reserpine	Treatment of high blood pressure
Belladonna	Atropine	Dilates pupils for eye exams
Pacific yew	Taxol	Treatment of ovarian and breast cancer
Rosy periwinkle	Vincristine	Treatment of Hodgkin's disease (a type of cancer)
	Vinblastine	Treatment of leukemia

Table 1.A

NOTES

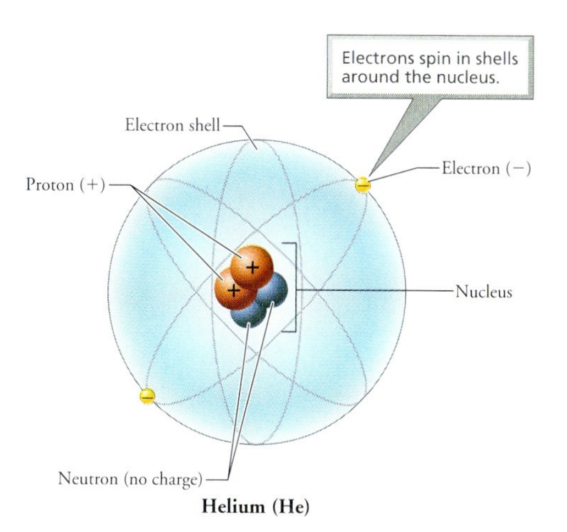

Fig. 2.1

SUBATOMIC PARTICLES			
PARTICLE	LOCATION	CHARGE	MASS
Proton	Nucleus	1 positive unit	1 atomic mass unit
Neutron	Nucleus	None	1 atomic mass unit
Electron	Outside the nucleus	1 negative unit	Negligible

Table 2.1

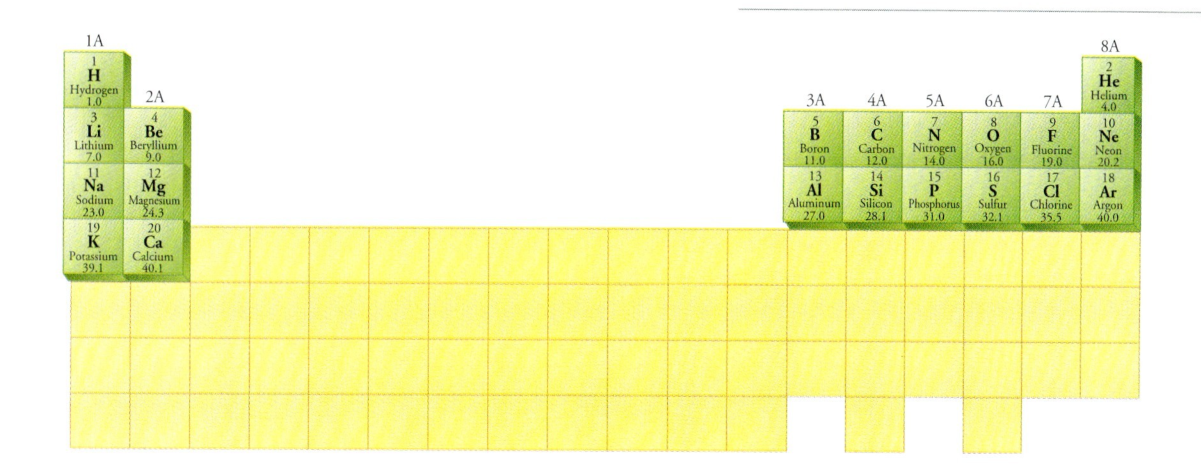

Fig. 2.2

NOTES

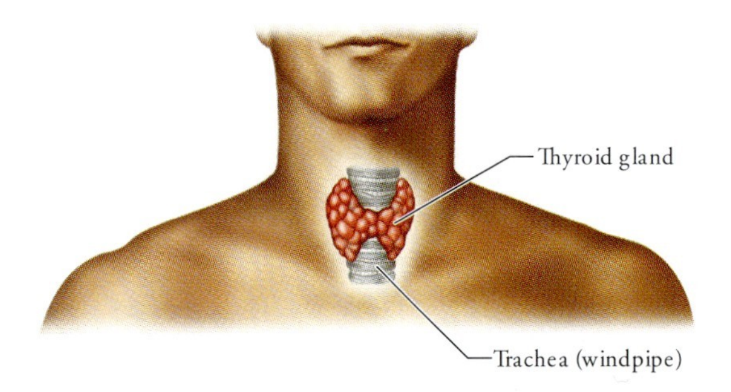

Fig. 2.4

Fig. 2.6

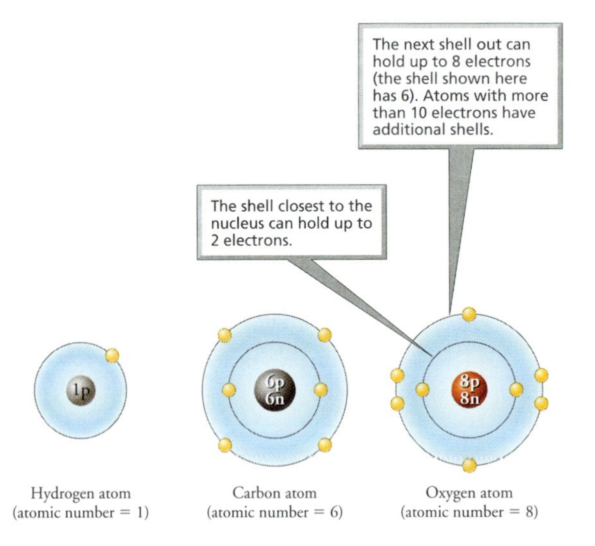

Fig. 2.8

NOTES

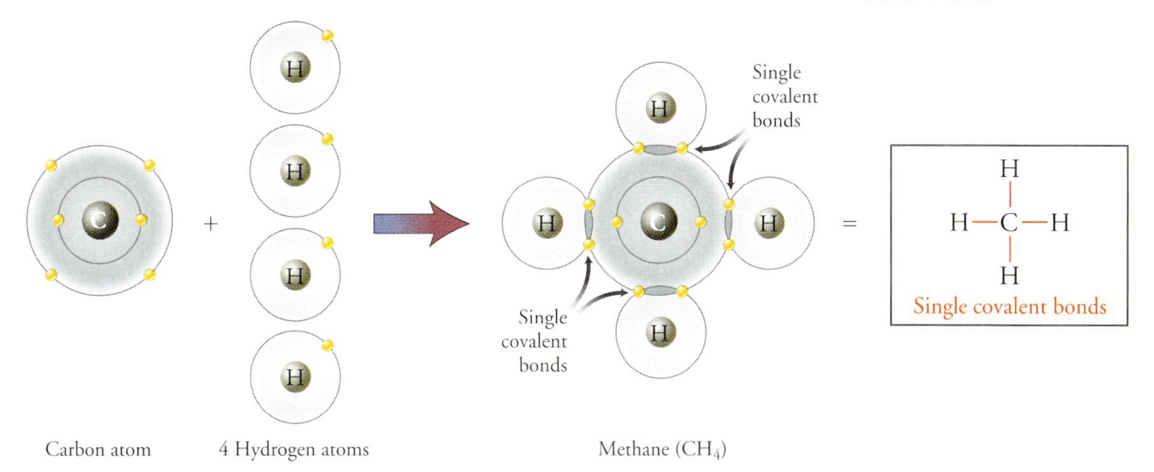

(a) The molecule methane (CH_4) is formed by the sharing of electrons between one carbon atom and four hydrogen atoms. Because, in each case one pair of electrons is shared, the bonds formed are single covalent bonds.

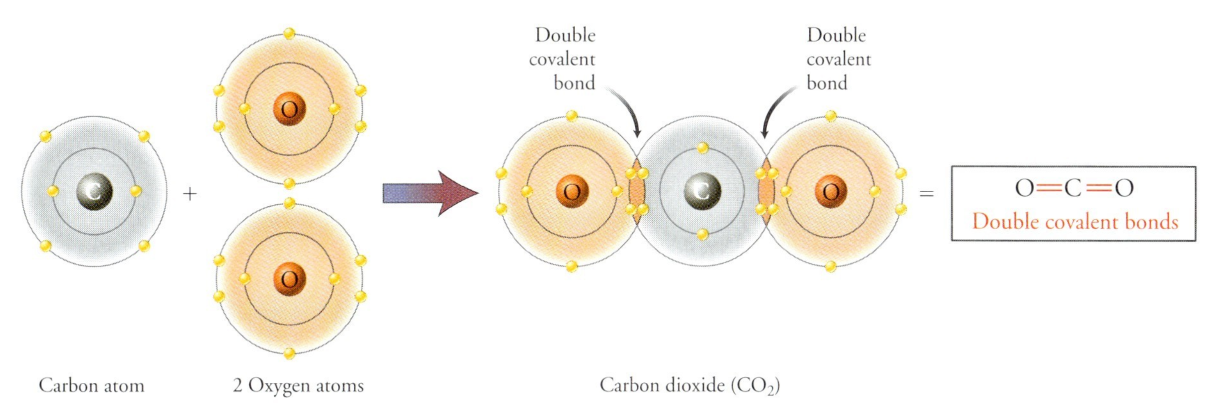

(b) The oxygen atoms in a molecule of carbon dioxide (CO_2) form double covalent bonds with the carbon atom. In double bonds, two pairs of electrons are shared.

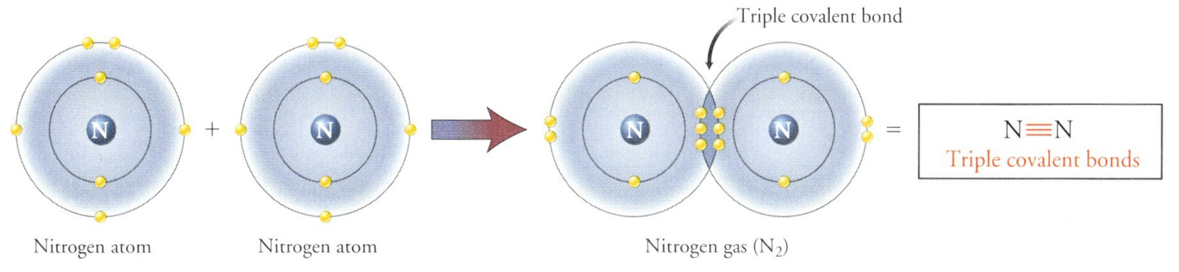

(c) The nitrogen atoms in nitrogen gas (N_2) form a triple covalent bond in which three pairs of electrons are shared.

Fig, 2.9

NOTES

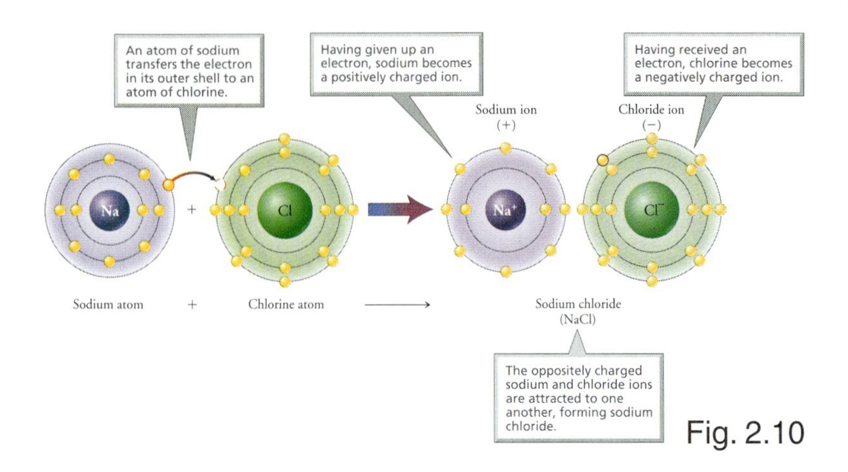

Fig. 2.10

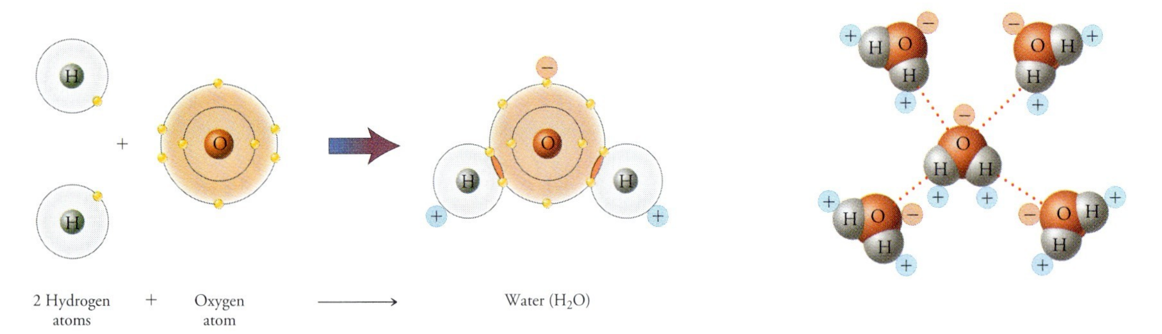

(a) Water is formed when an oxygen atom covalently bonds (shares electrons) with two hydrogen atoms. As a result of unequal sharing of electrons, oxygen carries a slight negative charge and the hydrogen atoms a slight positive charge.

(b) The hydrogen atoms from one water molecule are attracted to the oxygen atoms of other water molecules. This relatively weak attraction (shown by dotted lines) is called a hydrogen bond.

Fig.2.11

CHEMICAL BONDS			
TYPE	BASIS FOR ATTRACTION	STRENGTH	EXAMPLE
Covalent	Sharing of electrons between atoms; the sharing between atoms may be equal or unequal	Strongest	CH_4 (methane)
Ionic	Transfer of electrons between atoms creates oppositely charged ions that are attracted to one another	Strong	NaCl (table salt)
Hydrogen	Attraction between a hydrogen atom with a slight positive charge and another atom (often oxygen) with a slight negative charge	Weak	Between a hydrogen atom on one water molecule and an oxygen atom on another water molecule

Table 2.2

NOTES

ETHICS IN RADIOACTIVITY EXPERIMENTS ON HUMANS

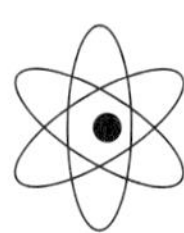

In some of the radiation studies cited in newspapers in late 1993 and in an earlier congressional report, known as the Markey Report, subjects did not freely consent to the experiments. In other studies it is doubtful whether informed consent was obtained. But in some of the studies informed consent was truly given. Here are examples from each category.

Date	Experiment
Possible Infliction of Harm or No Informed Consent	
1945–47	18 supposedly terminal patients were injected with high doses of plutonium to learn whether the body absorbed it.
1946–47	6 hospital patients were injected with uranium salts to determine the dose that produced injury to the kidneys.
1963–70	64 prison inmates had their testicles exposed to x-rays to relate radiation damage to sperm production.
1963–71	67 prison inmates had their testicles exposed to x-rays to measure radiation damage to sperm production.
Questionable Consent	
1946	17 mentally delayed teenagers at the Fernald School in Waltham, Massachusetts, ate meals with trace amounts of radioactive iron to learn about iron absorption in the body.
1953–57	11 comatose brain cancer patients were injected with uranium to learn whether it is absorbed by brain tumors.
1954–56	32 mentally delayed teenagers at the Fernald School drank milk with trace amounts of radioactive calcium to learn whether oatmeal impeded its absorption by the body.
Informed Consent	
1945	10 researchers and workers at Clinton Laboratory, in Oak Ridge, Tennessee, voluntarily exposed patches of their skin to radioactive phosphorus.
1951	14 researchers at Hanford Nuclear Reservation voluntarily exposed patches of their skin to gaseous tritium.
1963	54 hospital patients volunteered to take trace amounts of radioactive lanthanum to measure effects on the large intestine.
1965	Trace doses of radioactive technetium were given to 8 healthy volunteers to determine its utility as a medical diagnostic tool.

Fig.2.A

NOTES

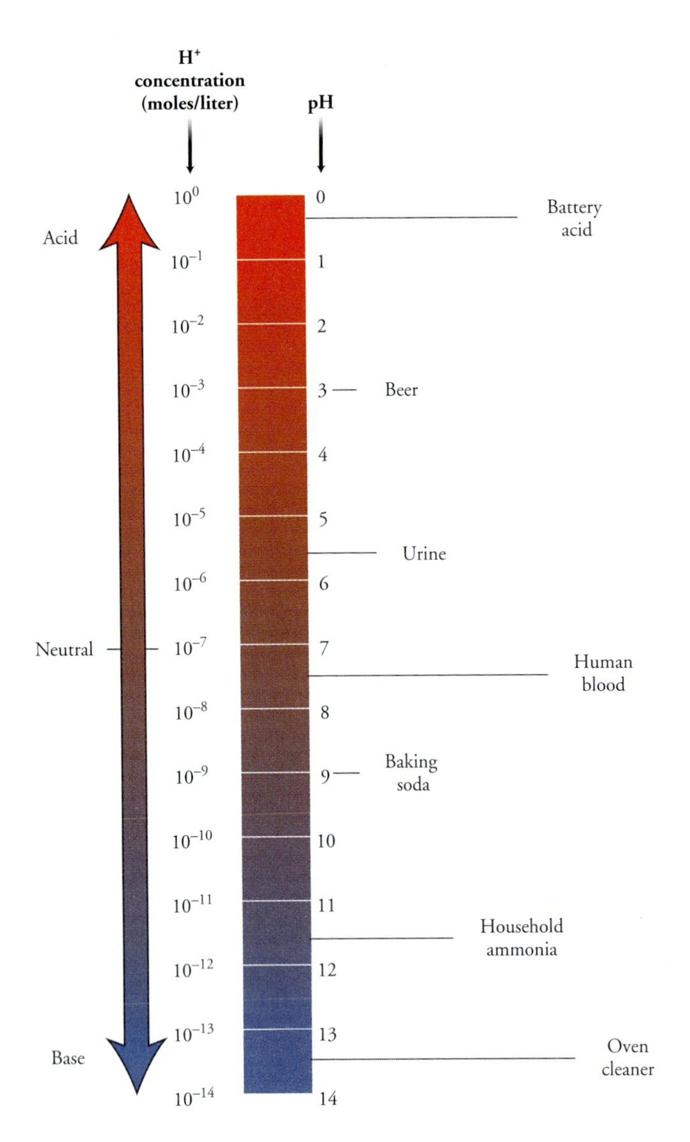

Fig. 2.13

ACIDS AND BASES COMPARED		
CHARACTERISTIC	ACID	BASE
Dissociation in Water	Releases H^+	Releases OH^-
pH	Less than 7	Greater than 7
Example	HCl (hydrochloric acid)	NaOH (sodium hydroxide)

Table 2.3

NOTES

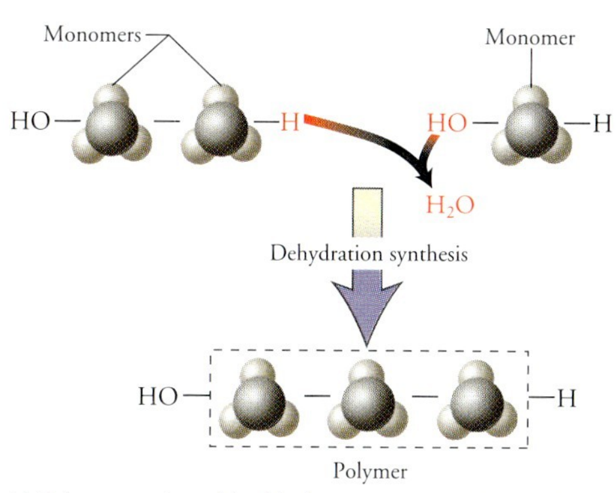

(a) Polymers are formed by dehydration synthesis, the removal of a water molecule, and the joining of two monomers.

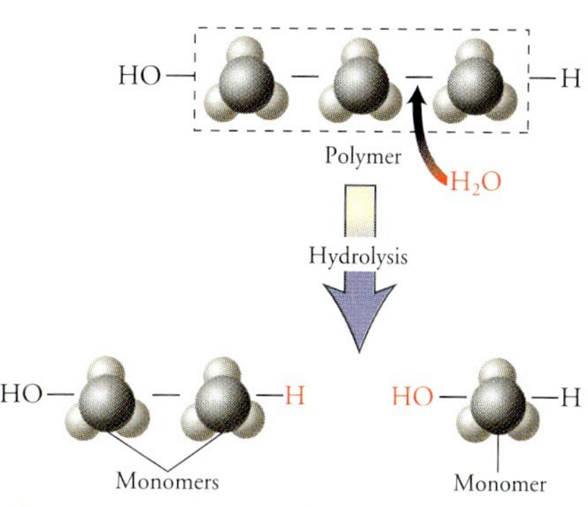

(b) Polymers are broken apart by hydrolysis, the addition of a water molecule, and the breaking of bonds between monomers.

Fig. 2.14

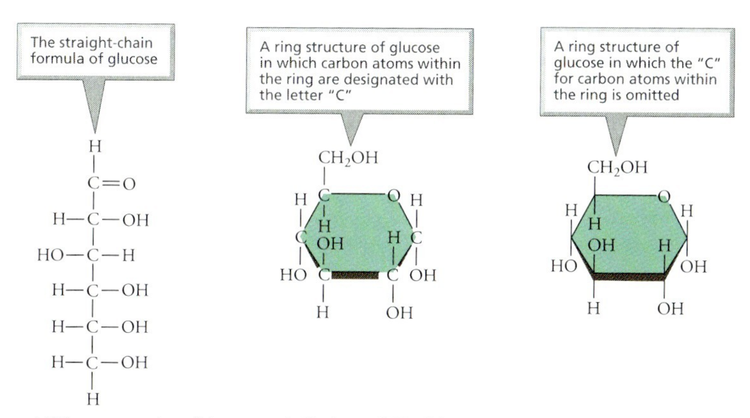

(a) Three representations of the monosaccharide glucose ($C_6H_{12}O_6$)

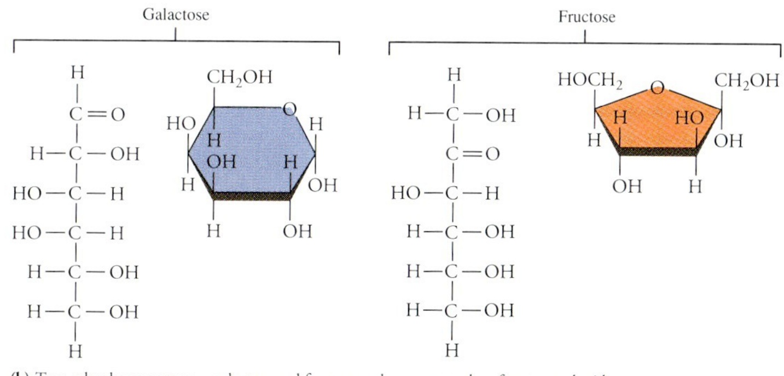

(b) Two other hexose sugars—galactose and fructose—also are examples of monosaccharides. Galactose and fructose have the same molecular formula as glucose ($C_6H_{12}O_6$), but slightly different structures.

Fig. 2.15

NOTES

Common Carbohydrates

Carbohydrate	Molecular Formula	Source	Component Monosaccharides
Monosaccharides			
Glucose	$C_6H_{12}O_6$	Blood, fruit, honey	
Fructose	$C_6H_{12}O_6$	Fruit, honey	
Galactose	$C_6H_{12}O_6$	From hydrolysis of lactose (milk sugar)	
Disaccharides			
Sucrose	$C_{12}H_{22}O_{11}$	Sugar cane, maple syrup	Glucose, fructose
Maltose	$C_{12}H_{22}O_{11}$	From hydrolysis of starch; ingredient in beer	Glucose
Lactose	$C_{12}H_{22}O_{11}$	Component of milk	Glucose, galactose
Polysaccharides			
Starch	*	Potatoes, corn, some grains	Glucose
Glycogen	*	Stored in muscle and liver cells	Glucose
Cellulose	*	Cell walls of plants	Glucose
Chitin	*	Outer coverings of insects, crustaceans	Glucose

* These complex carbohydrates consist of chains containing hundreds of glucose molecules joined to each other in long strings.

Table 2.4

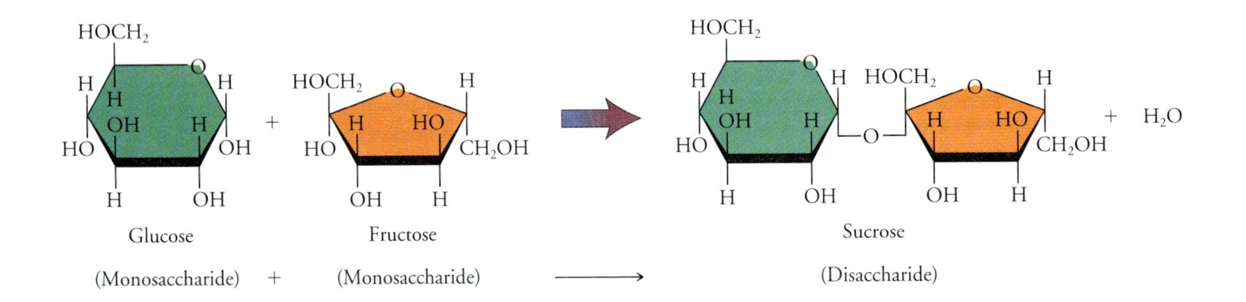

Fig. 2.16

NOTES

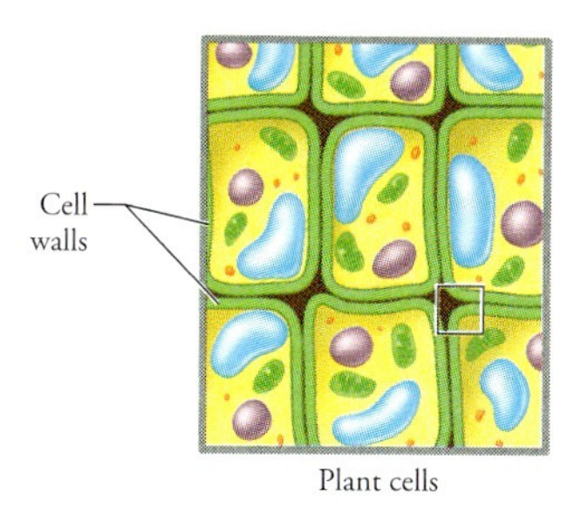

Fig. 2.17

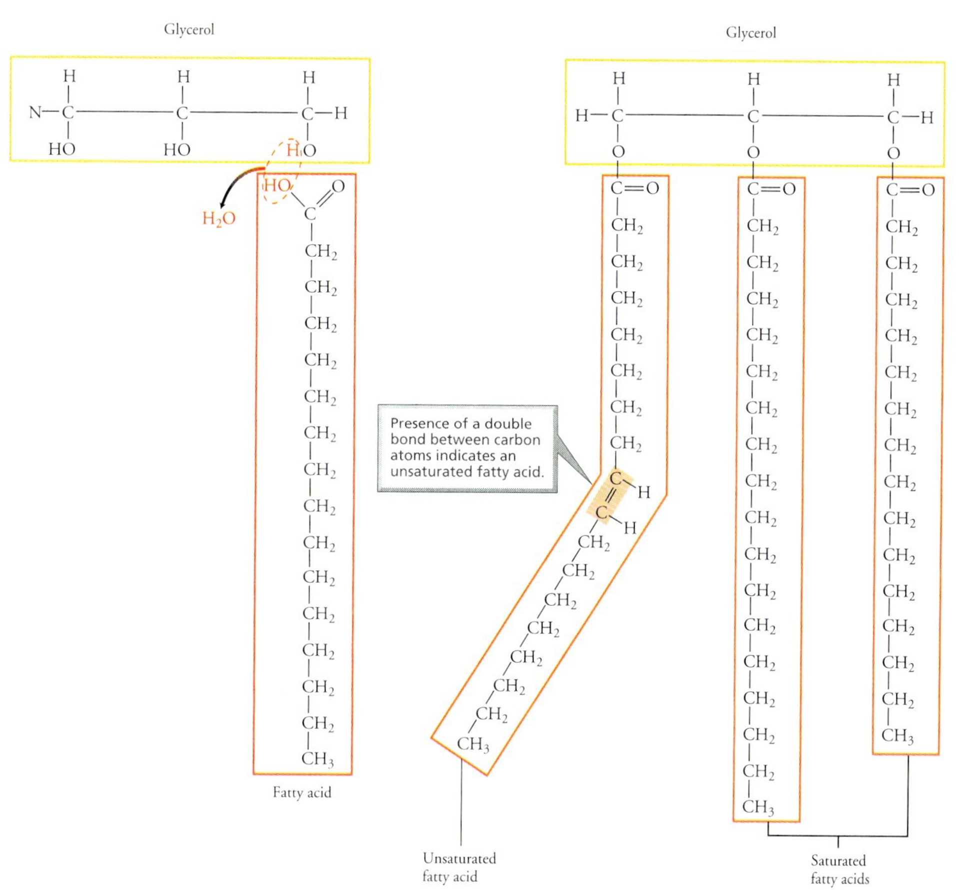

(a) A fatty acid bonds to glycerol through dehydration synthesis.

(b) This triglyceride contains one unsaturated fatty acid (note the presence of a double bond between the carbon atoms) and two saturated fatty acids (note the absence of any double bonds between the carbon atoms).

Fig. 2.18

NOTES

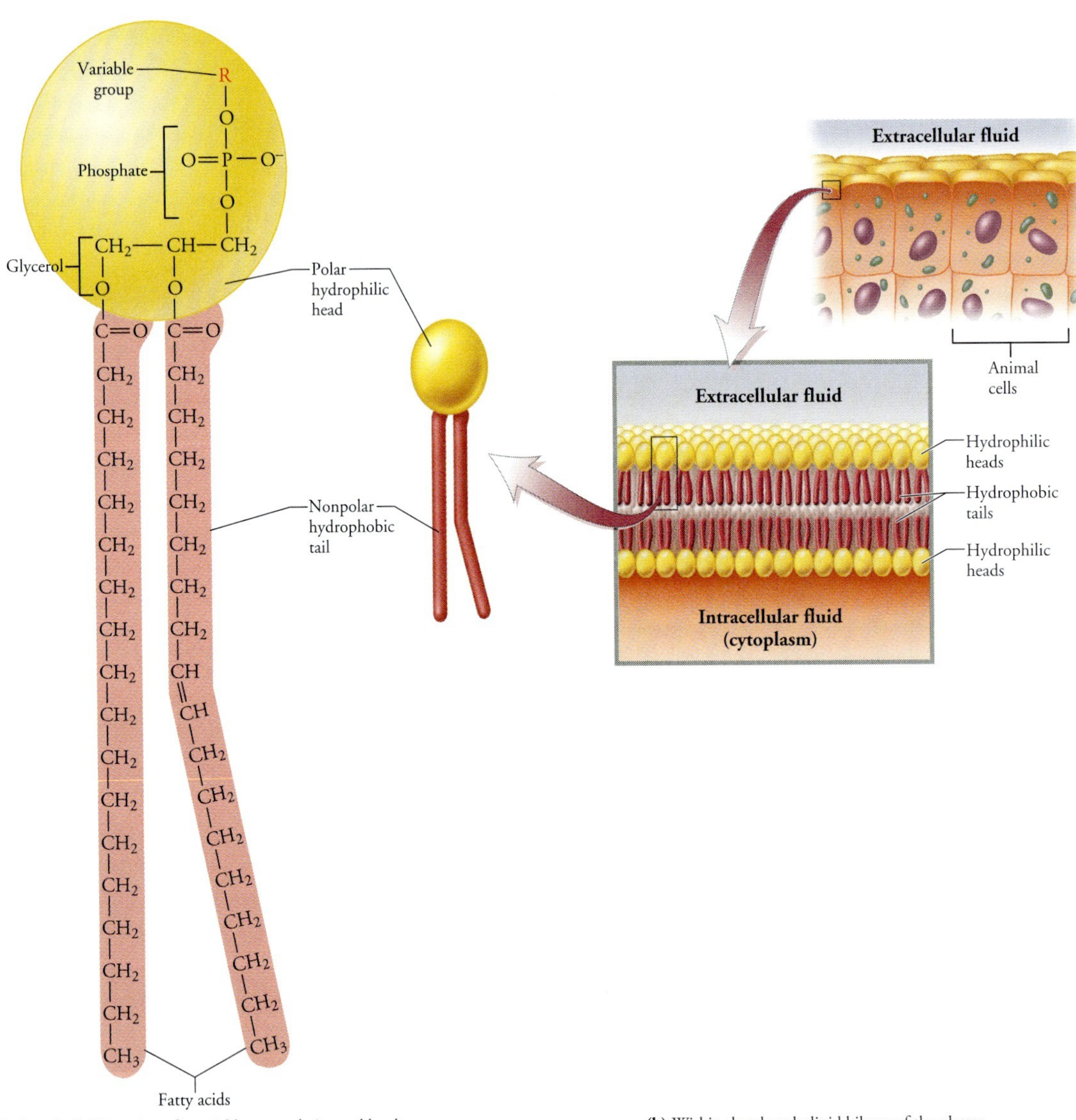

(a) A phospholipid consists of a variable group designated by the letter R, a phosphate, a glycerol, and two fatty acids. Because the variable group is often polar and the fatty acids nonpolar, phospholipids have a polar hydrophilic (water-loving) head and a nonpolar hydrophobic (water-fearing) tail.

(b) Within the phospholipid bilayer of the plasma membrane, the hydrophobic tails point inward and help hold the membrane together. The outward-pointing hydrophilic heads mix with the watery environments inside and outside the cell.

Fig. 2.19

NOTES

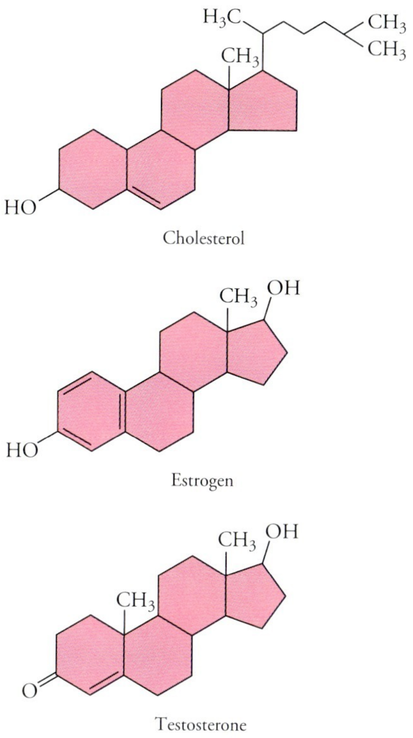

Cholesterol

Estrogen

Testosterone

Fig. 2.20

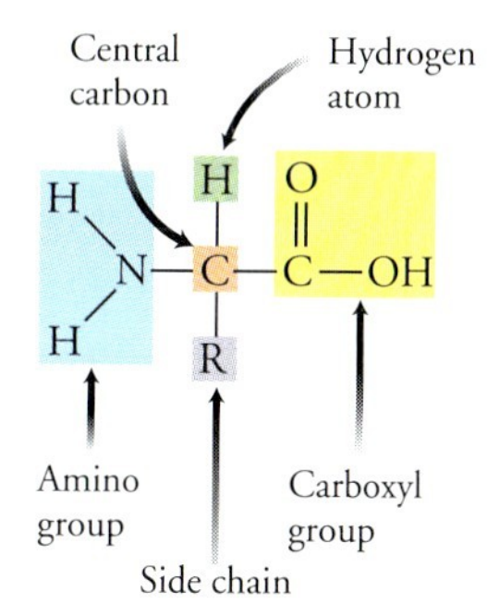

Fig. 2.21

NOTES

Peptide bond

Amino acid glycine + Amino acid alanine → Dehydration synthesis → Dipeptide glycylalanine + H_2O

Fig. 2.22

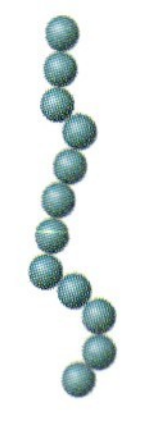

(a) Primary structure is the specific sequence of amino acids. Each amino acid is depicted here as a bead within the polypeptide chain.

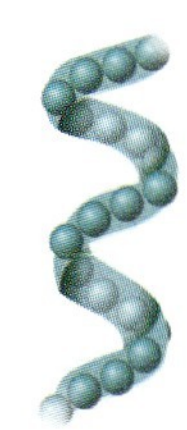

(b) Secondary structure, such as the helix shown here, results from the bending and coiling of the chain of amino acids.

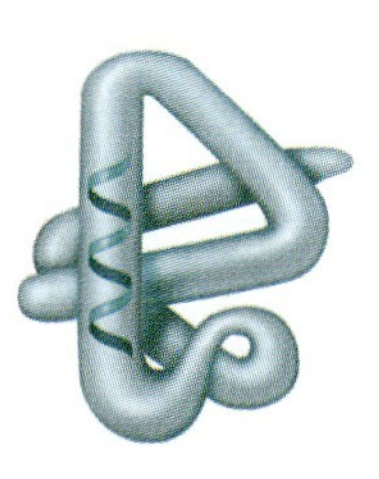

(c) Tertiary structure is the three-dimensional shape of proteins.

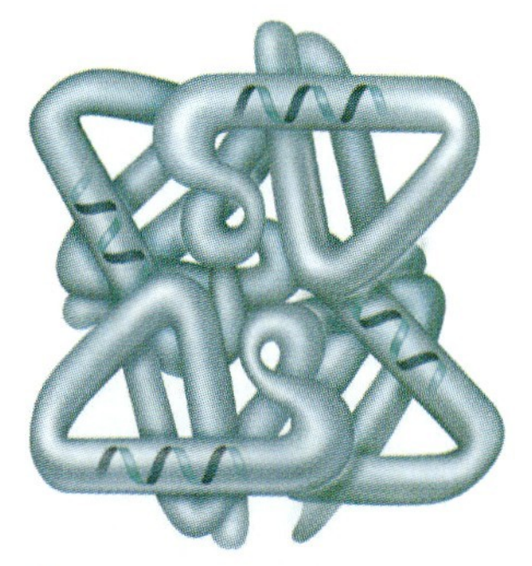

(d) Some proteins have two or more polypeptide chains, each chain forming a subunit. Quaternary structure results from the attractive forces between two or more subunits.

Fig. 2.23

NOTES

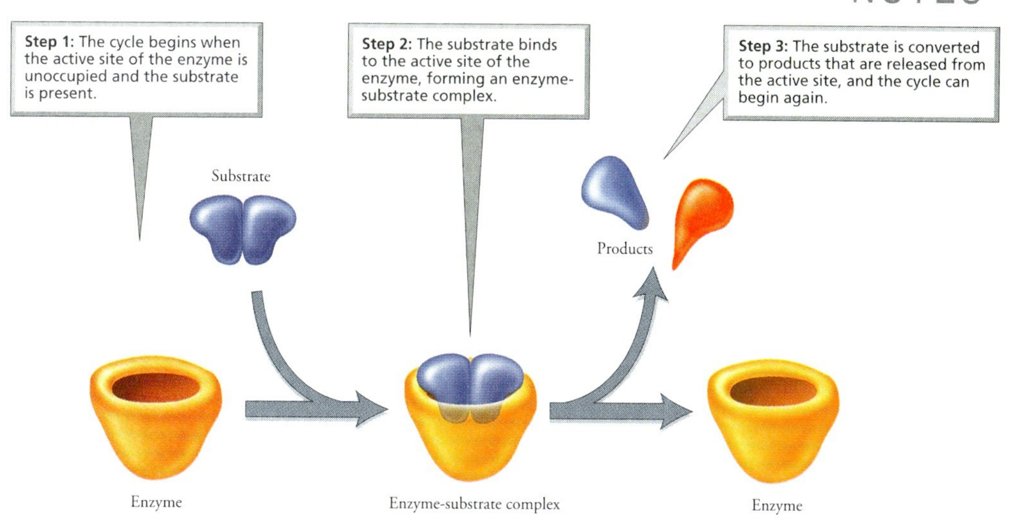

(a) A decomposition reaction involving an enzyme

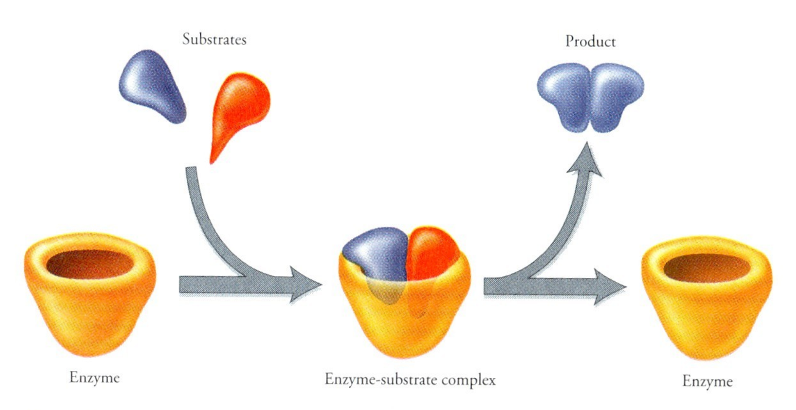

(b) A synthesis reaction involving an enzyme

Fig. 2.24

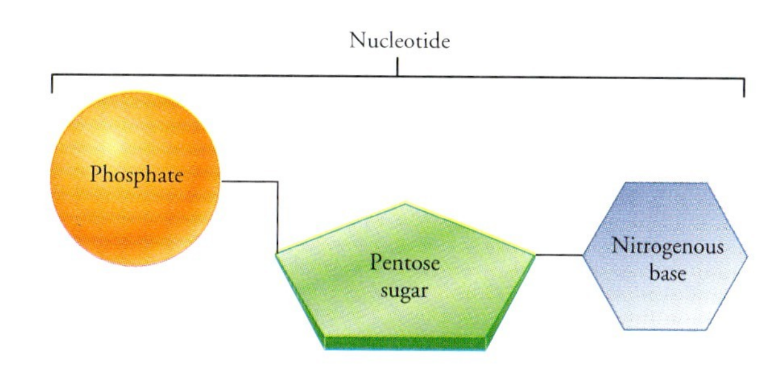

Fig. 2.25

NOTES

STRUCTURAL DIFFERENCES BETWEEN RNA AND DNA		
CHARACTERISTIC	RNA	DNA
Sugar	Ribose	Deoxyribose
Bases	Adenine, guanine, cytosine, uracil	Adenine, guanine, cytosine, thymine
Number of strands	One	Two; twisted to form double helix

Table 2.5

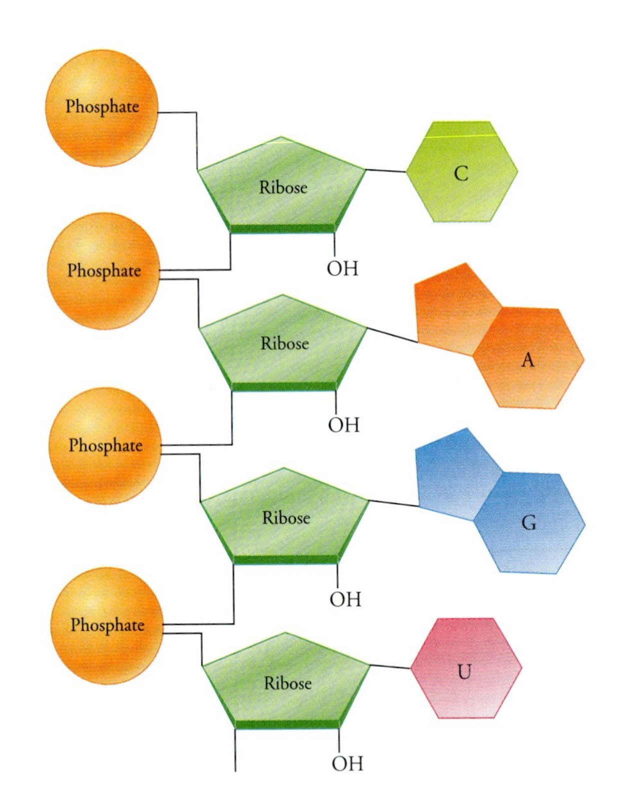

Fig. 2.26

NOTES

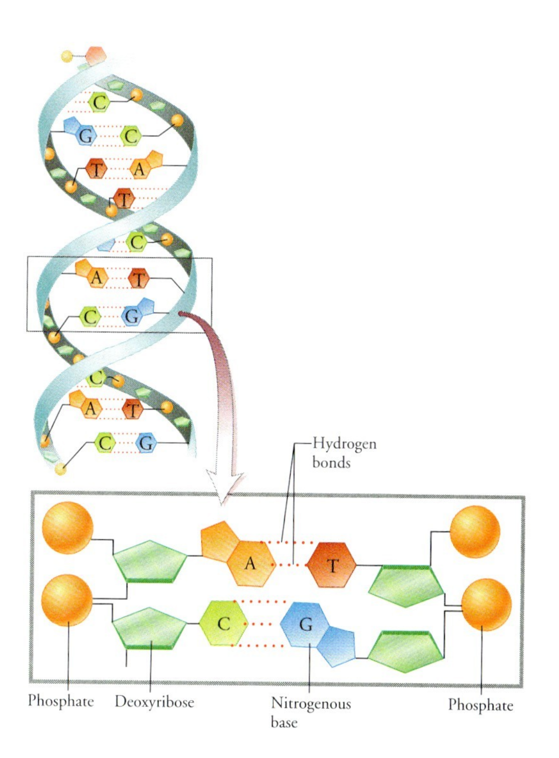

Fig. 2.27

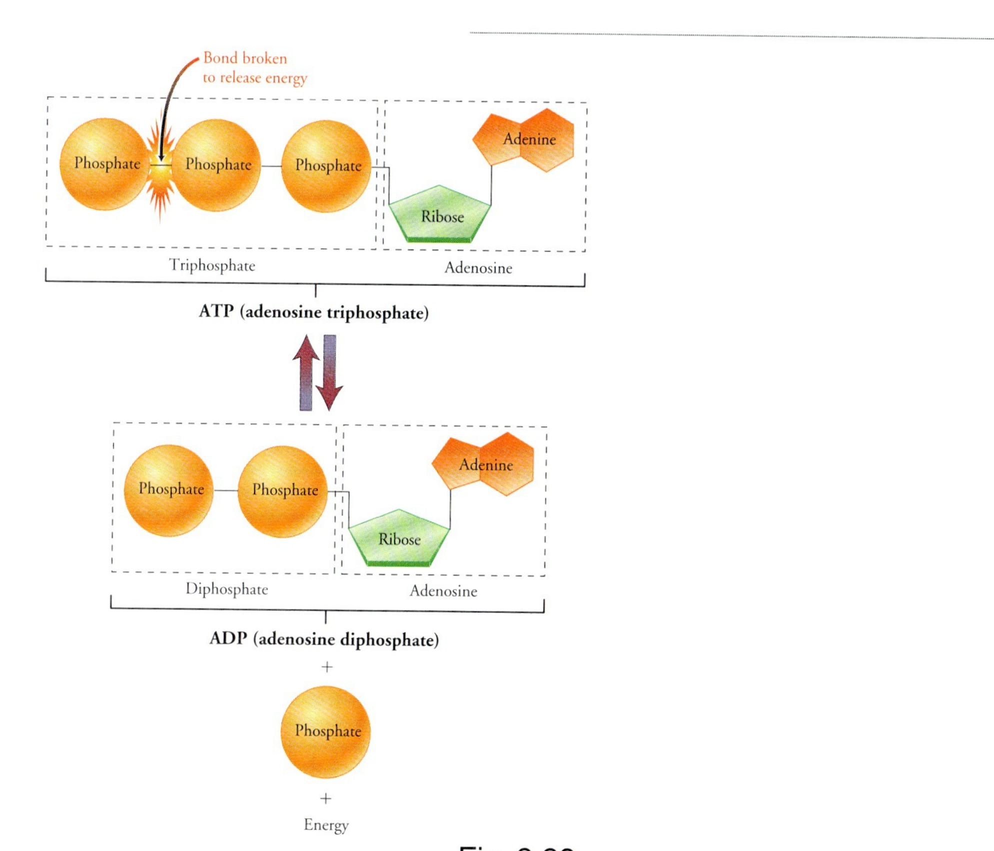

Fig. 2.28

NOTES

NOTES

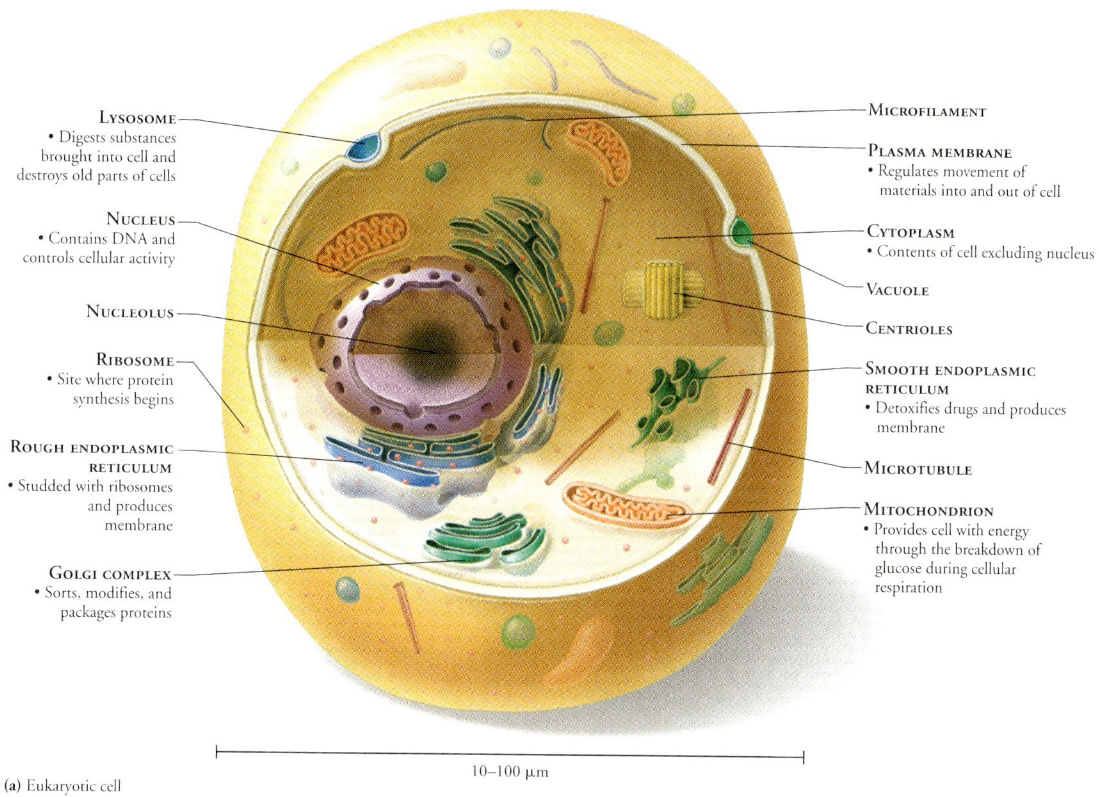

Lysosome
• Digests substances brought into cell and destroys old parts of cells

Nucleus
• Contains DNA and controls cellular activity

Nucleolus

Ribosome
• Site where protein synthesis begins

Rough endoplasmic reticulum
• Studded with ribosomes and produces membrane

Golgi complex
• Sorts, modifies, and packages proteins

Microfilament

Plasma membrane
• Regulates movement of materials into and out of cell

Cytoplasm
• Contents of cell excluding nucleus

Vacuole

Centrioles

Smooth endoplasmic reticulum
• Detoxifies drugs and produces membrane

Microtubule

Mitochondrion
• Provides cell with energy through the breakdown of glucose during cellular respiration

10–100 μm

(a) Eukaryotic cell

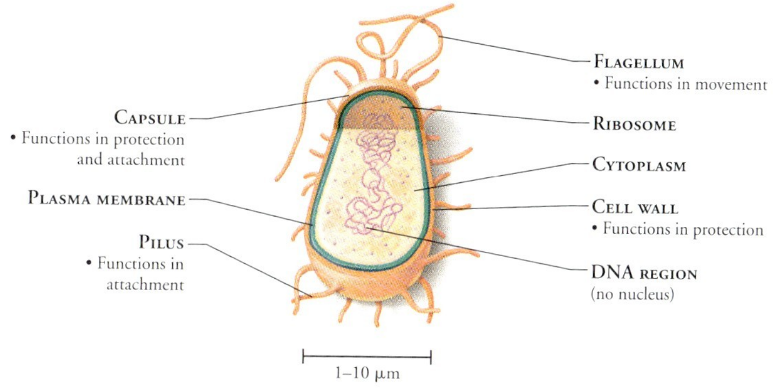

Capsule
• Functions in protection and attachment

Plasma membrane

Pilus
• Functions in attachment

Flagellum
• Functions in movement

Ribosome

Cytoplasm

Cell wall
• Functions in protection

DNA region
(no nucleus)

1–10 μm

(b) Prokaryotic cell

Fig. 3.1

NOTES

Comparisons of Eukaryotic and Prokaryotic Cells		
FEATURE	**EUKARYOTIC CELLS**	**PROKARYOTIC CELLS**
Organisms	Plants, animals, fungi, protists	Bacteria, archaea
Size	10–100 μm across	1–10 μm across
Membrane-bound organelles	Present	Absent
DNA form	Coiled, linear strands	Circular
DNA location	Nucleus	Cytoplasm
Internal membranes	Many	Rare
Cytoskeleton	Present	Absent

Table 3.1

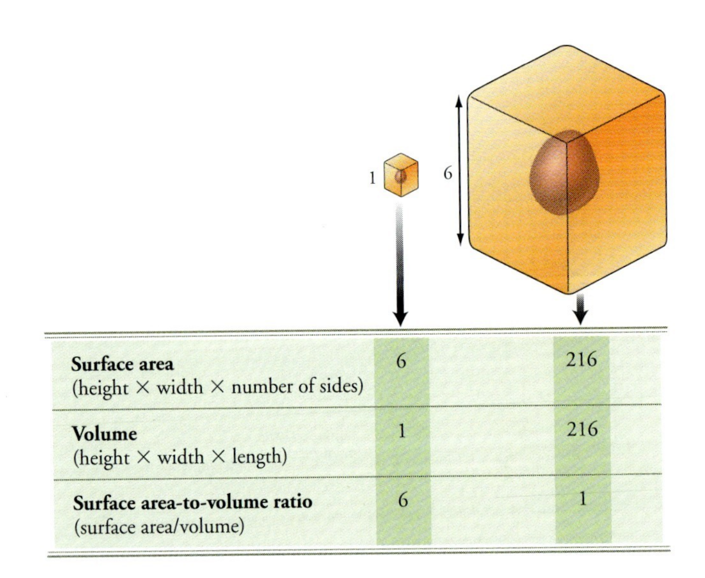

Surface area (height × width × number of sides)	6	216
Volume (height × width × length)	1	216
Surface area-to-volume ratio (surface area/volume)	6	1

Fig. 3.2

NOTES

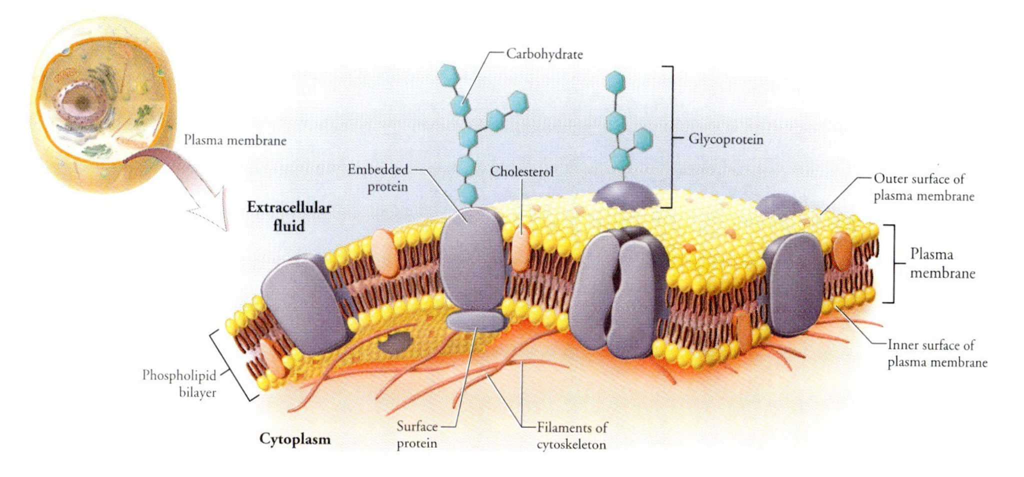

Fig. 3.3

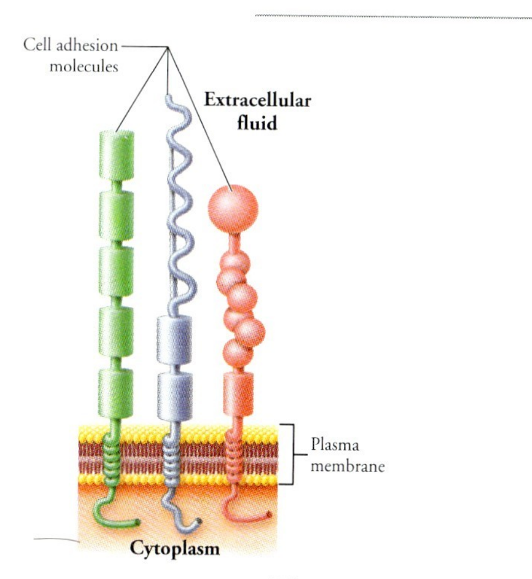

Fig. 3.4

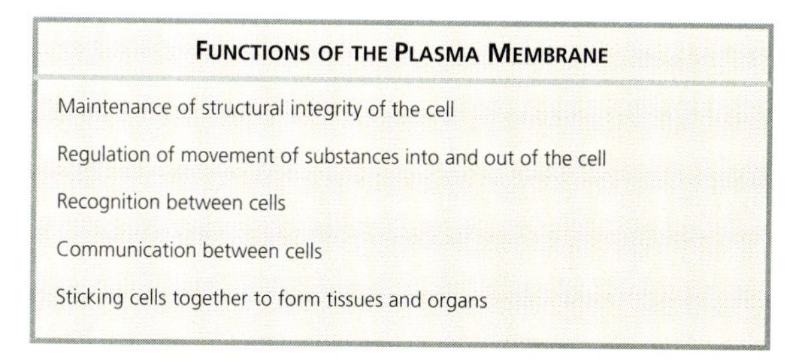

FUNCTIONS OF THE PLASMA MEMBRANE
Maintenance of structural integrity of the cell
Regulation of movement of substances into and out of the cell
Recognition between cells
Communication between cells
Sticking cells together to form tissues and organs

Table 3.2

NOTES

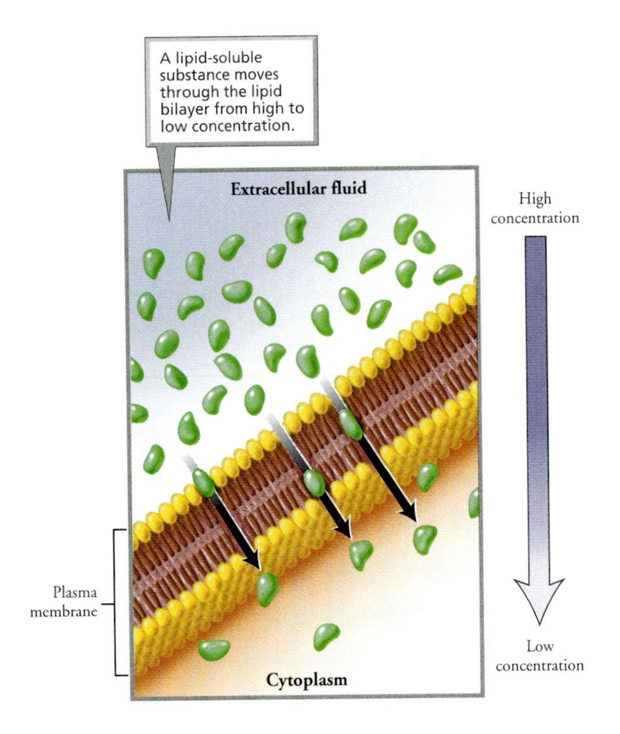

Fig. 3.5

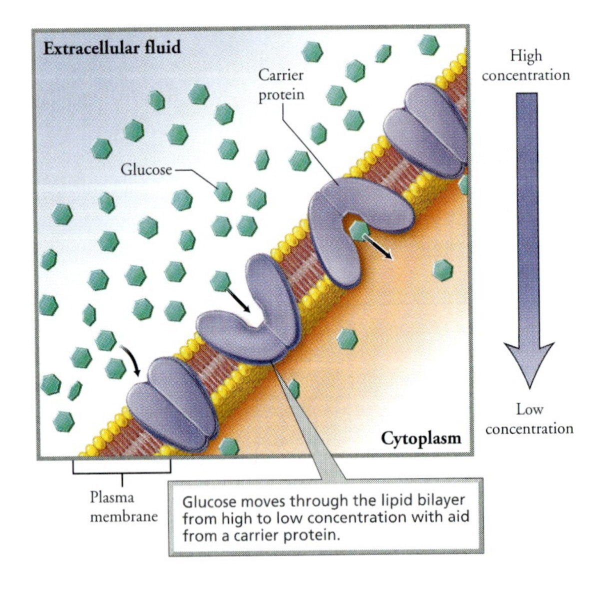

Fig. 3.6

NOTES

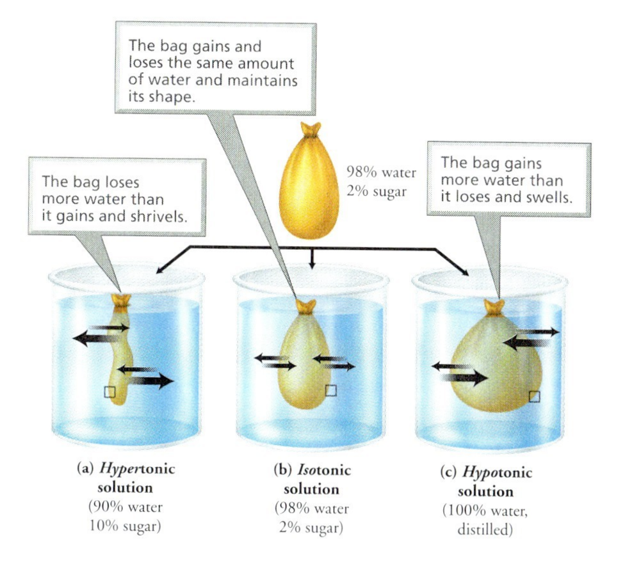

Fig. 3.7

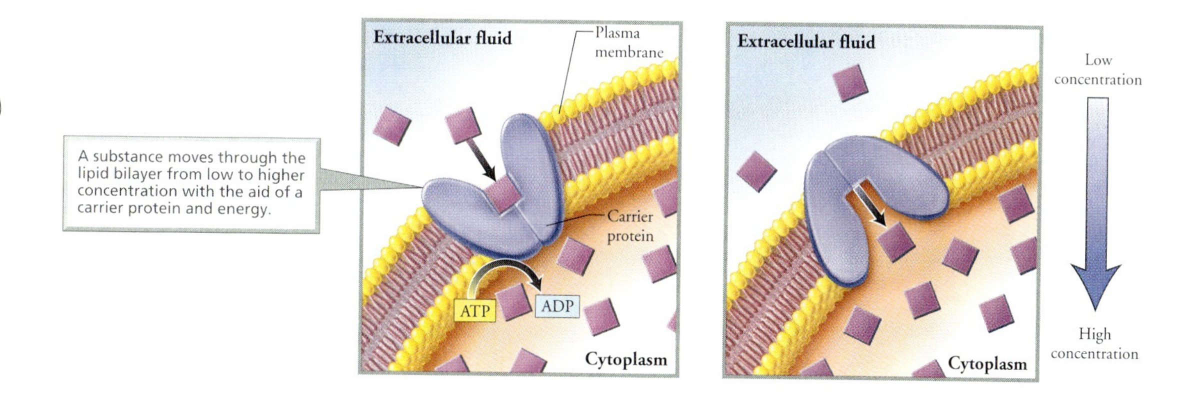

Fig. 3.8

NOTES

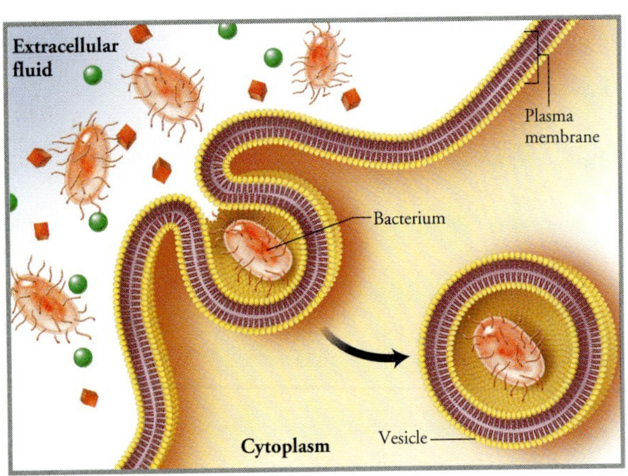

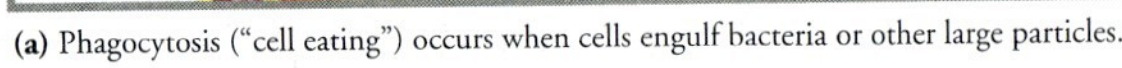

(a) Phagocytosis ("cell eating") occurs when cells engulf bacteria or other large particles.

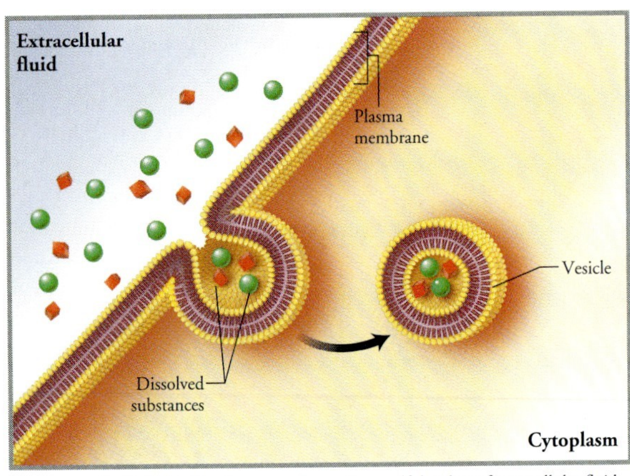

(b) Pinocytosis ("cell drinking") occurs when cells engulf droplets of extracellular fluid and the dissolved substances therein.

Fig. 3.9

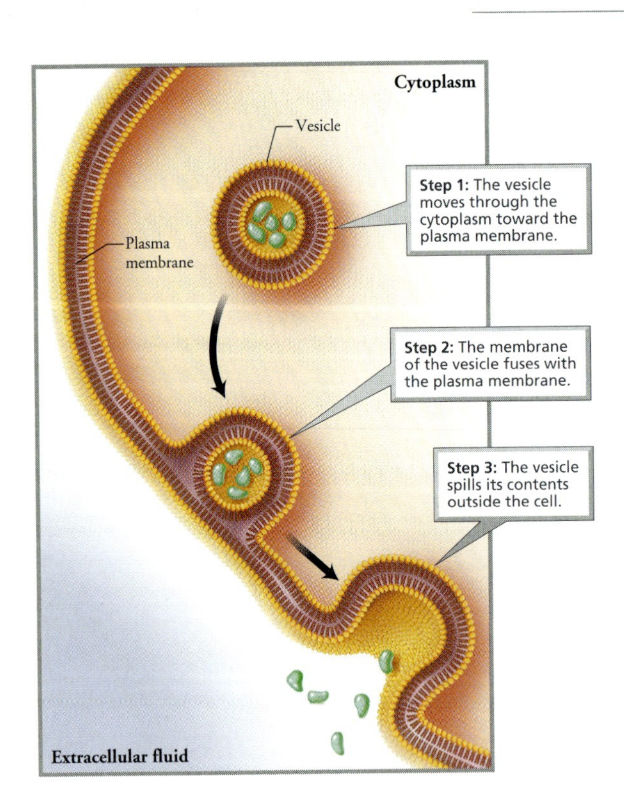

Fig. 3.10

NOTES

Mechanisms of Transport across the Plasma Membrane	
MECHANISM	**DESCRIPTION**
Simple diffusion	Random movement from region of high to low concentration
Facilitated diffusion	Movement from region of high to low concentration with the aid of a carrier or channel protein
Osmosis	Movement of water from region of high water concentration (low solute concentration) to low water concentration (high solute concentration)
Active transport	Movement often from region of low to high concentration with the aid of a carrier protein and energy usually from ATP
Endocytosis	Materials are engulfed by plasma membrane and drawn into cell in a vesicle
Exocytosis	Membrane-bound vesicle from inside the cell fuses with the plasma membrane and spills contents outside of cell

Table 3.3

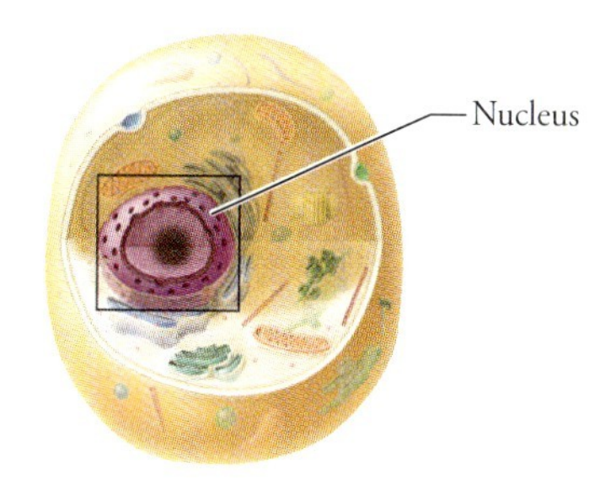

Fig. 3.11

NOTES

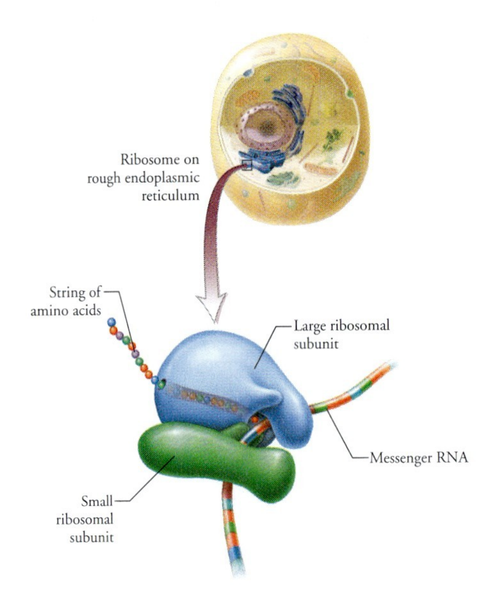

Diagram of a ribosome. A ribosome consists of RNA and protein organized into two subunits, one large and one small.

Fig. 3.13

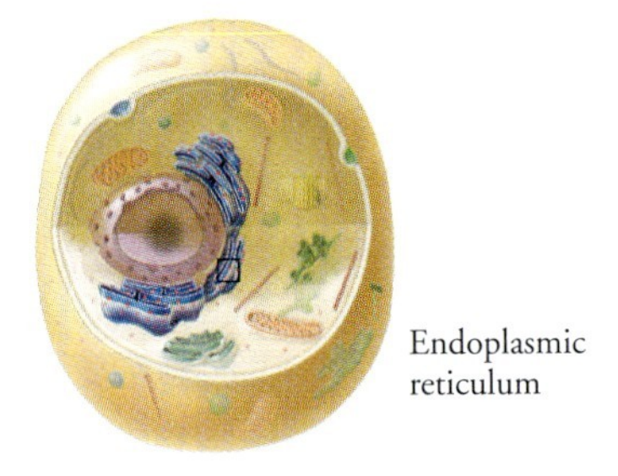

Fig. 3.14

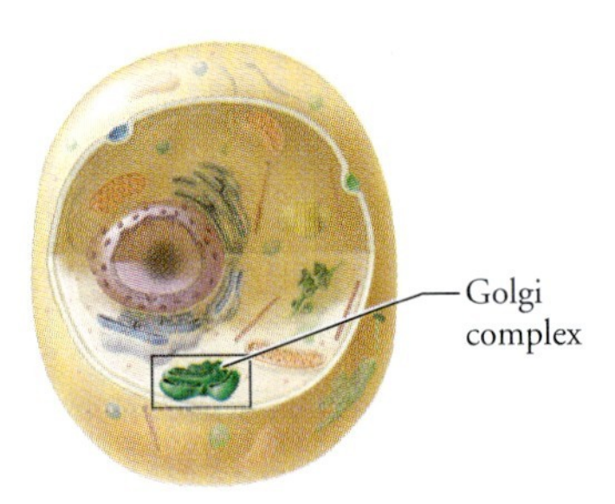

Fig. 3.15

NOTES

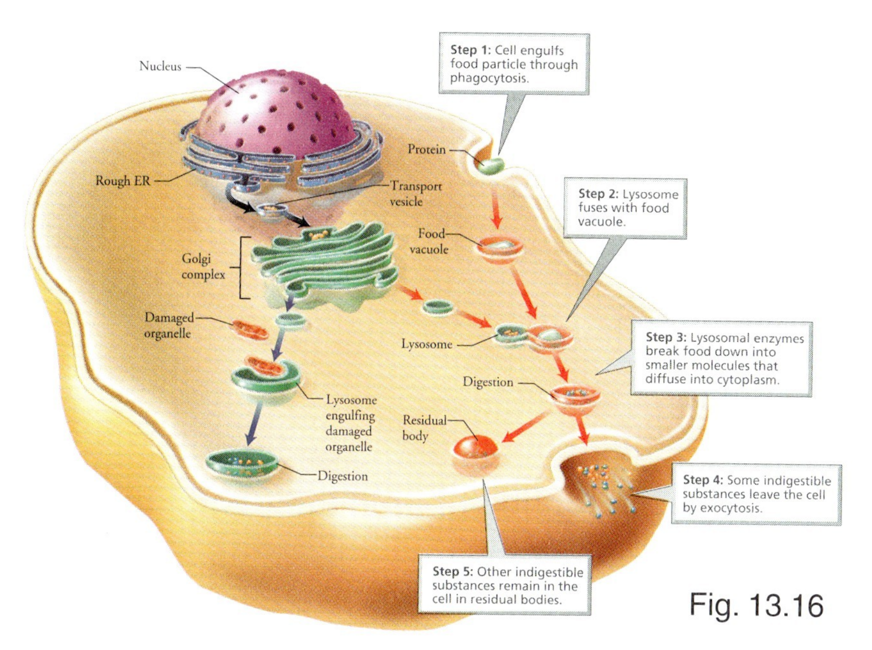

Fig. 13.16

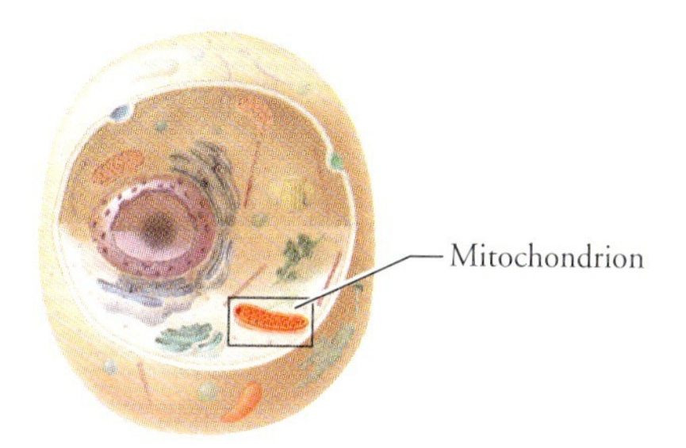

Fig. 13.18

SUMMARY OF MAJOR ORGANELLES AND THEIR FUNCTIONS

ORGANELLE	FUNCTION
Nucleus	Contains almost all the genetic information and influences cellular structure and function
Rough endoplasmic reticulum (RER)	Studded with ribosomes (sites where the synthesis of proteins begins); produces membrane
Smooth endoplasmic reticulum (SER)	Detoxifies drugs; produces membrane
Golgi complex	Sorts, modifies, and packages products of RER
Lysosomes	Digest substances imported from outside the cell; destroy old or defective cell parts or cells
Mitochondria	Provide cell with energy through the breakdown of glucose during cellular respiration

Table 3.4

NOTES

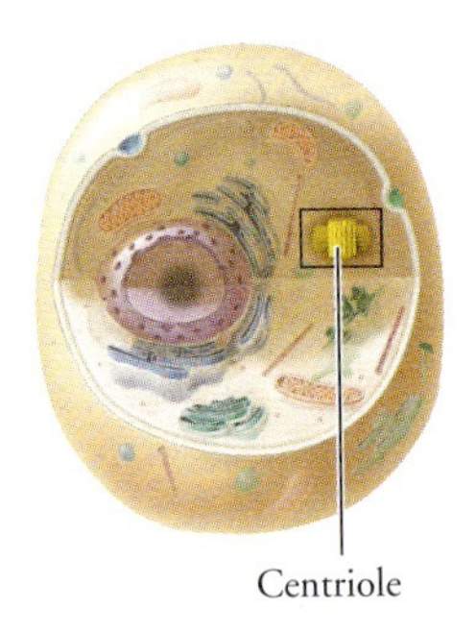

Fig. 13.19

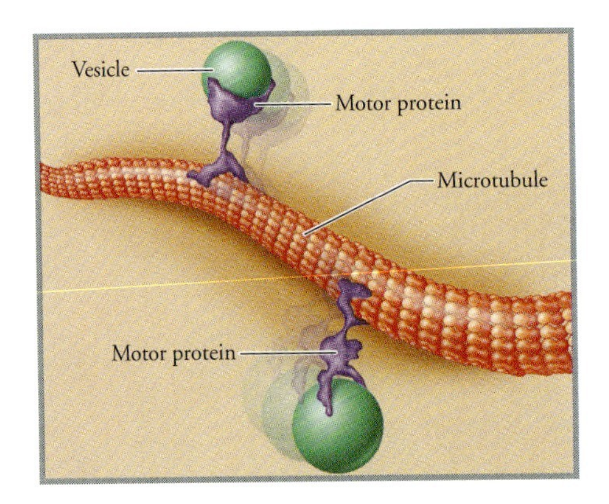

Fig. 13.20

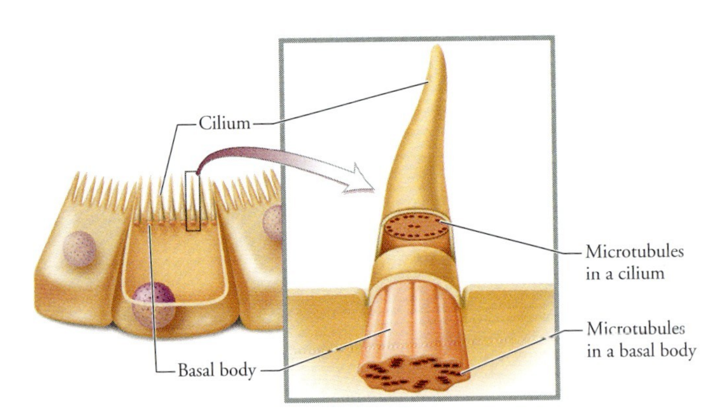

Fig.3.24

NOTES

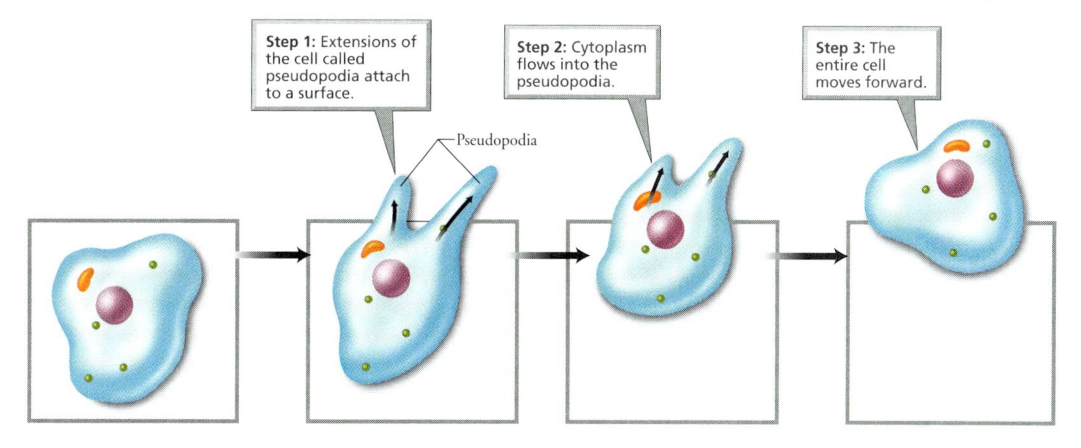

Fig.3.26

STRUCTURES AND FUNCTIONS OF CYTOSKELETAL ELEMENTS

ELEMENT	STRUCTURE	PROTEIN	FUNCTIONS
Microtubules	Hollow rods	Tubulin	• Cell movement (cilia, flagella) • Support • Tracks for organelles and vesicles • Chromosome movements in cell division
Microfilaments	Solid rods	Actin	• Muscle contraction • Support • Cell movement (pseudopodia) • Pinch cell in two in cell division
Intermediate filaments	Ropelike fibers	Different proteins; depends on cell type	• Maintain cell shape • Support • Anchor organelles

Table 3.5

NOTES

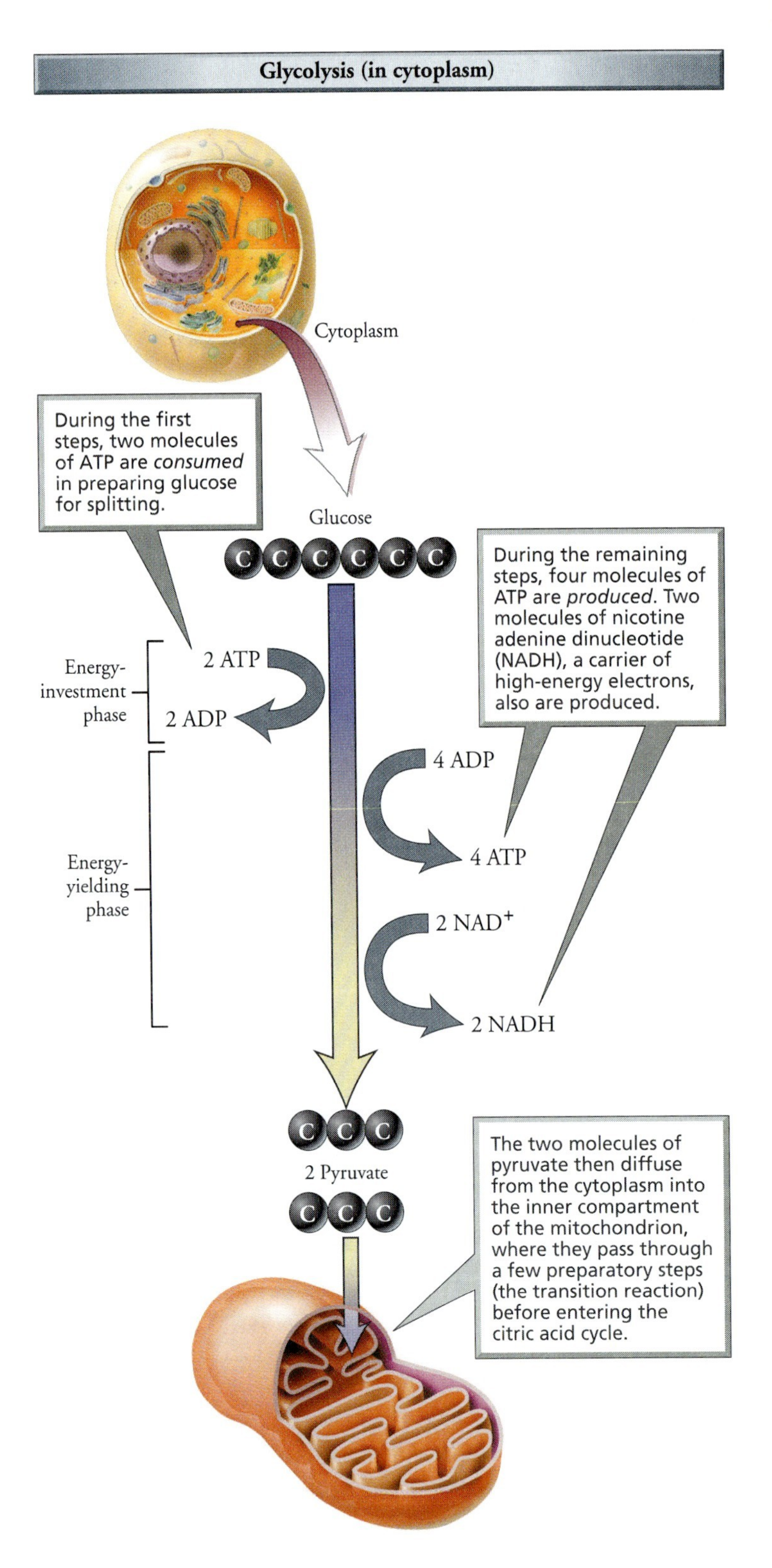

Fig. 3.27

NOTES

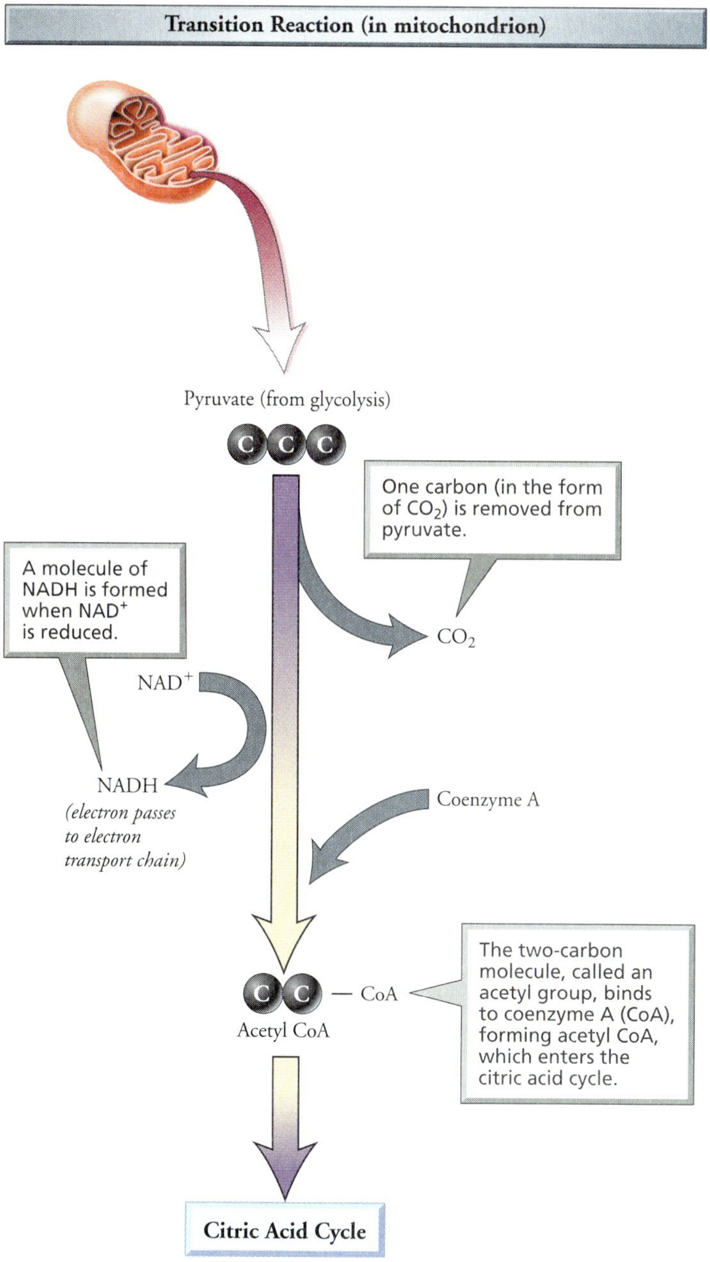

Transition Reaction (in mitochondrion)

Pyruvate (from glycolysis)

C C C

One carbon (in the form of CO_2) is removed from pyruvate.

CO_2

A molecule of NADH is formed when NAD^+ is reduced.

NAD^+

NADH
(electron passes to electron transport chain)

Coenzyme A

C C — CoA
Acetyl CoA

The two-carbon molecule, called an acetyl group, binds to coenzyme A (CoA), forming acetyl CoA, which enters the citric acid cycle.

Citric Acid Cycle

Fig. 3.28

NOTES

Citric Acid Cycle (in mitochondrion)

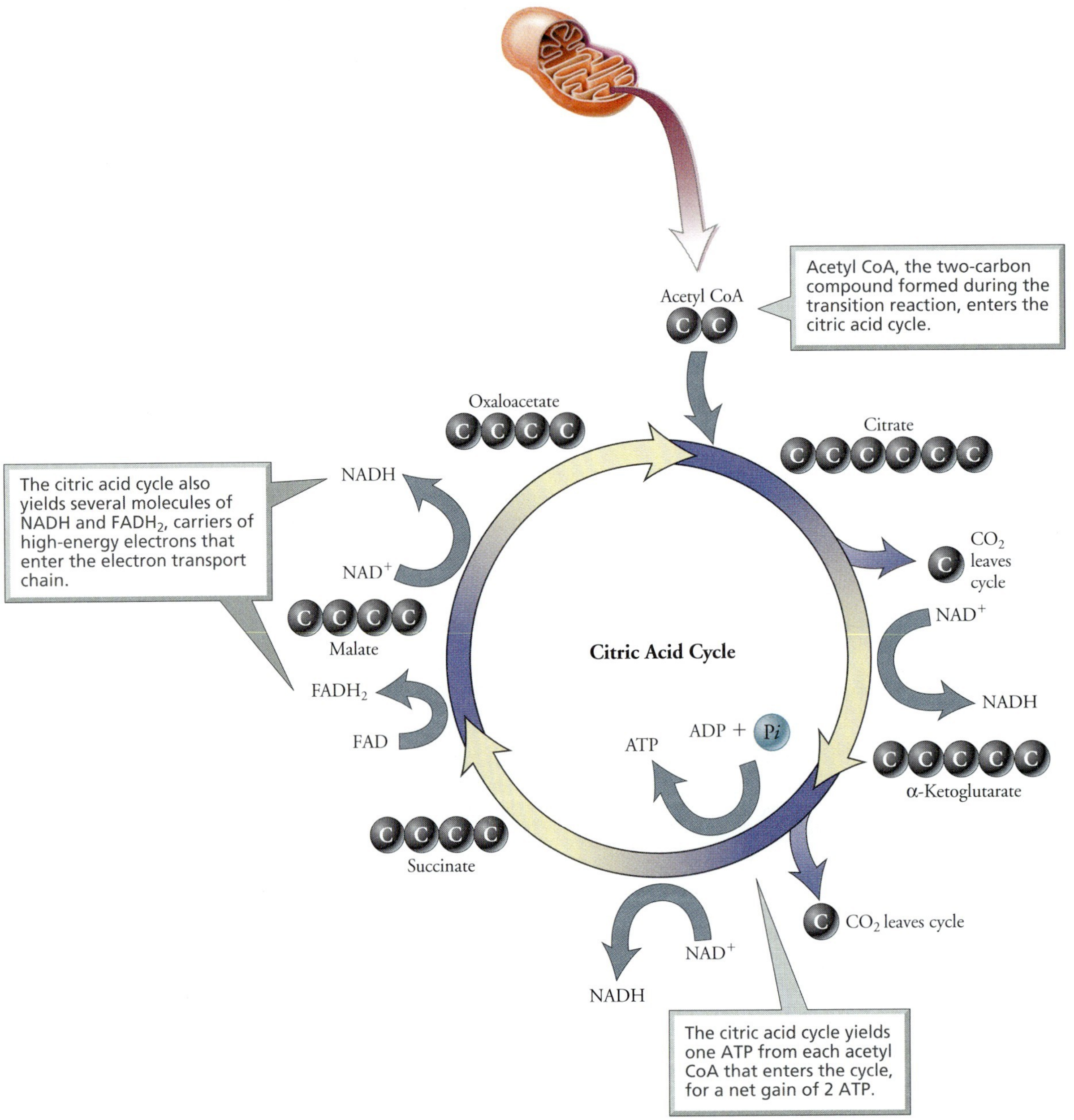
Acetyl CoA
C C
Acetyl CoA, the two-carbon compound formed during the transition reaction, enters the citric acid cycle.
Oxaloacetate
C C C C
Citrate
C C C C C C
C
CO_2 leaves cycle
NAD+
NADH
The citric acid cycle also yields several molecules of NADH and $FADH_2$, carriers of high-energy electrons that enter the electron transport chain.
NADH
NAD+
C C C C
Malate
Citric Acid Cycle
$FADH_2$
FAD
ADP + Pi
ATP
C C C C C
α-Ketoglutarate
C C C C
Succinate
C
CO_2 leaves cycle
NAD+
NADH
The citric acid cycle yields one ATP from each acetyl CoA that enters the cycle, for a net gain of 2 ATP.

Fig. 3.29

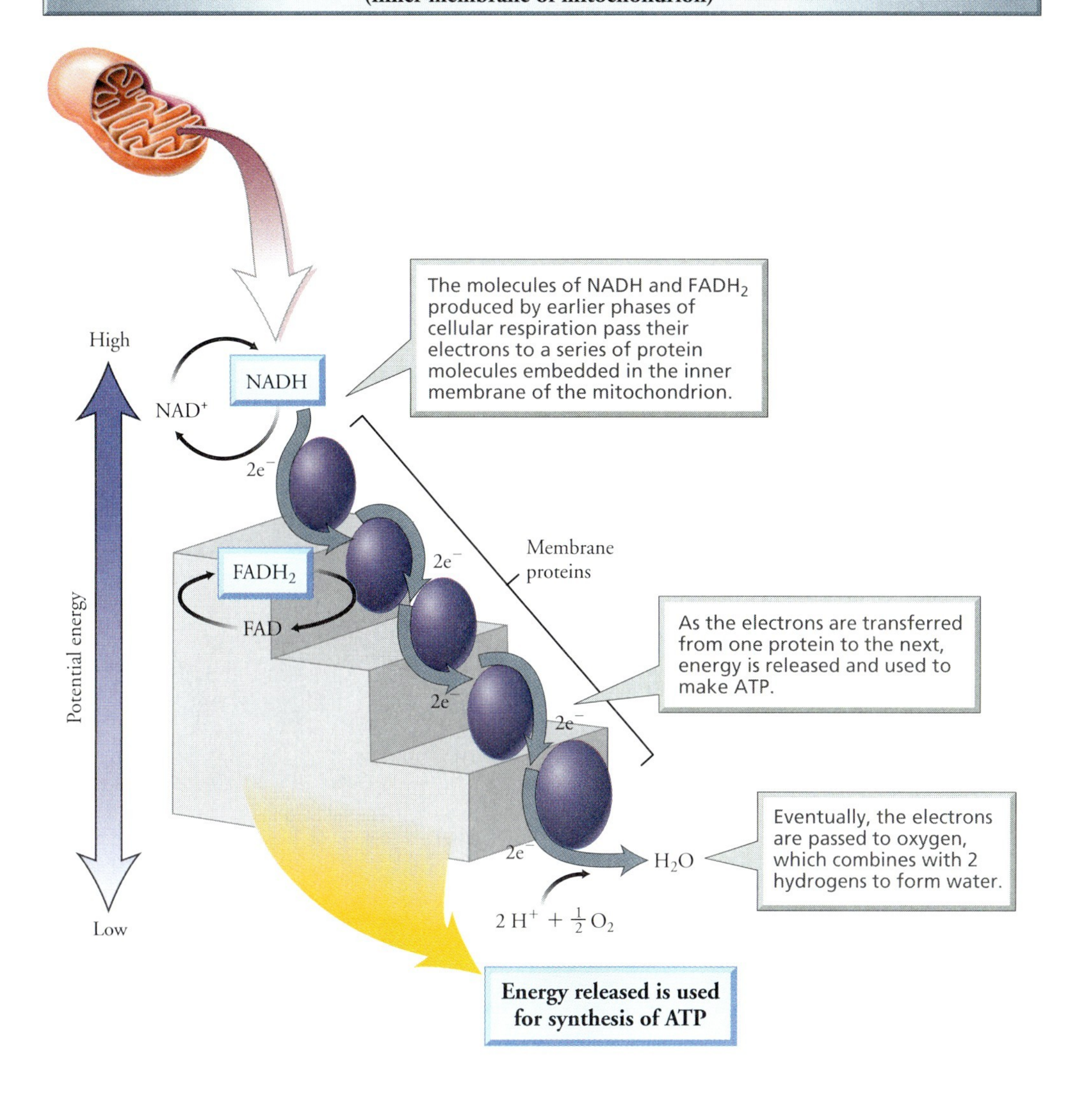
Electron Transport Chain
(inner membrane of mitochondrion)
High
NADH
NAD+
2e−
FADH2
FAD
2e−
Membrane proteins
2e−
2e−
2e−
H2O
2 H+ + ½ O2
Low
Potential energy
The molecules of NADH and FADH2 produced by earlier phases of cellular respiration pass their electrons to a series of protein molecules embedded in the inner membrane of the mitochondrion.
As the electrons are transferred from one protein to the next, energy is released and used to make ATP.
Eventually, the electrons are passed to oxygen, which combines with 2 hydrogens to form water.
Energy released is used for synthesis of ATP

Fig. 3.30

NOTES

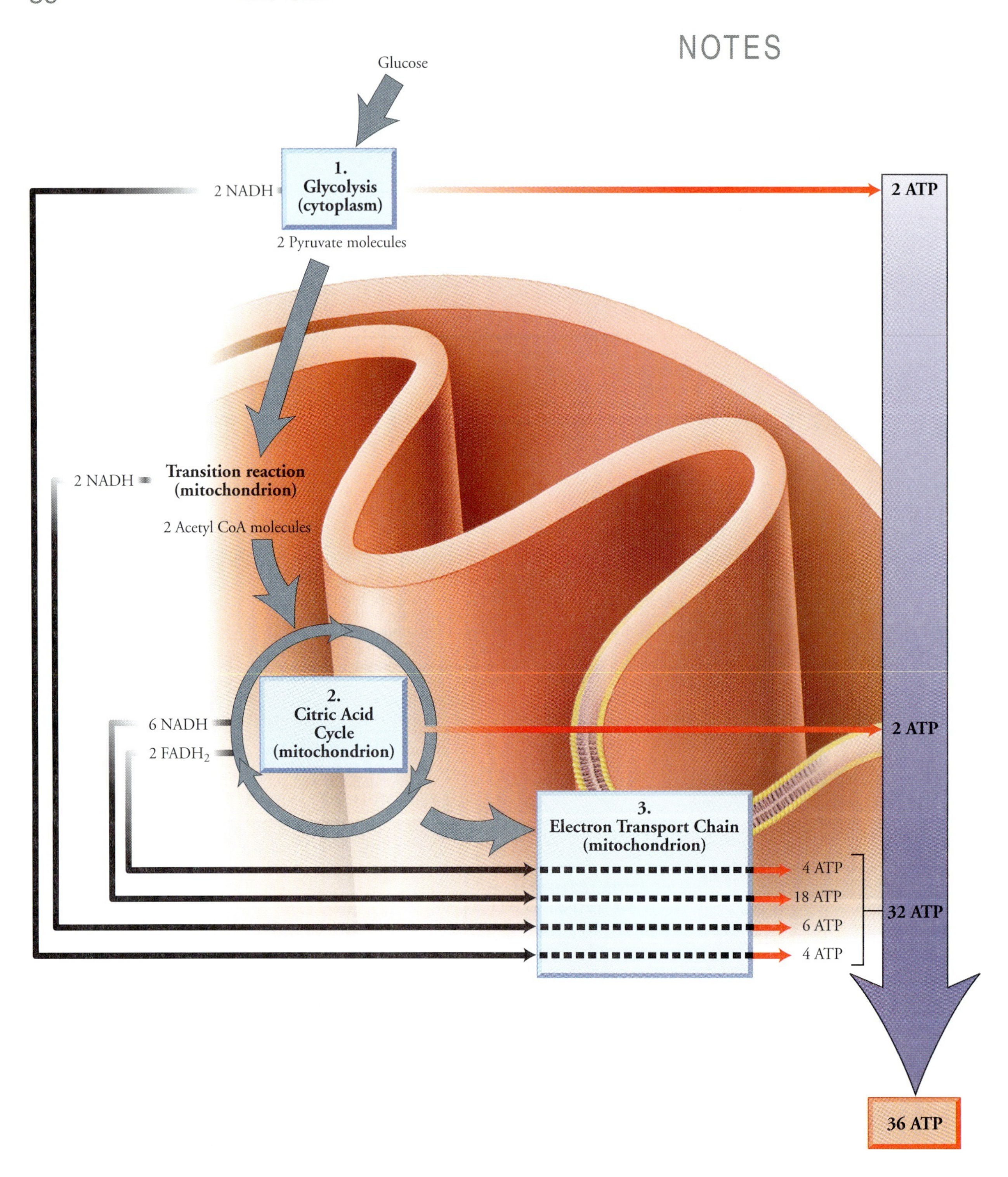

Fig. 3.31

NOTES

THE PHASES OF CELLULAR RESPIRATION			
PHASE	LOCATION	DESCRIPTION	MAIN PRODUCTS
Glycolysis	Cytoplasm	Several-step process by which glucose is split into 2 pyruvate	2 pyruvate 2 ATP 2 NADH
Transition reaction	Mitochondria	One CO_2 is removed from each pyruvate; the resulting molecules bind to CoA, forming 2 acetyl CoA	2 acetyl CoA 2 NADH
and			
Citric acid cycle	Mitochondria	Cyclic series of eight chemical reactions by which acetyl CoA is broken down	2 ATP 2 $FADH_2$ 6 NADH
Electron transport chain	Mitochondria	Electrons from NADH and $FADH_2$ are passed from one protein to the next, releasing energy for ATP synthesis	32 ATP H_2O

Table 3.6

NOTES

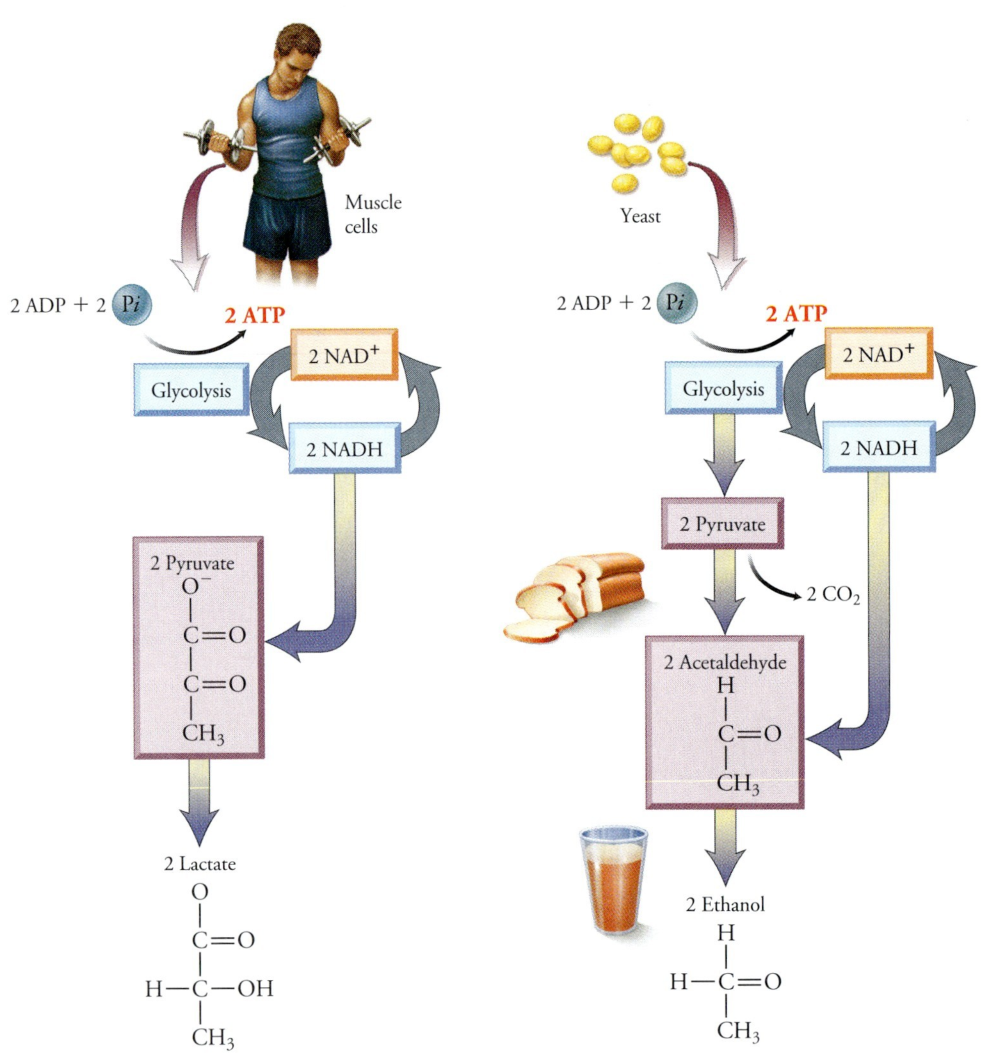

(a) Lactic acid fermentation occurs in our muscle cells during strenuous exercise. It consists of glycolysis plus chemical reactions in which NADH passes electrons directly to pyruvate, forming the waste product lactate (lactic acid).

(b) Alcohol fermentation occurs in yeast, a fungus used in making bread and beer. It consists of glycolysis plus a two-step chemical process. In the first step, CO_2 is released from pyruvate, leaving a two-carbon derivative of pyruvate called acetaldehyde. In the second step, NADH passes electrons to acetaldehyde, generating ethanol (ethyl alcohol).

Fig. 3.32

NOTES

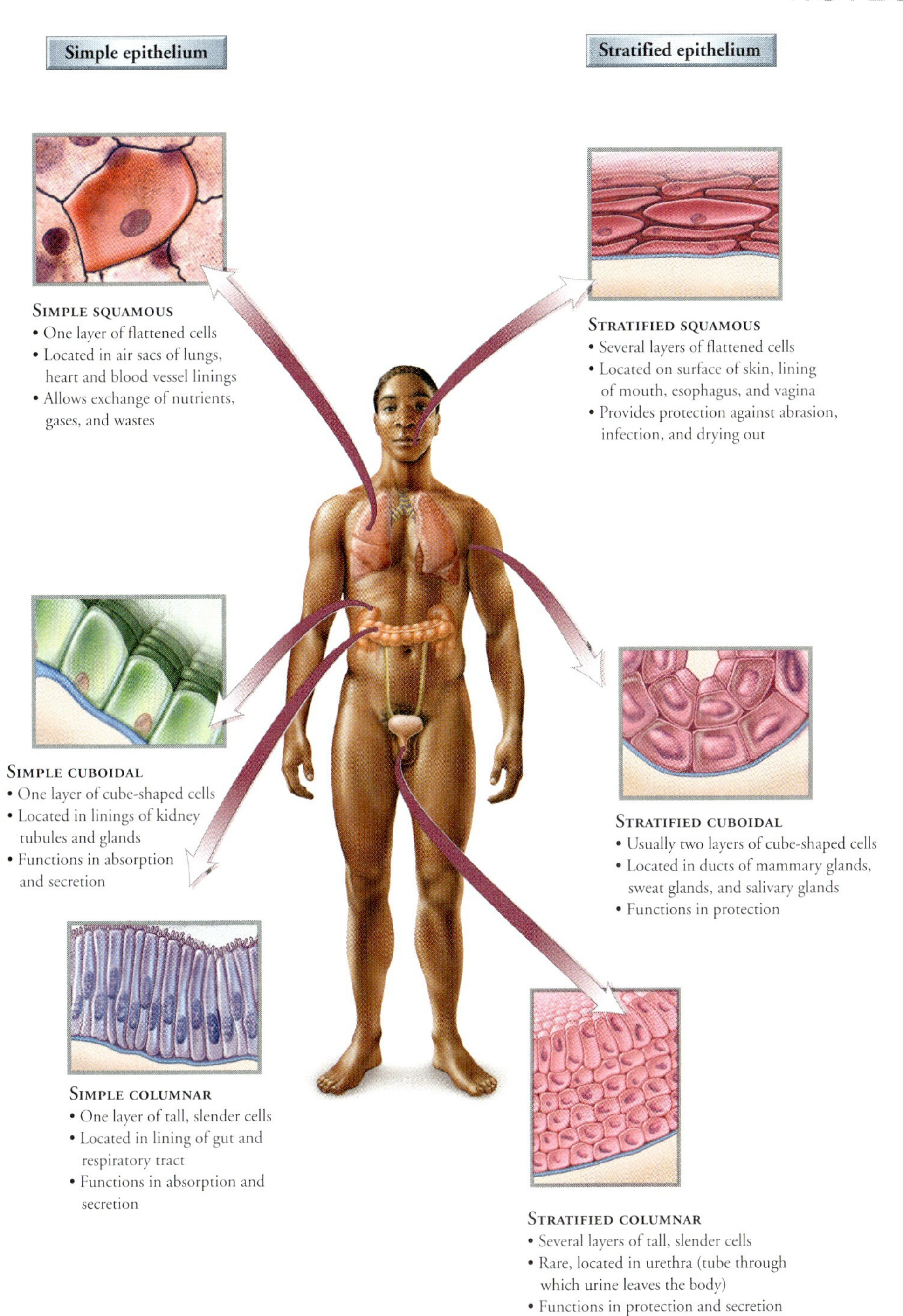
Simple epithelium
Stratified epithelium
Simple squamous
• One layer of flattened cells
• Located in air sacs of lungs, heart and blood vessel linings
• Allows exchange of nutrients, gases, and wastes
Stratified squamous
• Several layers of flattened cells
• Located on surface of skin, lining of mouth, esophagus, and vagina
• Provides protection against abrasion, infection, and drying out
Simple cuboidal
• One layer of cube-shaped cells
• Located in linings of kidney tubules and glands
• Functions in absorption and secretion
Stratified cuboidal
• Usually two layers of cube-shaped cells
• Located in ducts of mammary glands, sweat glands, and salivary glands
• Functions in protection
Simple columnar
• One layer of tall, slender cells
• Located in lining of gut and respiratory tract
• Functions in absorption and secretion
Stratified columnar
• Several layers of tall, slender cells
• Rare, located in urethra (tube through which urine leaves the body)
• Functions in protection and secretion

Fig. 4.1

NOTES

TYPES OF EPITHELIAL TISSUE			
SHAPE	NUMBER OF LAYERS	EXAMPLE LOCATIONS	FUNCTIONS
Squamous (flat, scale-like cells)	Simple (single layer)	Lining of heart and blood vessels, air sacs of lungs	Allows passage of materials by diffusion
	Stratified (more than one layer)	Linings of mouth, esophagus, and vagina; outer layer of skin	Protects underlying areas
Cuboidal (cube-shaped cells)	Simple	Kidney tubules, secretory portion of glands and their ducts	Secretion; absorption
	Stratified	Ducts of sweat glands, mammary glands, and salivary glands	Protects underlying areas
Columnar	Simple	Most of digestive tract (stomach to anus), air tubes of lungs (bronchi), excretory ducts of some glands, uterus	Absorbs; secretes mucus, enzymes, and other substances
	Stratified	Rare; urethra, junction of esophagus and stomach	Protects underlying areas, secretes

Table 4.1

TYPES OF CONNECTIVE TISSUE		
TYPE	EXAMPLE LOCATIONS	FUNCTIONS
Connective tissue proper		
Loose, areolar	Between muscles, surrounding glands, wrapping small blood vessels and nerves	Wraps and cushions organs
Loose, adipose (fat)	Under skin, around kidneys and heart	Stores energy, insulates, cushions organs
Dense	Tendons; ligaments	Attaches bone to bone (ligaments) or bone to muscle (tendons)
Specialized connective tissue		
Cartilage (semisolid)	Nose (tip); rings in respiratory air tubules; external ear	Provides flexible support, cushions
Bone (solid)	Skeleton	Provides support and protection (by enclosing), and levers for muscles to act on
Blood (fluid)	Within blood vessels	Transports oxygen and carbon dioxide, nutrients, hormones, and wastes; helps fight infections

Table 4.2

NOTES

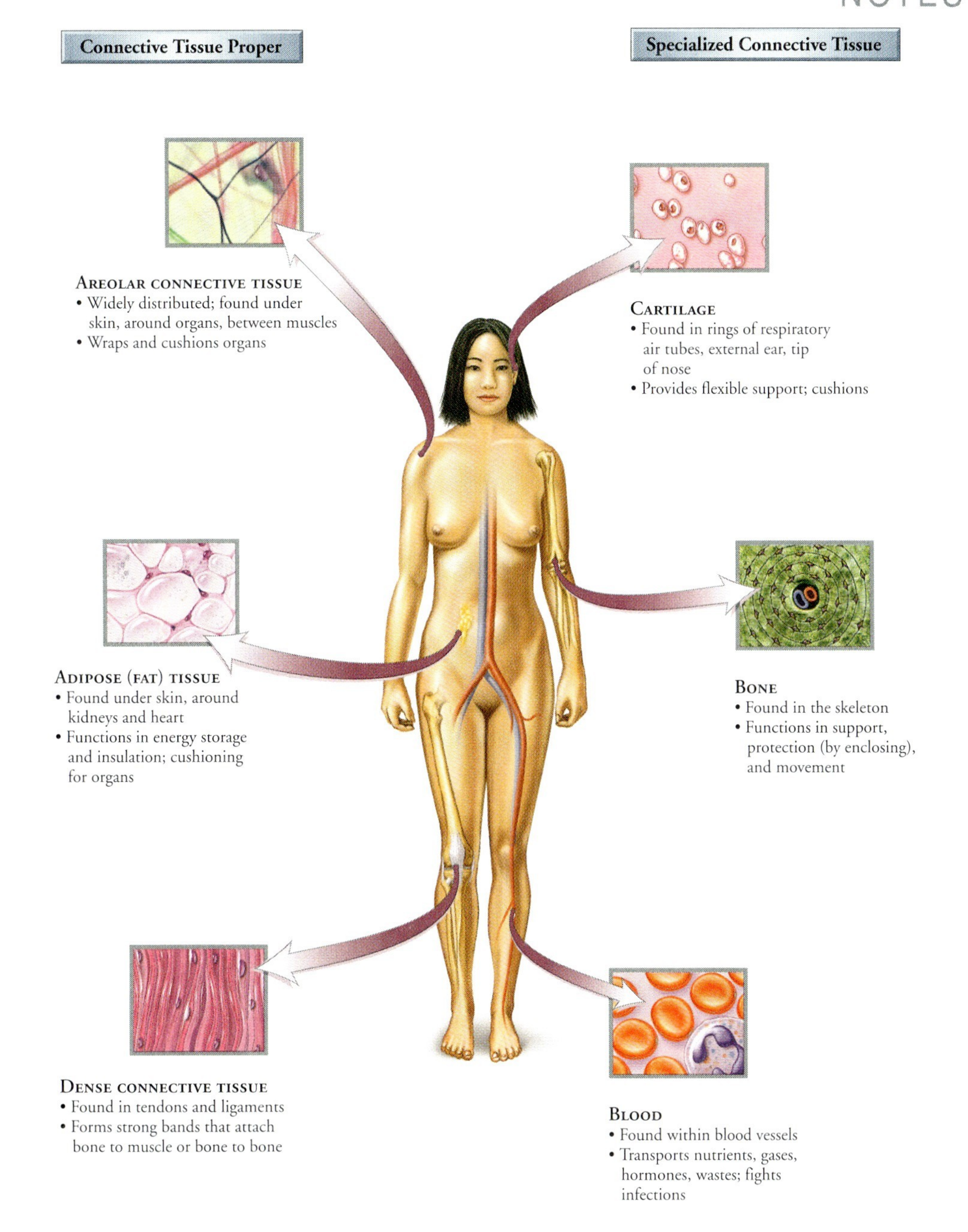
Connective Tissue Proper
Specialized Connective Tissue
Areolar connective tissue
• Widely distributed; found under skin, around organs, between muscles
• Wraps and cushions organs
Cartilage
• Found in rings of respiratory air tubes, external ear, tip of nose
• Provides flexible support; cushions
Adipose (fat) tissue
• Found under skin, around kidneys and heart
• Functions in energy storage and insulation; cushioning for organs
Bone
• Found in the skeleton
• Functions in support, protection (by enclosing), and movement
Dense connective tissue
• Found in tendons and ligaments
• Forms strong bands that attach bone to muscle or bone to bone
Blood
• Found within blood vessels
• Transports nutrients, gases, hormones, wastes; fights infections

Fig. 4.2

NOTES

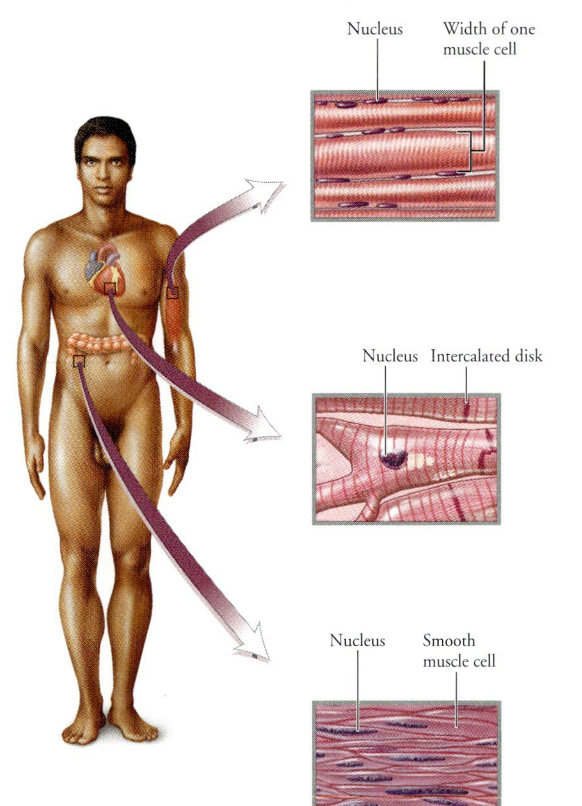

Skeletal muscle
- Long cylindrical striated cells with many nuclei
- Voluntary contraction
- Most are found attached to the skeleton
- Responsible for voluntary movement

Cardiac muscle
- Branching striated cells, one nucleus
- Involuntary contraction
- Found in wall of heart
- Pumps blood through the body

Smooth muscle
- Cells tapered at each end, one nucleus
- Involuntary contraction
- Found in walls of hollow internal organs, such as the intestines, and tubes, such as blood vessels
- Contractions in digestive system move food along
- When arranged in circle, controls diameter of tube

Fig. 4.3

Types of Muscle Tissue

TYPE	DESCRIPTION	EXAMPLE LOCATIONS	FUNCTIONS
Skeletal	Long, cylindrical cells; multiple nuclei per cell; obvious striations	Muscles attached to bones	Provides voluntary movement
Cardiac	Branching striated cells; one nucleus; specialized junctions between cells	Wall of heart	Contracts and propels blood through the circulatory system
Smooth	Cells taper at each end; single nucleus; arranged in sheets; no striations	Walls of digestive system, blood vessels, and tubules of urinary system	Propels substances or objects through internal passageways

Table 4.3

NOTES

(a) Tight junction
- Creates an impermeable junction that prevents the exchange of materials between cells
- Found between epithelial cells of the digestive tract, where they prevent digestive enzymes and microorganisms from entering the blood

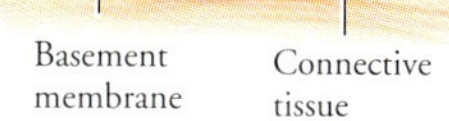

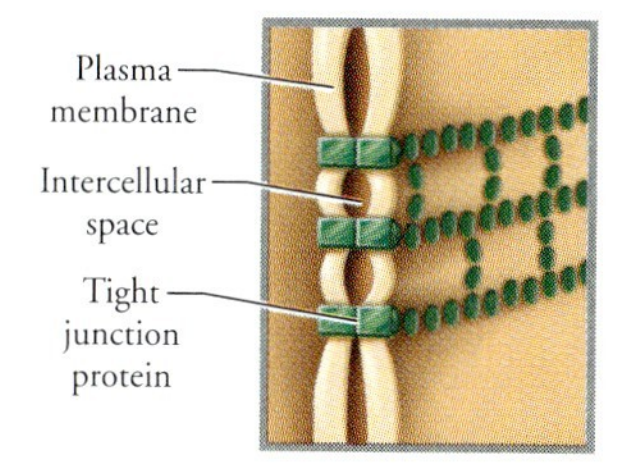

(b) Adhesion junction
- Holds cells together despite stretching
- Found in tissues that are often stretched, such as the skin and the opening of the uterus

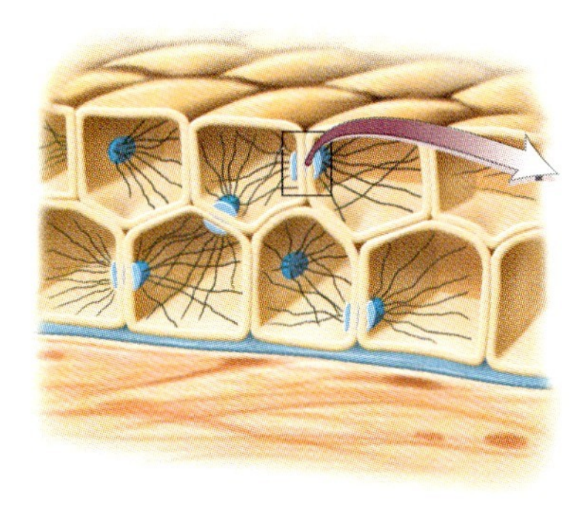

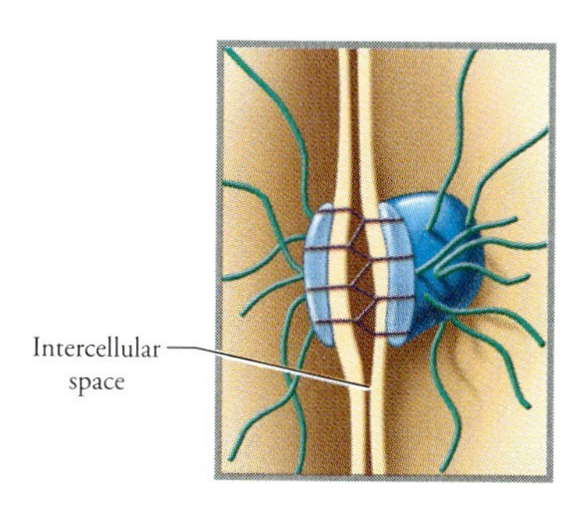

(c) Gap junction
- Allows cells to communicate by allowing small molecules and ions to pass from cell to cell
- Found in epithelia in which the movement of ions coordinates functions, such as the beating of cilia; found in excitable tissue such as heart and smooth muscle

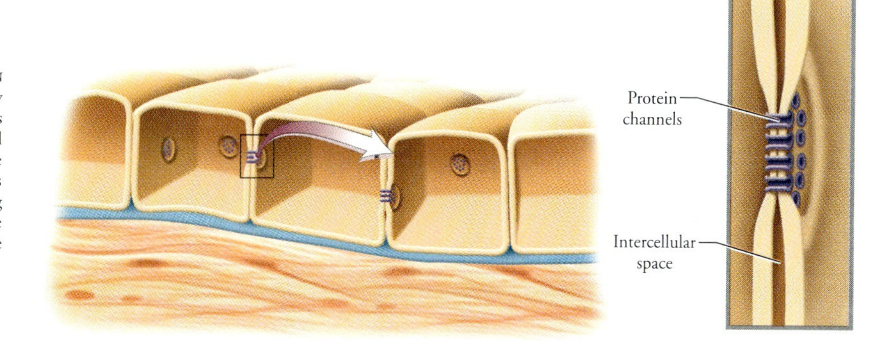

Fig. 4.5

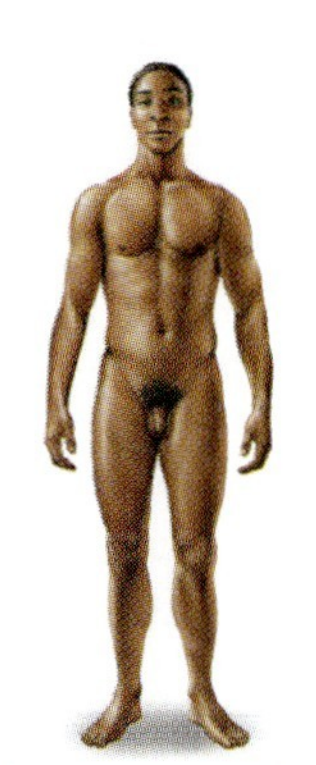

INTEGUMENTARY SYSTEM
- Protects underlying tissues from abrasion and dehydration
- Provides cutaneous sensation
- Regulates body temperature
- Immune function
- Synthesizes vitamin D
- Excretion

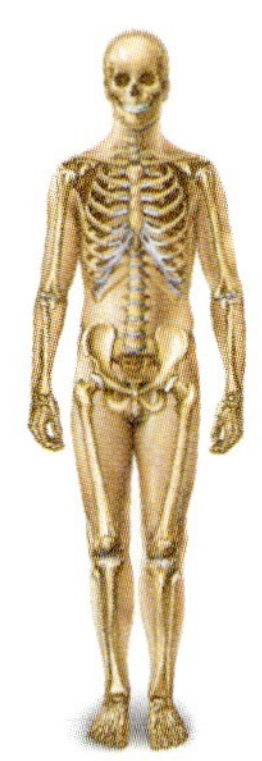

SKELETAL SYSTEM
- Attachment for muscles
- Encloses and protects organs
- Stores calcium and phosphorus
- Produces blood cells
- Stores fat

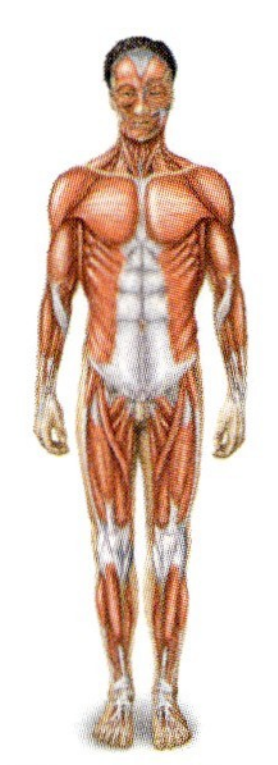

MUSCULAR SYSTEM
- Moves body and maintains posture
- Internal transport of fluids
- Generation of heat

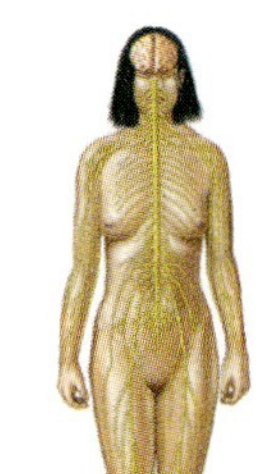

NERVOUS SYSTEM
- Regulates and integrates body functions via neurons

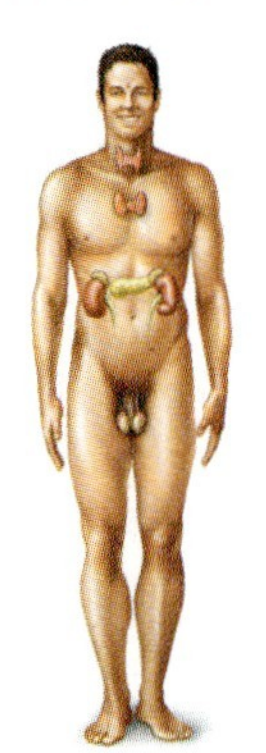

ENDOCRINE SYSTEM
- Regulates and integrates body functions via hormones

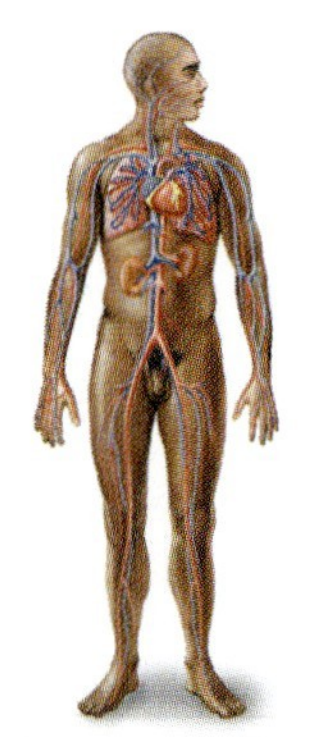

CIRCULATORY SYSTEM
- Transports nutrients, respiratory gases, wastes, and heat
- Transports cells and antibodies for immune response
- Transports hormones
- Regulates pH through buffers

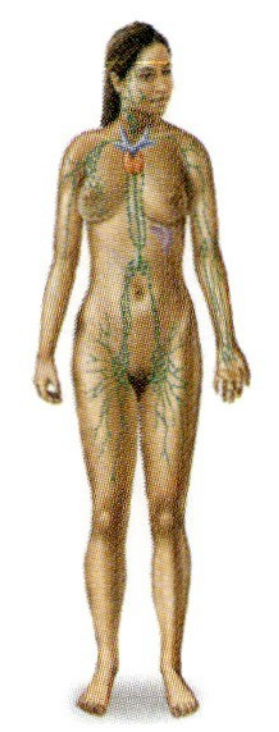

LYMPHATIC SYSTEM
- Returns tissue fluids to bloodstream
- Protects against infection and disease

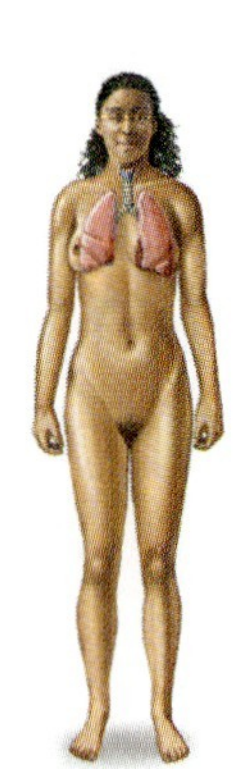

RESPIRATORY SYSTEM
- Exchanges respiratory gases with the environment

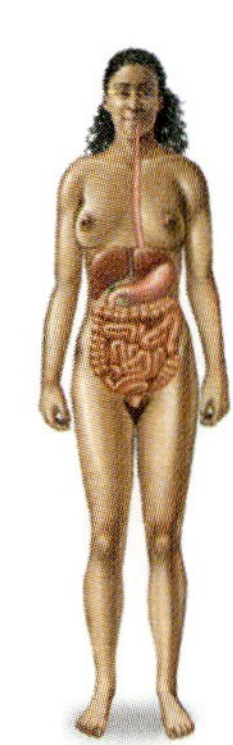

DIGESTIVE SYSTEM
- Physical and chemical breakdown of food
- Absorbs, processes, stores and controls release of digestive products

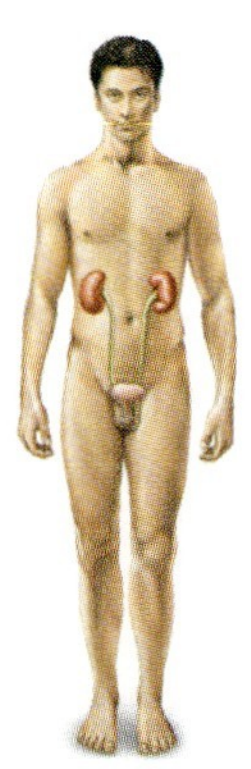

URINARY SYSTEM
- Maintains constant internal environment through the excretion of nitrogenous waste

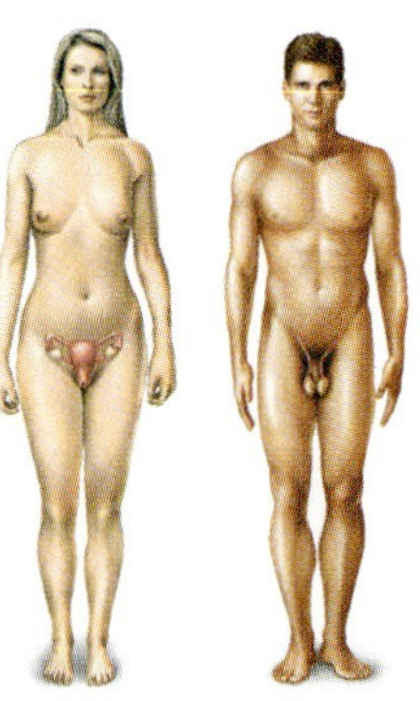

REPRODUCTIVE SYSTEM
- Produces and secretes hormones
- Produces and releases egg and sperm cells and accessory secretions
- Forms placental attachment with fetus (females only)

Fig. 4.6

NOTES

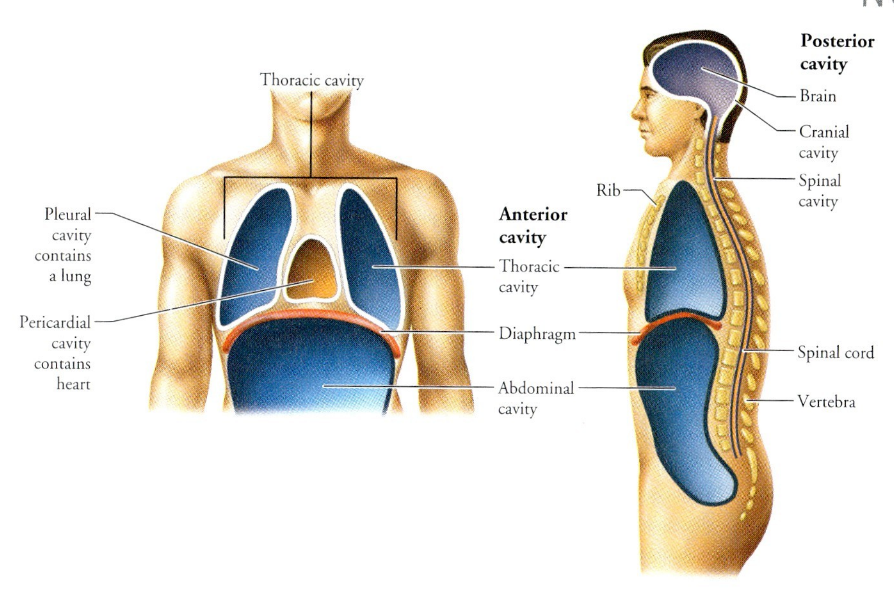

Fig. 4.7

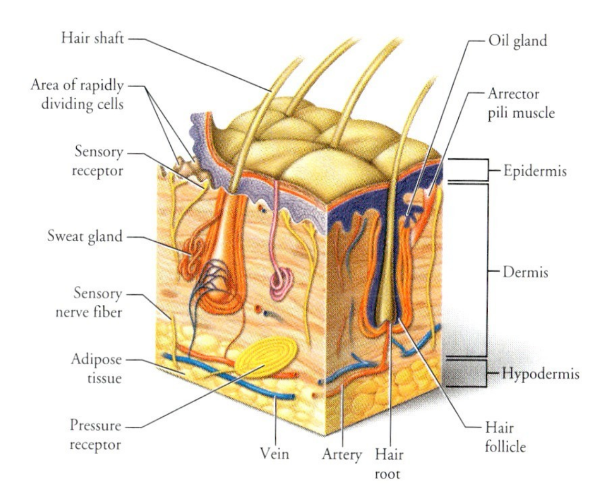

Fig. 4.8

NOTES

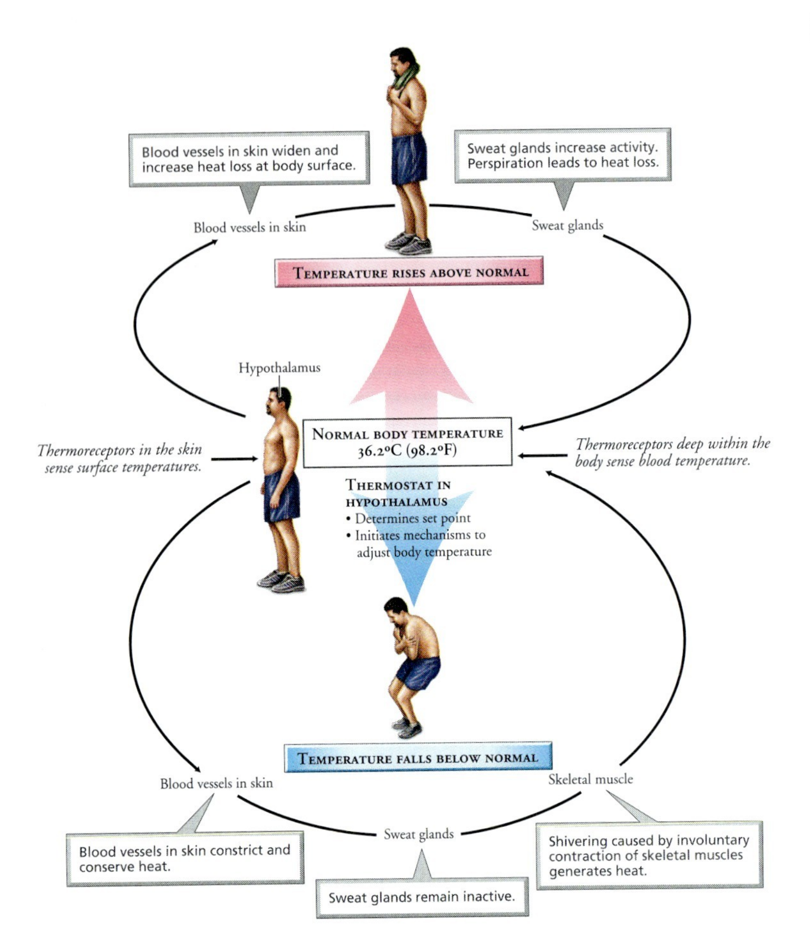

Fig. 4.12

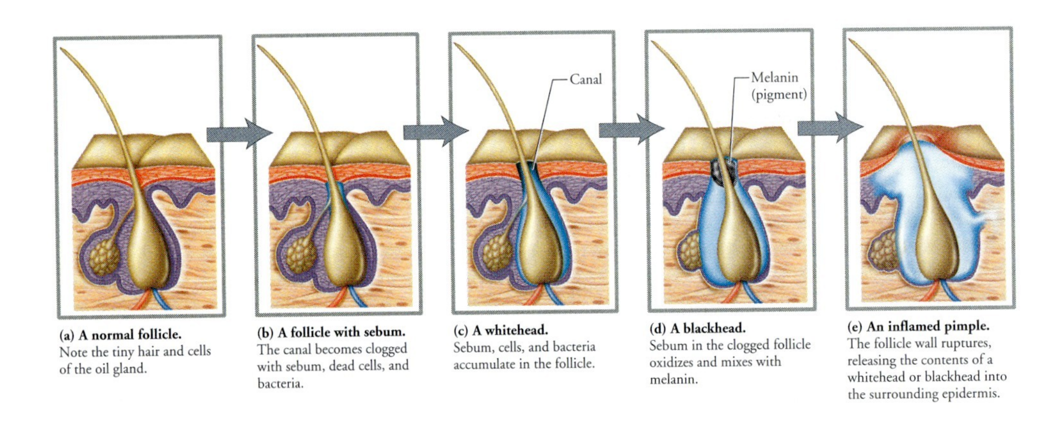

(a) A normal follicle. Note the tiny hair and cells of the oil gland.

(b) A follicle with sebum. The canal becomes clogged with sebum, dead cells, and bacteria.

(c) A whitehead. Sebum, cells, and bacteria accumulate in the follicle.

(d) A blackhead. Sebum in the clogged follicle oxidizes and mixes with melanin.

(e) An inflamed pimple. The follicle wall ruptures, releasing the contents of a whitehead or blackhead into the surrounding epidermis.

Fig. 4.B

NOTES

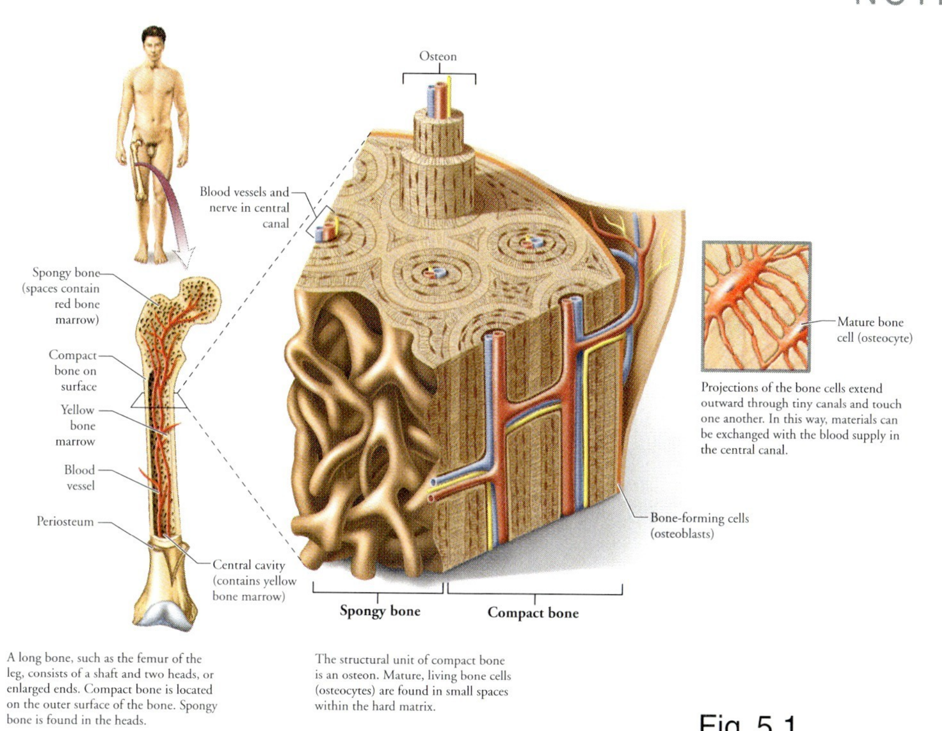

Fig. 5.1

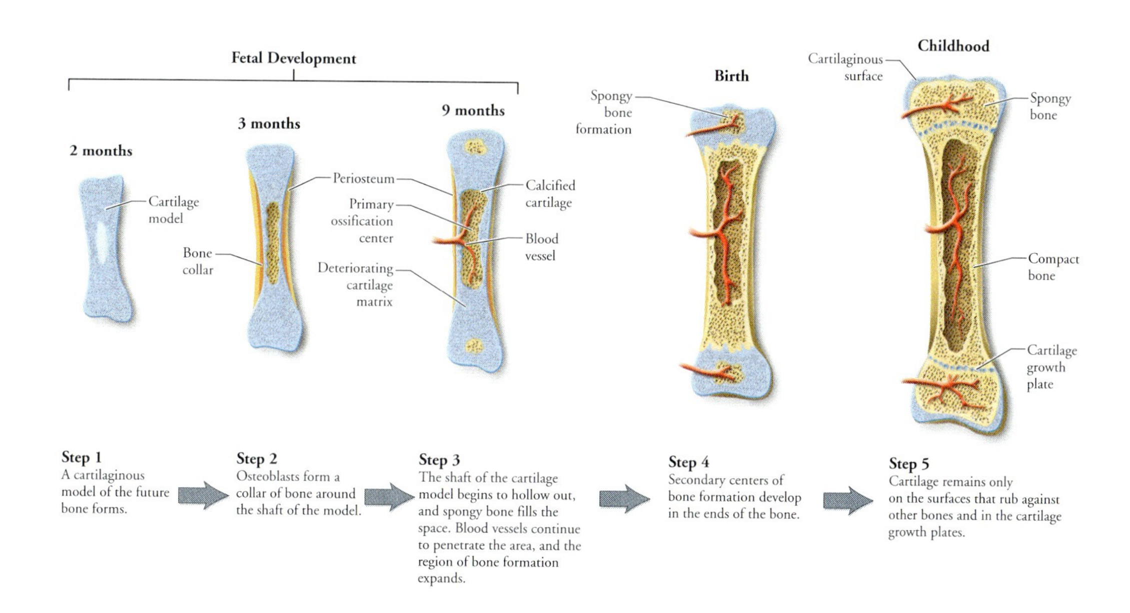

Fig. 5.4

NOTES

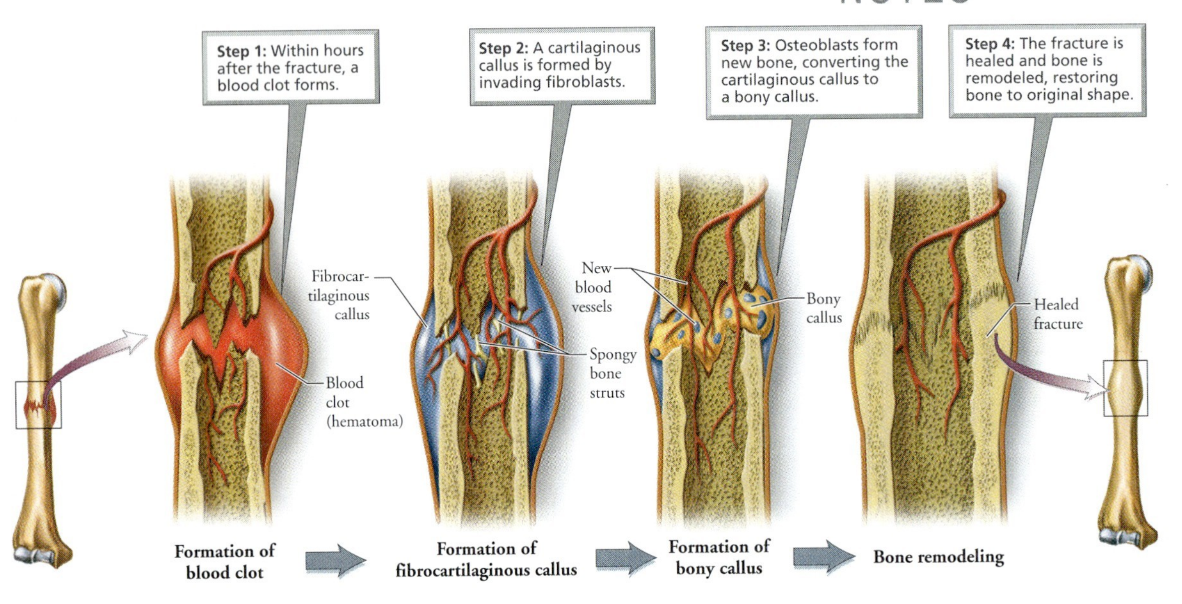

Fig. 5.5

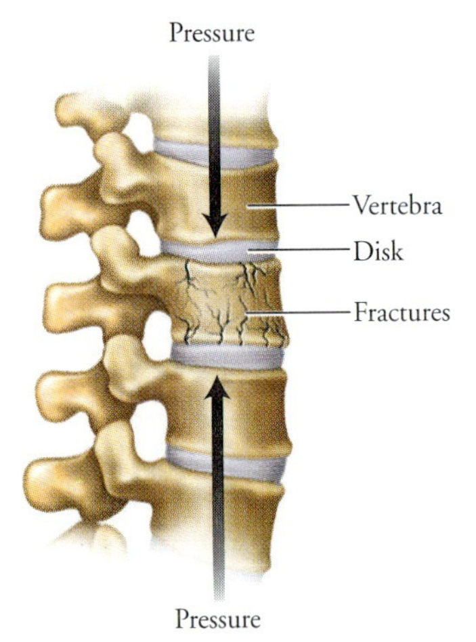

Normal bending can place pressure on vertebrae that can cause small fractures.

Fig. 5.A

NOTES

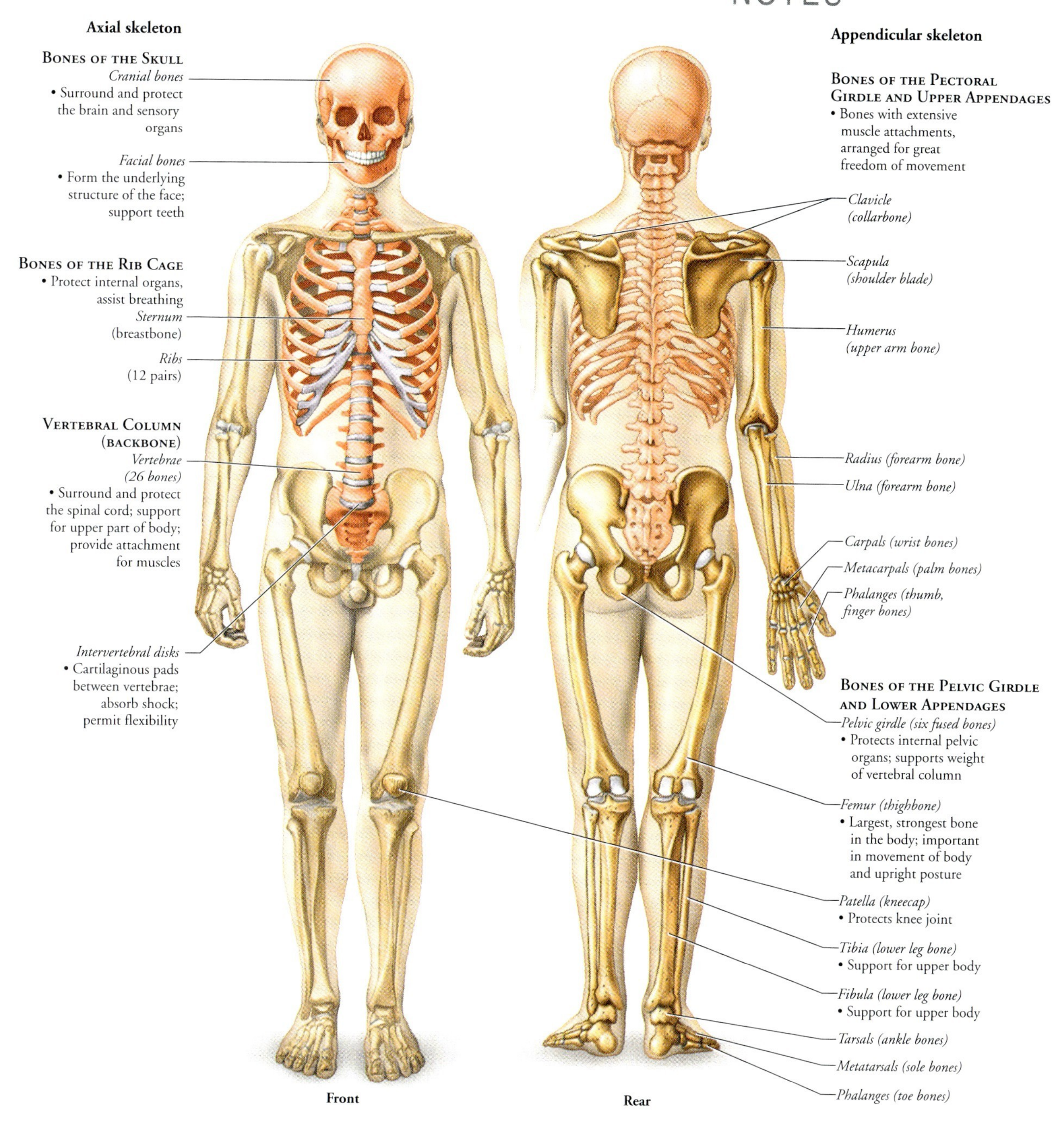
Axial skeleton
Bones of the Skull
Cranial bones
• Surround and protect the brain and sensory organs
Facial bones
• Form the underlying structure of the face; support teeth
Bones of the Rib Cage
• Protect internal organs, assist breathing
Sternum (breastbone)
Ribs (12 pairs)
Vertebral Column (backbone)
Vertebrae (26 bones)
• Surround and protect the spinal cord; support for upper part of body; provide attachment for muscles
Intervertebral disks
• Cartilaginous pads between vertebrae; absorb shock; permit flexibility
Appendicular skeleton
Bones of the Pectoral Girdle and Upper Appendages
• Bones with extensive muscle attachments, arranged for great freedom of movement
Clavicle (collarbone)
Scapula (shoulder blade)
Humerus (upper arm bone)
Radius (forearm bone)
Ulna (forearm bone)
Carpals (wrist bones)
Metacarpals (palm bones)
Phalanges (thumb, finger bones)
Bones of the Pelvic Girdle and Lower Appendages
Pelvic girdle (six fused bones)
• Protects internal pelvic organs; supports weight of vertebral column
Femur (thighbone)
• Largest, strongest bone in the body; important in movement of body and upright posture
Patella (kneecap)
• Protects knee joint
Tibia (lower leg bone)
• Support for upper body
Fibula (lower leg bone)
• Support for upper body
Tarsals (ankle bones)
Metatarsals (sole bones)
Phalanges (toe bones)
Front
Rear

Fig. 5.6

NOTES

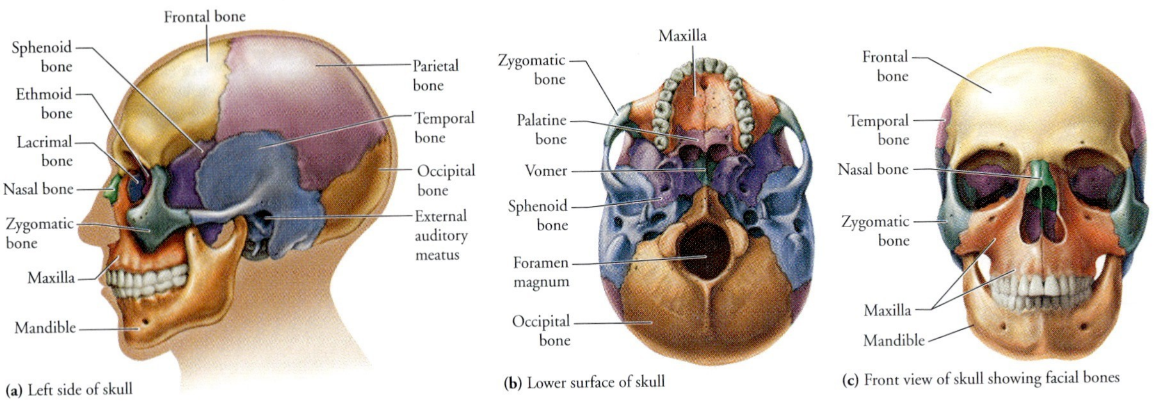

Frontal bone
Sphenoid bone
Ethmoid bone
Lacrimal bone
Nasal bone
Zygomatic bone
Maxilla
Mandible
Parietal bone
Temporal bone
Occipital bone
External auditory meatus

(a) Left side of skull

Maxilla
Zygomatic bone
Palatine bone
Vomer
Sphenoid bone
Foramen magnum
Occipital bone

(b) Lower surface of skull

Frontal bone
Temporal bone
Nasal bone
Zygomatic bone
Maxilla
Mandible

(c) Front view of skull showing facial bones

Fig. 5.7

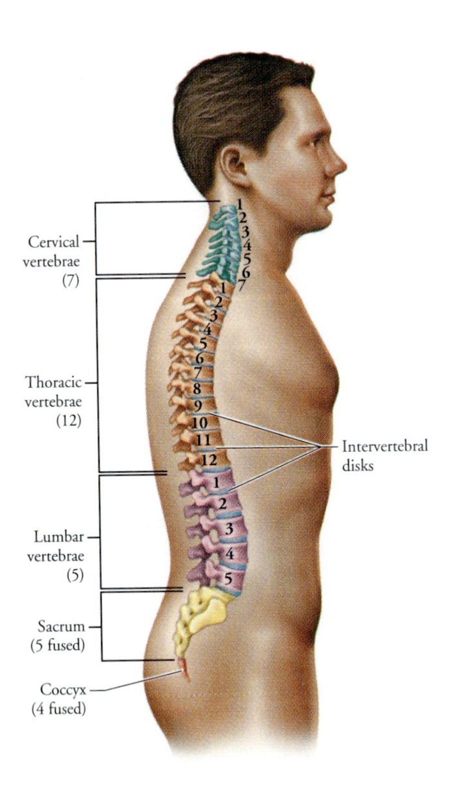

Fig. 5.9

NOTES

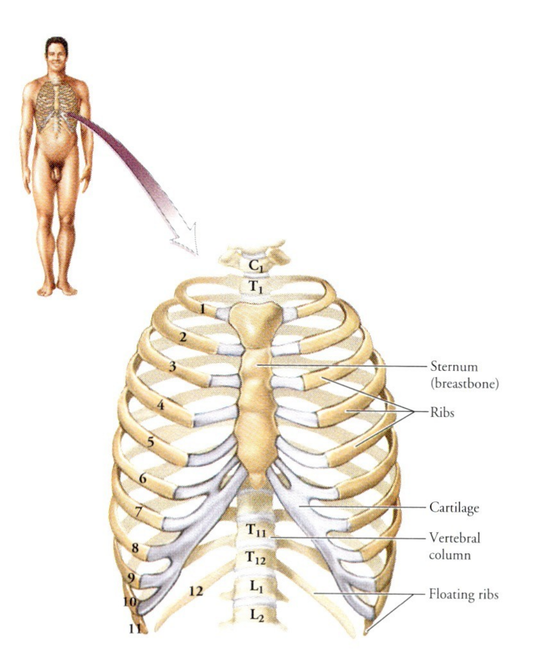

Fig. 5.10

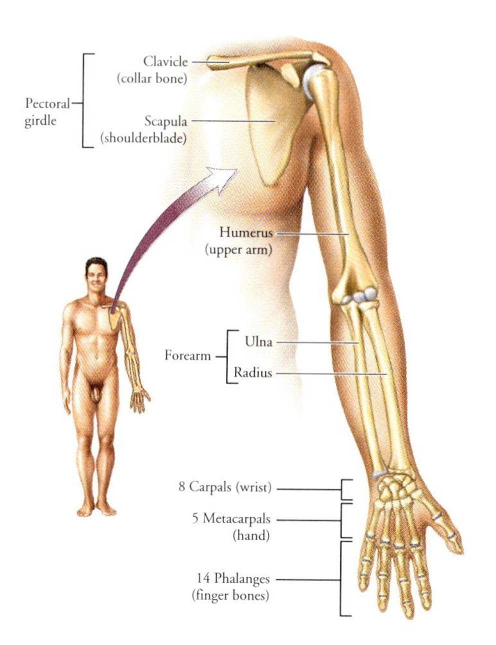

Fig. 5.11

NOTES

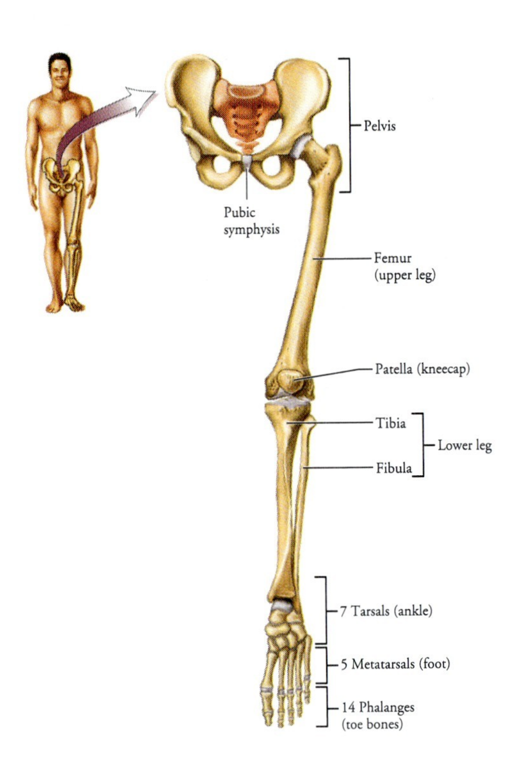

Fig. 5.12

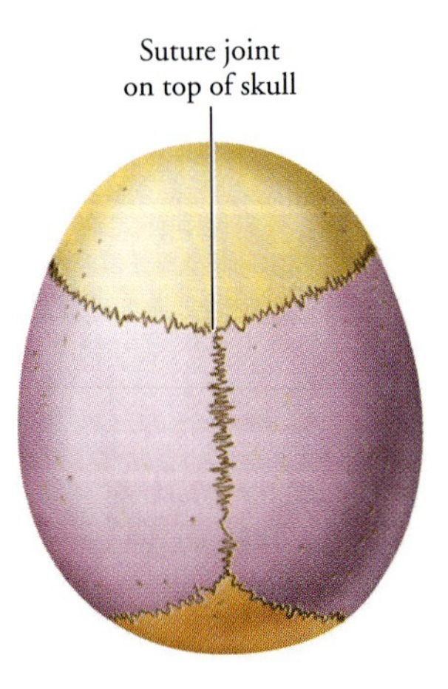

Fig. 5.13

NOTES

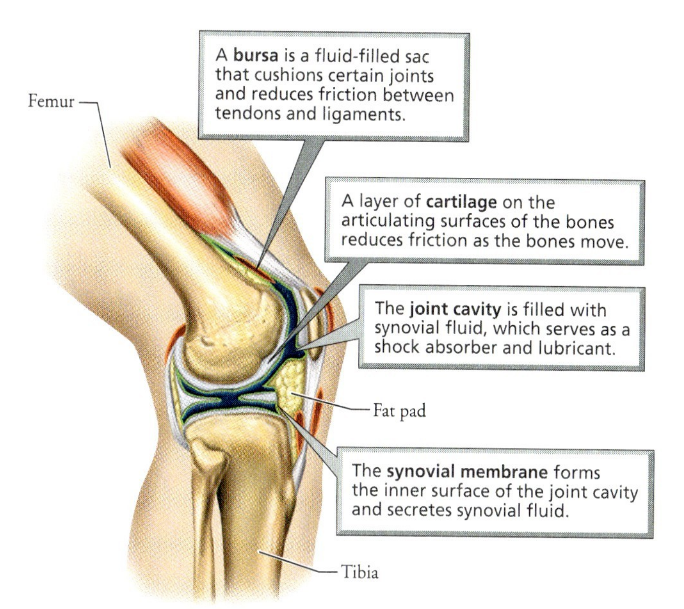

(a) Synovial joints, such as the knee shown here, permit a great range of movement.

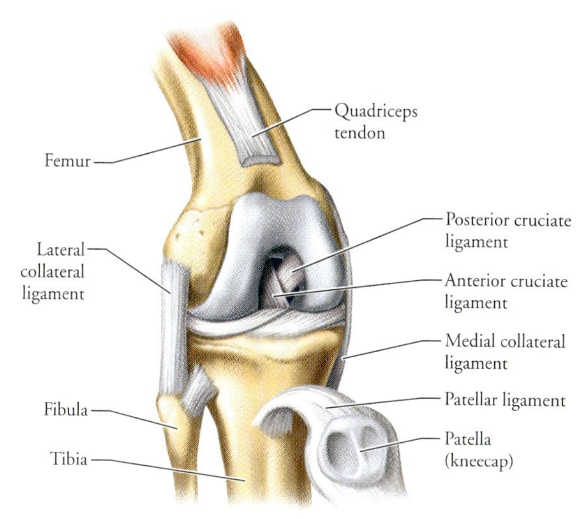

(b) Ligaments hold bones together, support the joint, and direct the movement of the bones.

Fig. 5.14

NOTES

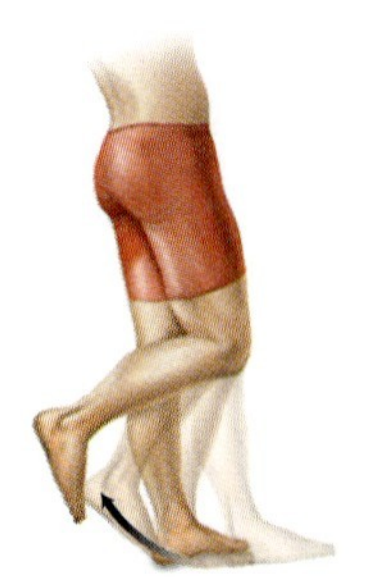

Flexion
Motion that *decreases* the angle between the bones of the joint, bringing the bones closer together

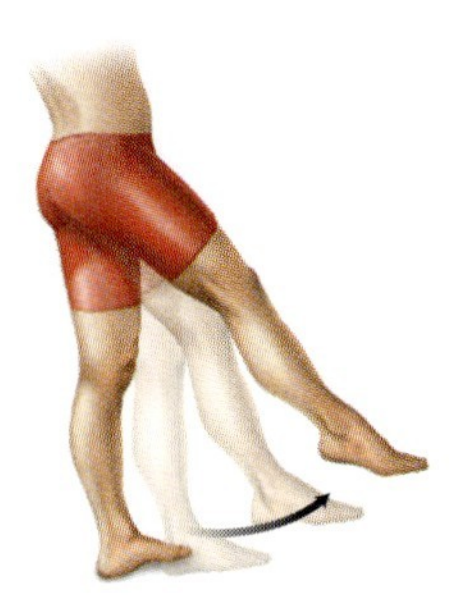

Extension
Motion that *increases* the angle between the bones of the joint

Adduction
Movement of a body part *toward* the body midline

Abduction
Movement of a body part *away from* the body midline

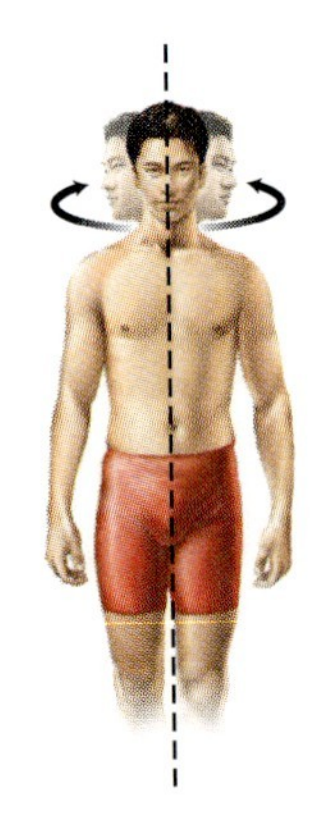

Rotation
Movement of a body part around it own axis

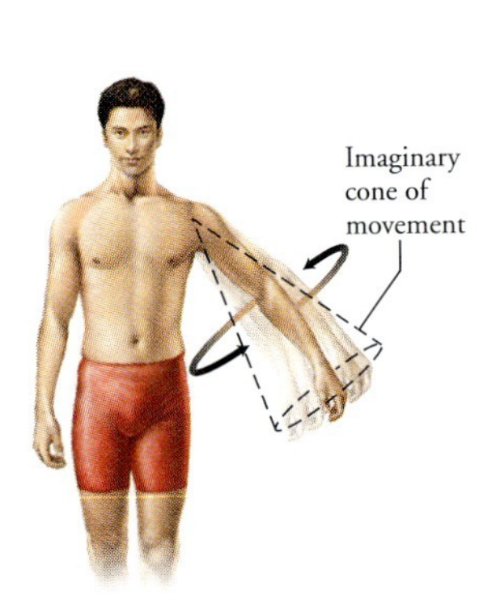

Circumduction
Movement of a body part in a wide circle so that the motion describes a cone

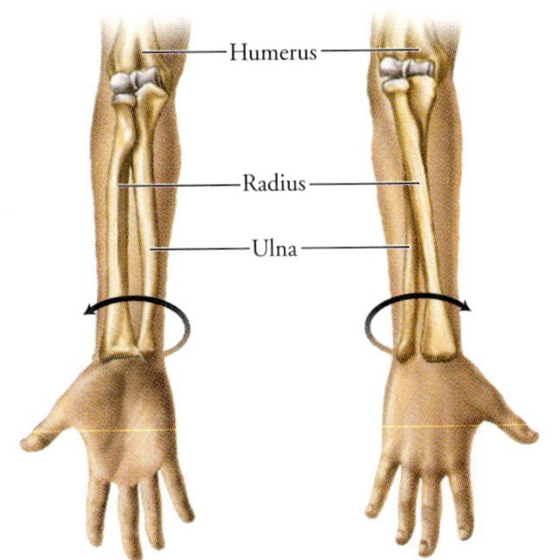

Supination
Rotation of the forearm so that the palm faces up

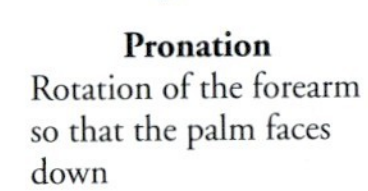

Pronation
Rotation of the forearm so that the palm faces down

Fig. 5.15

NOTES

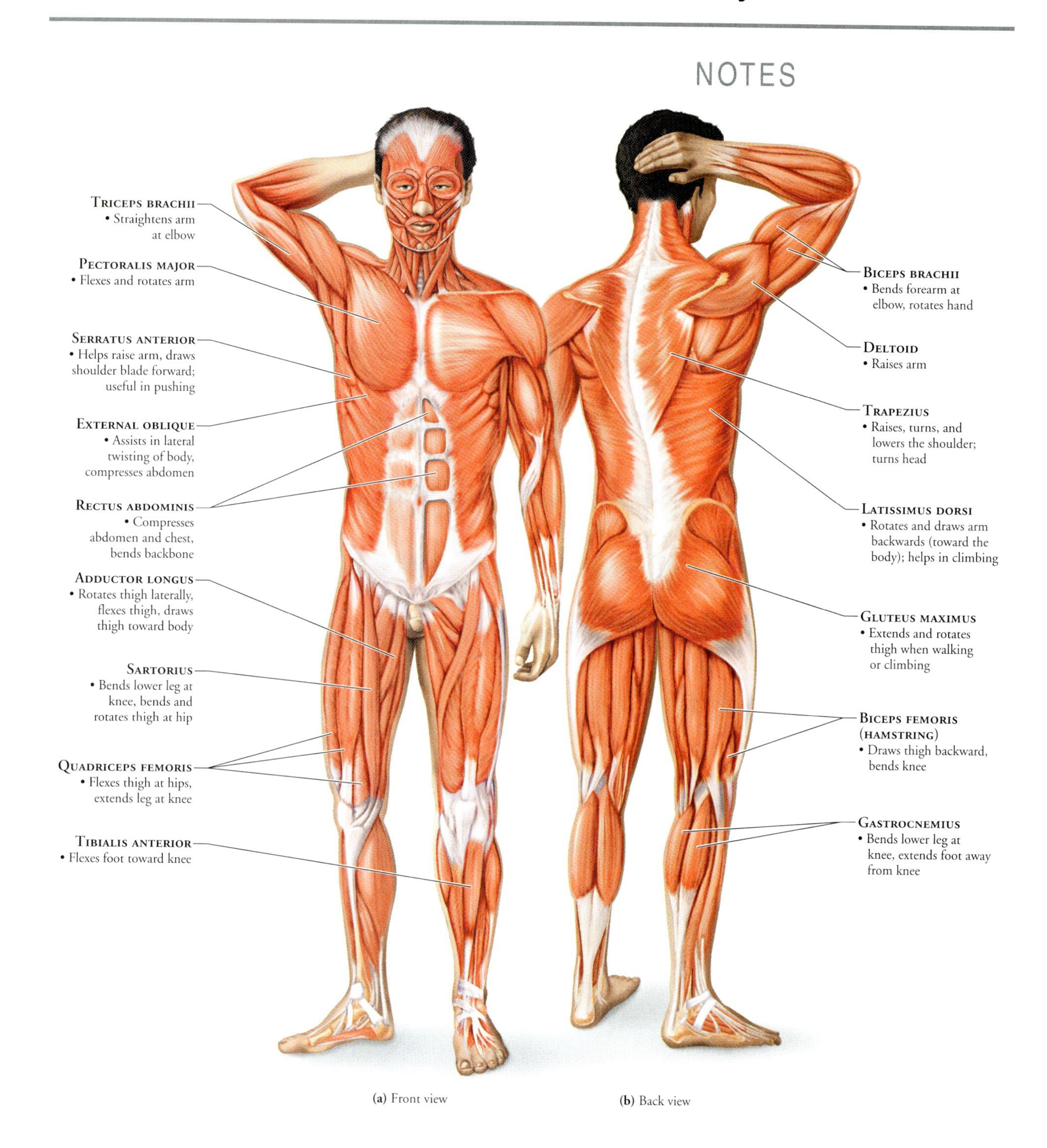
TRICEPS BRACHII
• Straightens arm at elbow
PECTORALIS MAJOR
• Flexes and rotates arm
SERRATUS ANTERIOR
• Helps raise arm, draws shoulder blade forward; useful in pushing
EXTERNAL OBLIQUE
• Assists in lateral twisting of body, compresses abdomen
RECTUS ABDOMINIS
• Compresses abdomen and chest, bends backbone
ADDUCTOR LONGUS
• Rotates thigh laterally, flexes thigh, draws thigh toward body
SARTORIUS
• Bends lower leg at knee, bends and rotates thigh at hip
QUADRICEPS FEMORIS
• Flexes thigh at hips, extends leg at knee
TIBIALIS ANTERIOR
• Flexes foot toward knee
BICEPS BRACHII
• Bends forearm at elbow, rotates hand
DELTOID
• Raises arm
TRAPEZIUS
• Raises, turns, and lowers the shoulder; turns head
LATISSIMUS DORSI
• Rotates and draws arm backwards (toward the body); helps in climbing
GLUTEUS MAXIMUS
• Extends and rotates thigh when walking or climbing
BICEPS FEMORIS (HAMSTRING)
• Draws thigh backward, bends knee
GASTROCNEMIUS
• Bends lower leg at knee, extends foot away from knee
(a) Front view
(b) Back view

Fig 6.1

NOTES

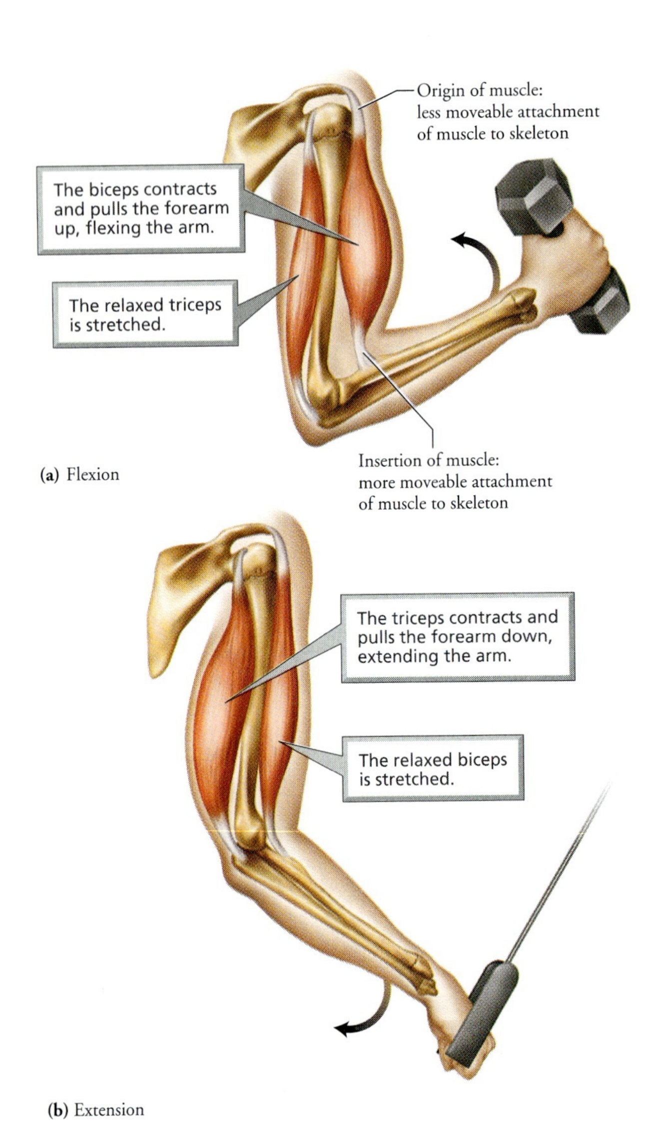
Origin of muscle:
less moveable attachment
of muscle to skeleton
The biceps contracts and pulls the forearm up, flexing the arm.
The relaxed triceps is stretched.
Insertion of muscle:
more moveable attachment
of muscle to skeleton
(a) Flexion
The triceps contracts and pulls the forearm down, extending the arm.
The relaxed biceps is stretched.
(b) Extension

Fig. 6.2

NOTES

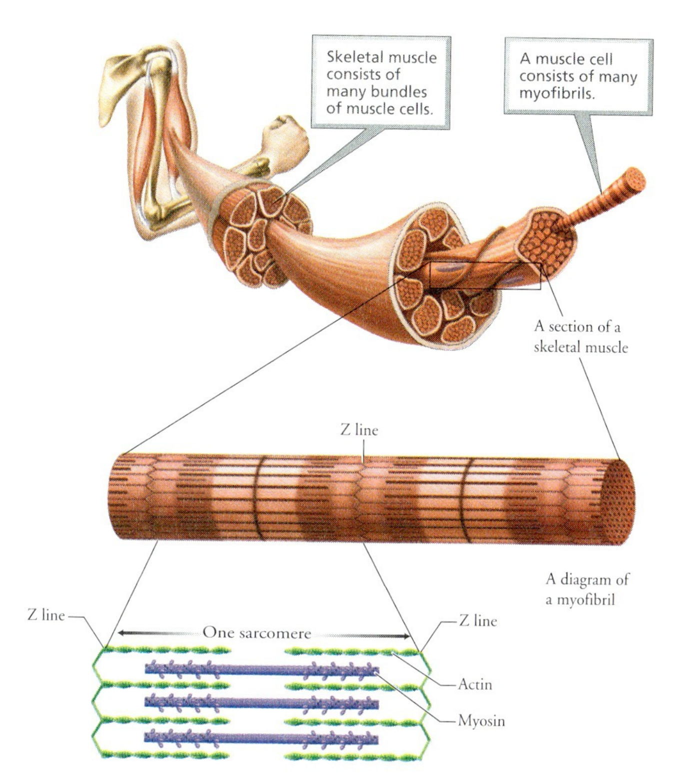

Fig. 6.3

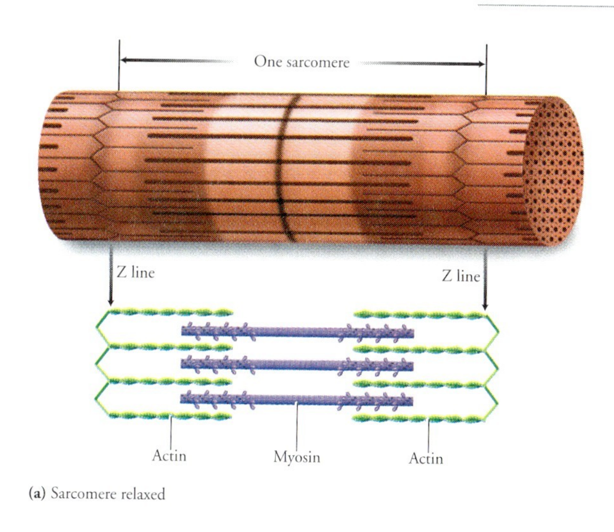

(a) Sarcomere relaxed

(b) Sarcomere contracted

Fig. 6.4

NOTES

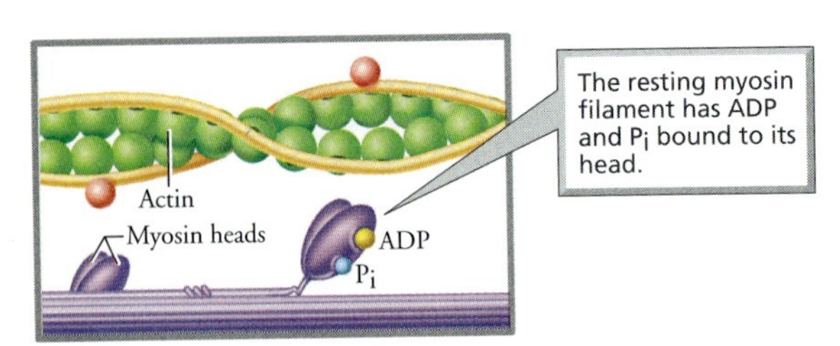

The resting myosin filament has ADP and P_i bound to its head.
Actin
Myosin heads
ADP
P_i

Resting sarcomere

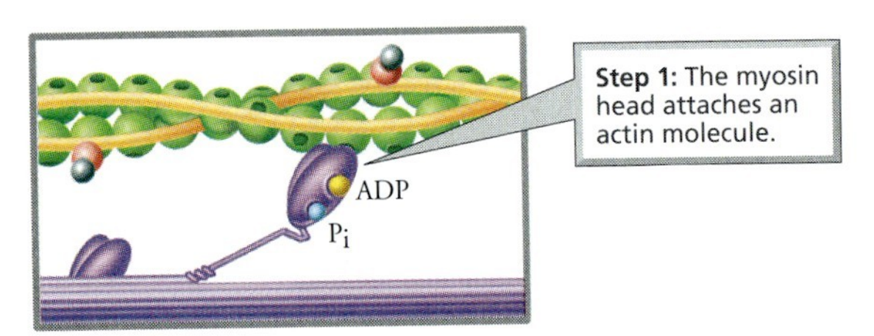

Step 1: The myosin head attaches an actin molecule.
ADP
P_i

Cross-bridge attachment

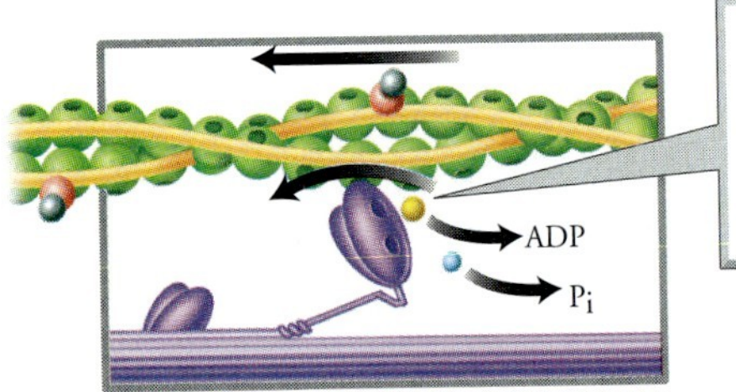

Step 2: The myosin head bends, causing the actin filament to slide across the myosin filament. The ADP and P_i are released from the myosin head.
ADP
P_i

Pivoting of myosin head

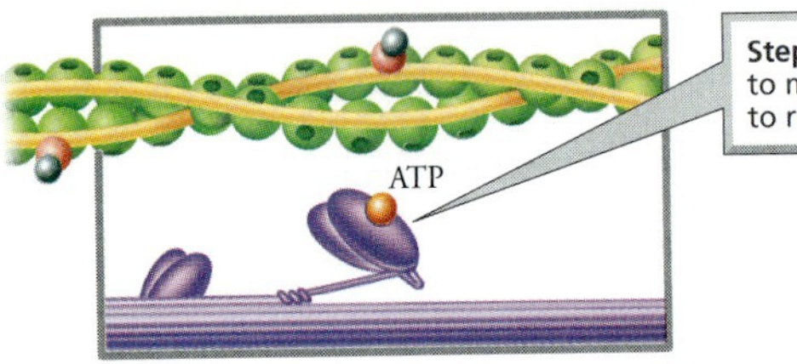

Step 3: Free ATP binds to myosin, causing it to release actin.
ATP

Cross-bridge detachment

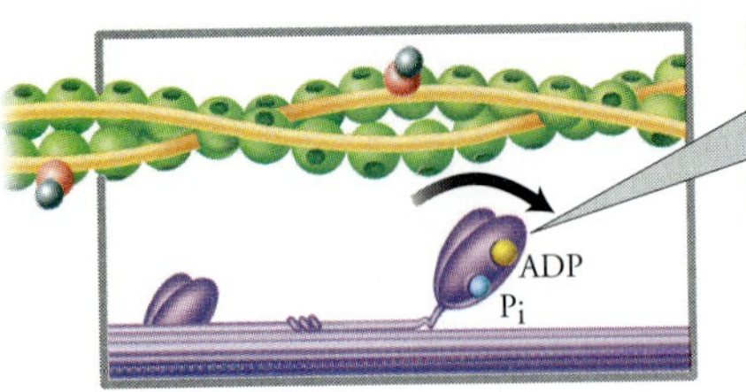

Step 4: ATP is split to ADP and P_i and myosin returns to its resting position.
ADP
P_i

Myosin reactivation

Fig. 6.5

NOTES

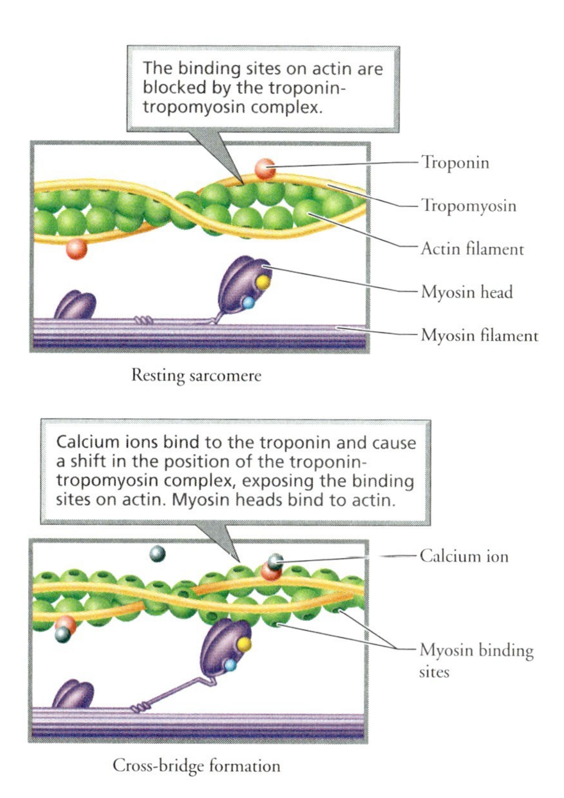
The binding sites on actin are blocked by the troponin-tropomyosin complex.
Troponin
Tropomyosin
Actin filament
Myosin head
Myosin filament
Resting sarcomere
Calcium ions bind to the troponin and cause a shift in the position of the troponin-tropomyosin complex, exposing the binding sites on actin. Myosin heads bind to actin.
Calcium ion
Myosin binding sites
Cross-bridge formation

Fig. 6.6

NOTES

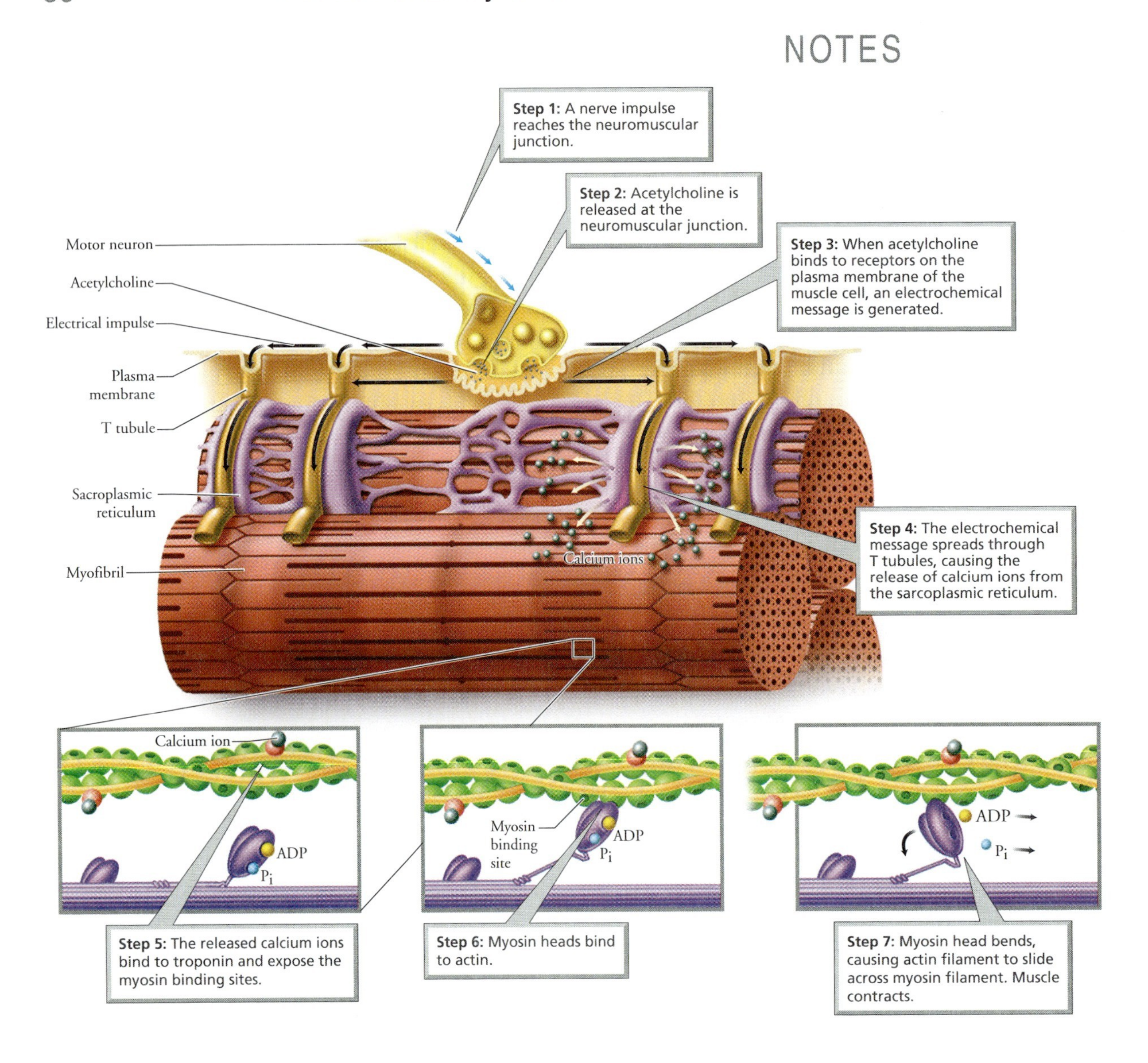
Step 1: A nerve impulse reaches the neuromuscular junction.
Step 2: Acetylcholine is released at the neuromuscular junction.
Step 3: When acetylcholine binds to receptors on the plasma membrane of the muscle cell, an electrochemical message is generated.
Motor neuron
Acetylcholine
Electrical impulse
Plasma membrane
T tubule
Sacroplasmic reticulum
Myofibril
Calcium ions
Step 4: The electrochemical message spreads through T tubules, causing the release of calcium ions from the sarcoplasmic reticulum.
Calcium ion
ADP
Pi
Myosin binding site
ADP
Pi
ADP
Pi
Step 5: The released calcium ions bind to troponin and expose the myosin binding sites.
Step 6: Myosin heads bind to actin.
Step 7: Myosin head bends, causing actin filament to slide across myosin filament. Muscle contracts.

Fig. 6.7

NOTES

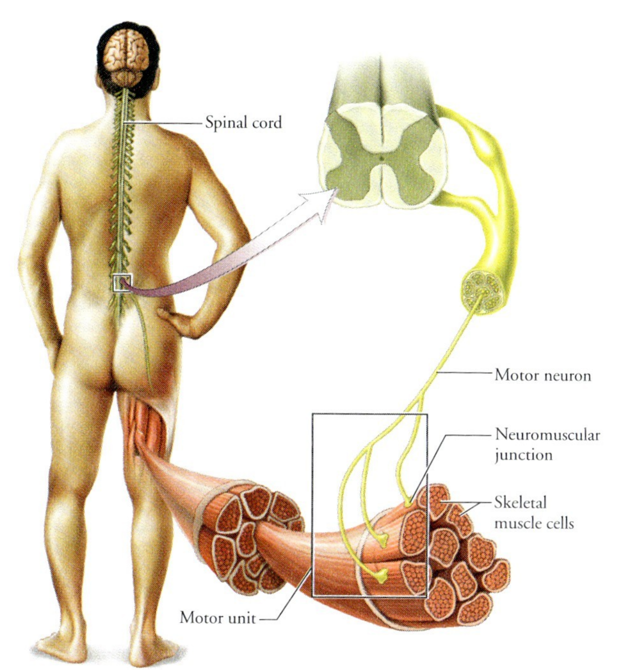

Fig. 6.8

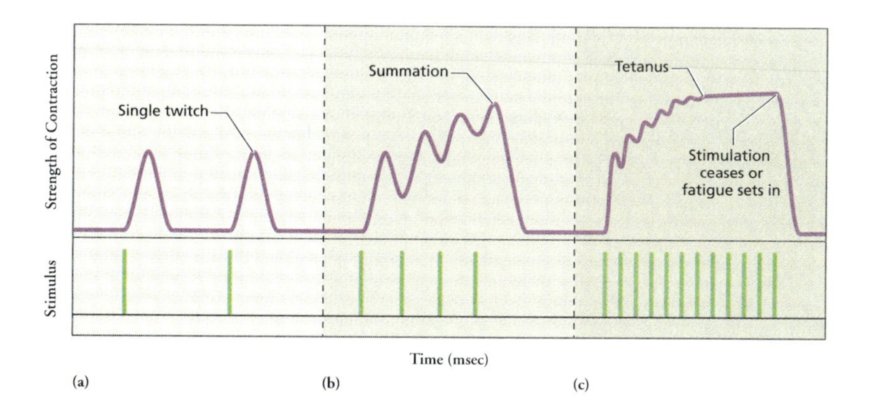

Fig. 6.9

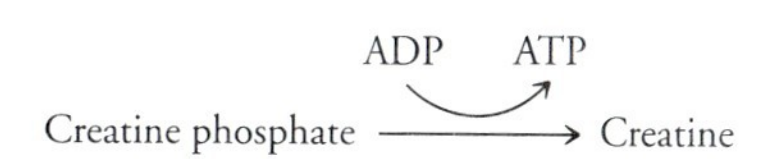

NOTES

6 seconds	25 seconds	10 minutes	End of exercise	After prolonged exercise
ATP stored in muscles	ATP formed from creatine phosphate and ADP	ATP generated from glycogen stored in muscles and broken down to form glucose		Oxygen debt paid back
		Oxygen limited • Glucose oxidized to lactic acid	Oxygen present • Heart beats faster to deliver oxygen more quickly • Myoglobin releases oxygen	Breathe heavily to deliver oxygen • Lactic acid used to produce ATP • Creatine phosphate restored • Oxygen restored to myoglobin • Glycogen reserves restored

Fig. 6.10

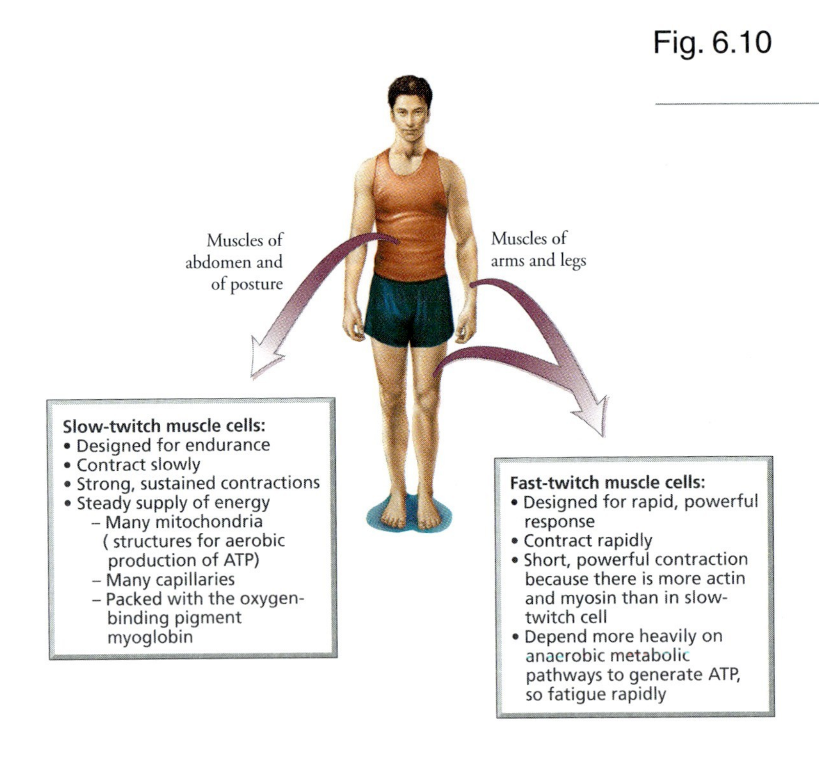
Muscles of abdomen and of posture
Muscles of arms and legs
Slow-twitch muscle cells:
• Designed for endurance
• Contract slowly
• Strong, sustained contractions
• Steady supply of energy
– Many mitochondria (structures for aerobic production of ATP)
– Many capillaries
– Packed with the oxygen-binding pigment myoglobin
Fast-twitch muscle cells:
• Designed for rapid, powerful response
• Contract rapidly
• Short, powerful contraction because there is more actin and myosin than in slow-twitch cell
• Depend more heavily on anaerobic metabolic pathways to generate ATP, so fatigue rapidly

Fig. 6.11

NOTES

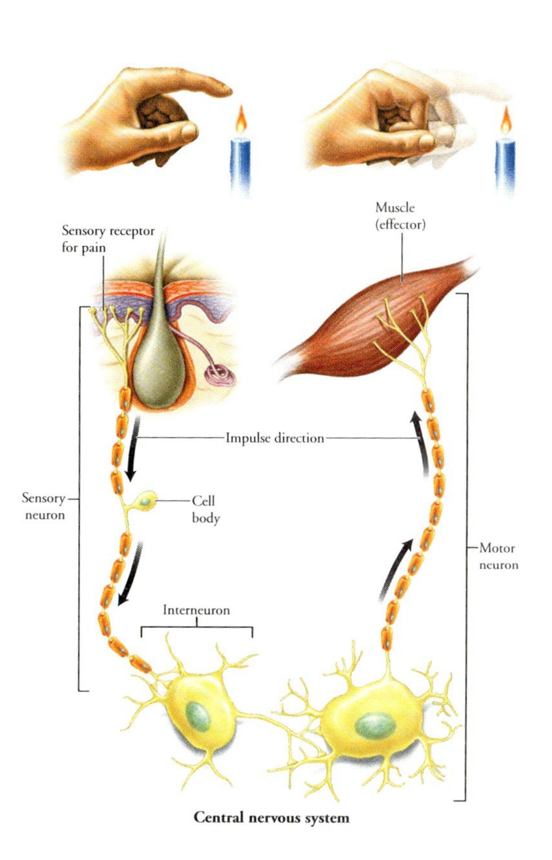

Fig. 7.1

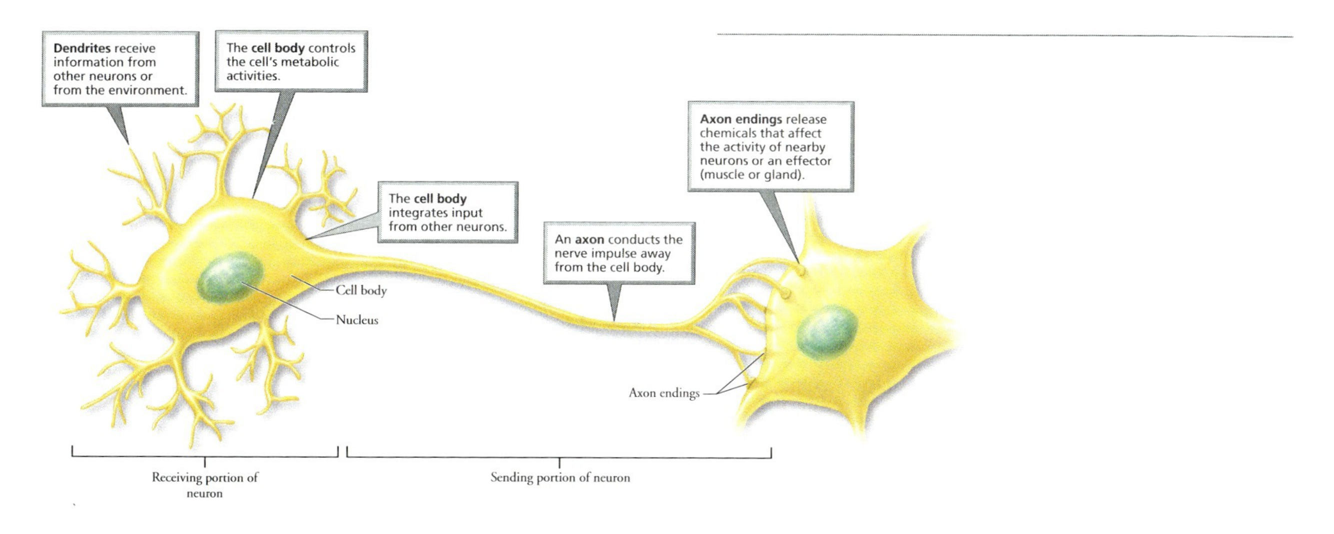

Fig. 7.2

NOTES

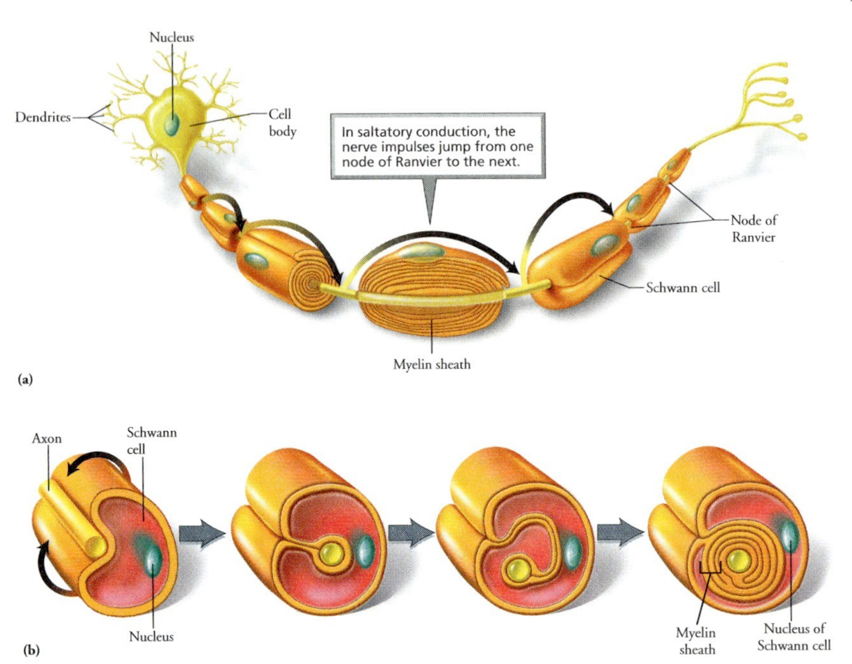

Fig. 7.3

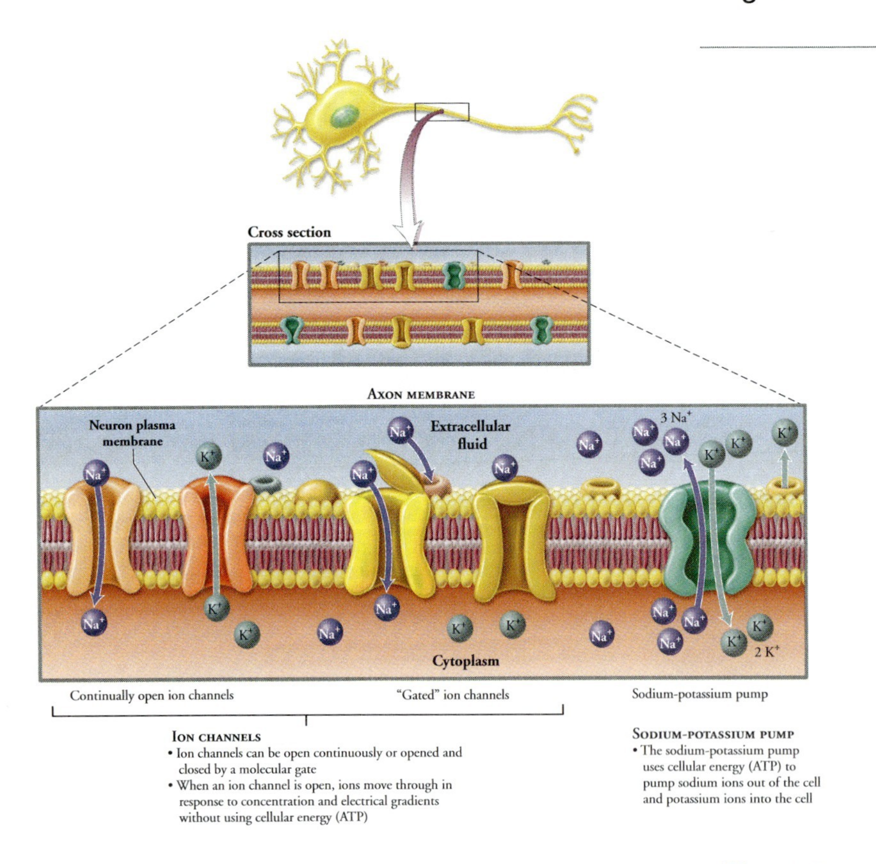

Fig. 7.4

NOTES

Resting Neuron
Plasma membrane is charged, with the inside negative relative to the outside.

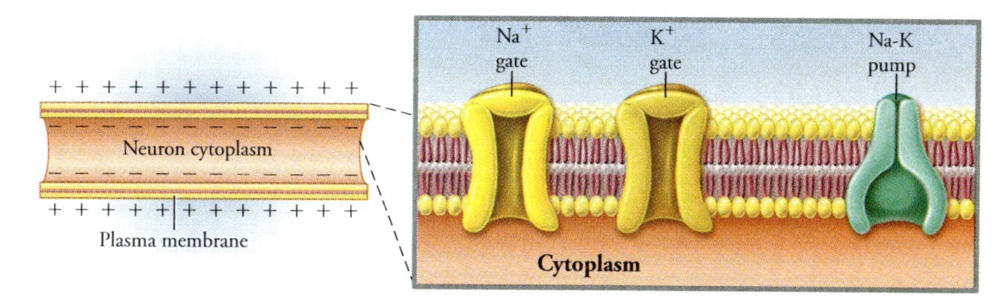

Action Potential
The charge difference across the membrane reverses and then is restored.

Step 1: The loss of the charge difference across the membrane (depolarization) occurs as sodium ions (Na^+) enter the axon. The inside of the membrane becomes positively charged.

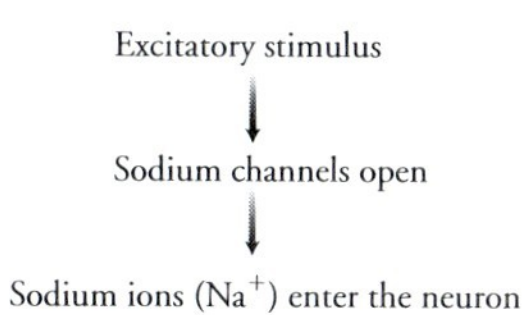

Step 2: The return of the membrane potential to near its resting value (repolarization) occurs as potassium (K^+) ions leave the axon.

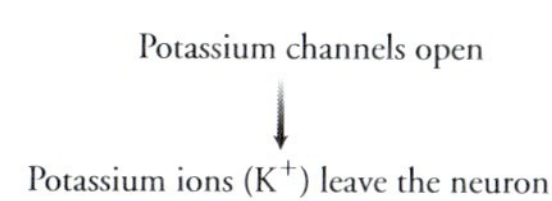

Charge restoration
The sodium-potassium pump restores the original distribution of ions.

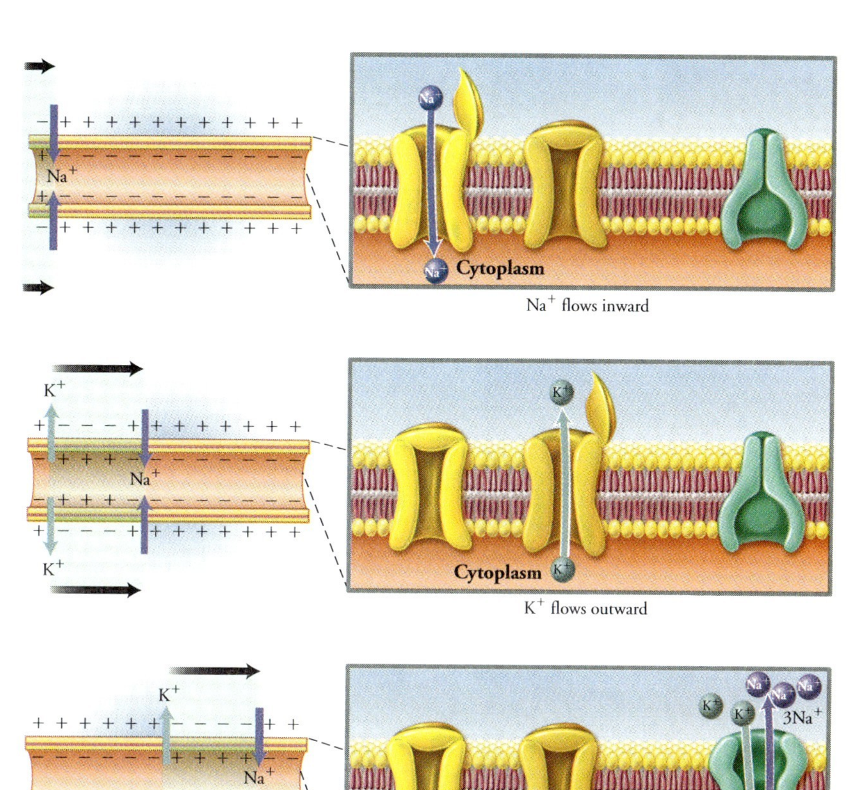

Fig. 7.5

NOTES

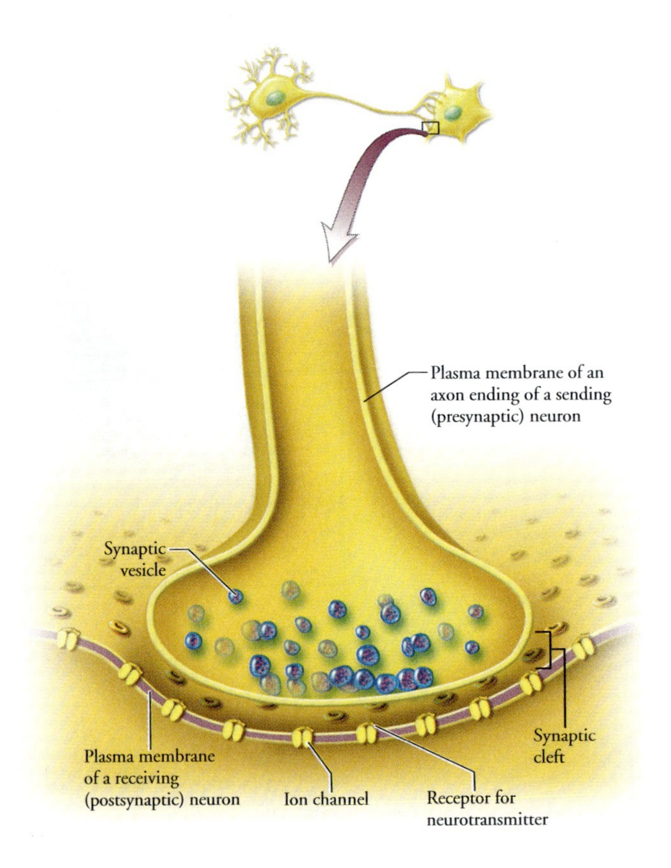
Plasma membrane of an
axon ending of a sending
(presynaptic) neuron
Synaptic
vesicle
Synaptic
cleft
Plasma membrane
of a receiving
(postsynaptic) neuron
Ion channel
Receptor for
neurotransmitter

Fig. 7.6

NOTES

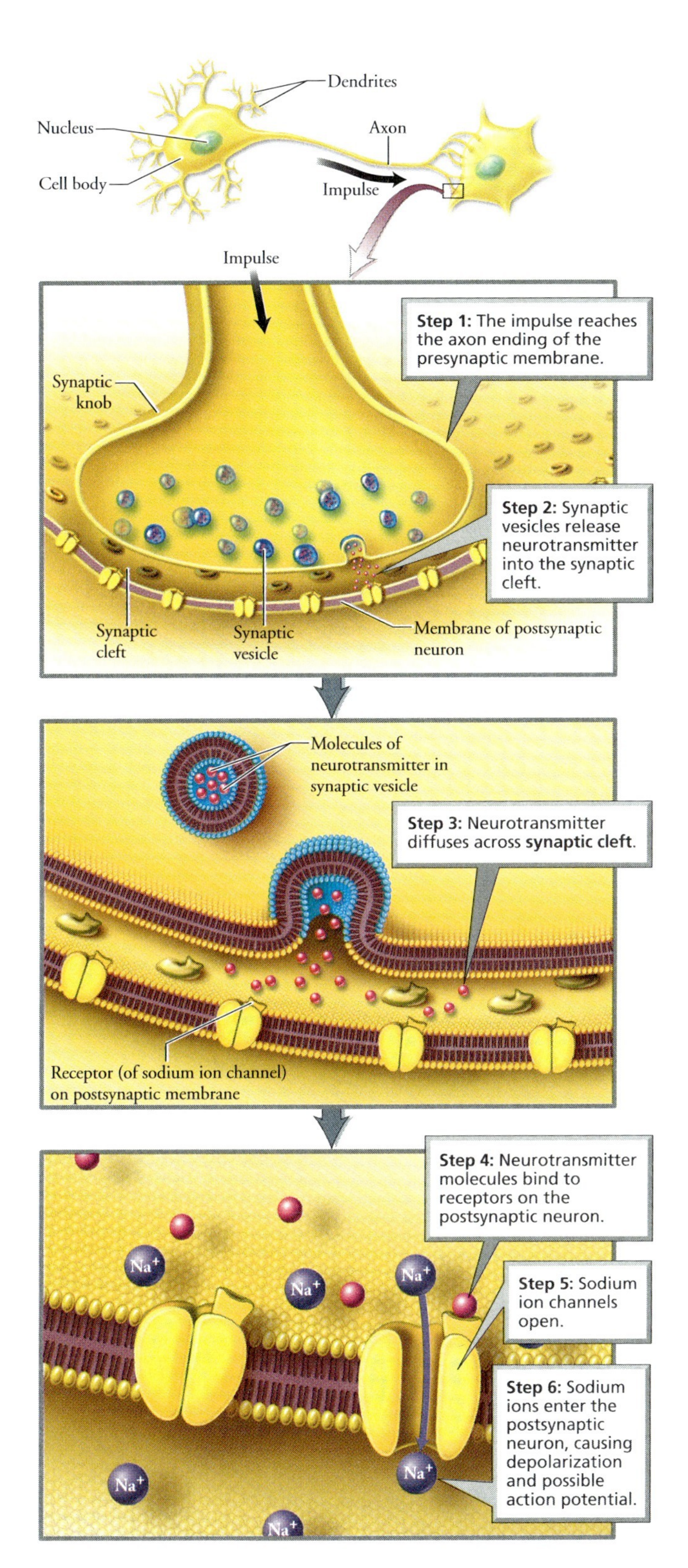
Dendrites
Nucleus
Axon
Cell body
Impulse
Impulse
Step 1: The impulse reaches the axon ending of the presynaptic membrane.
Synaptic knob
Step 2: Synaptic vesicles release neurotransmitter into the synaptic cleft.
Synaptic cleft
Synaptic vesicle
Membrane of postsynaptic neuron
Molecules of neurotransmitter in synaptic vesicle
Step 3: Neurotransmitter diffuses across synaptic cleft.
Receptor (of sodium ion channel) on postsynaptic membrane
Step 4: Neurotransmitter molecules bind to receptors on the postsynaptic neuron.
Na+
Step 5: Sodium ion channels open.
Step 6: Sodium ions enter the postsynaptic neuron, causing depolarization and possible action potential.

Fig. 7.7

NOTES

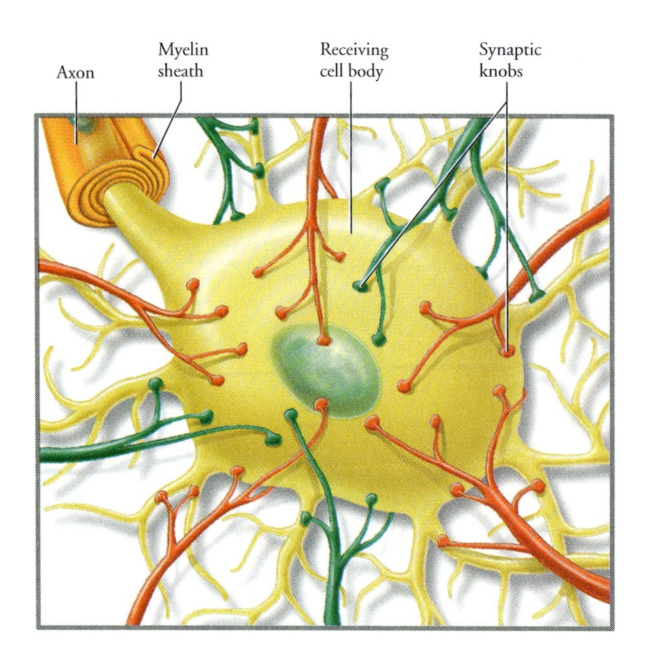

Fig. 7.8

NOTES

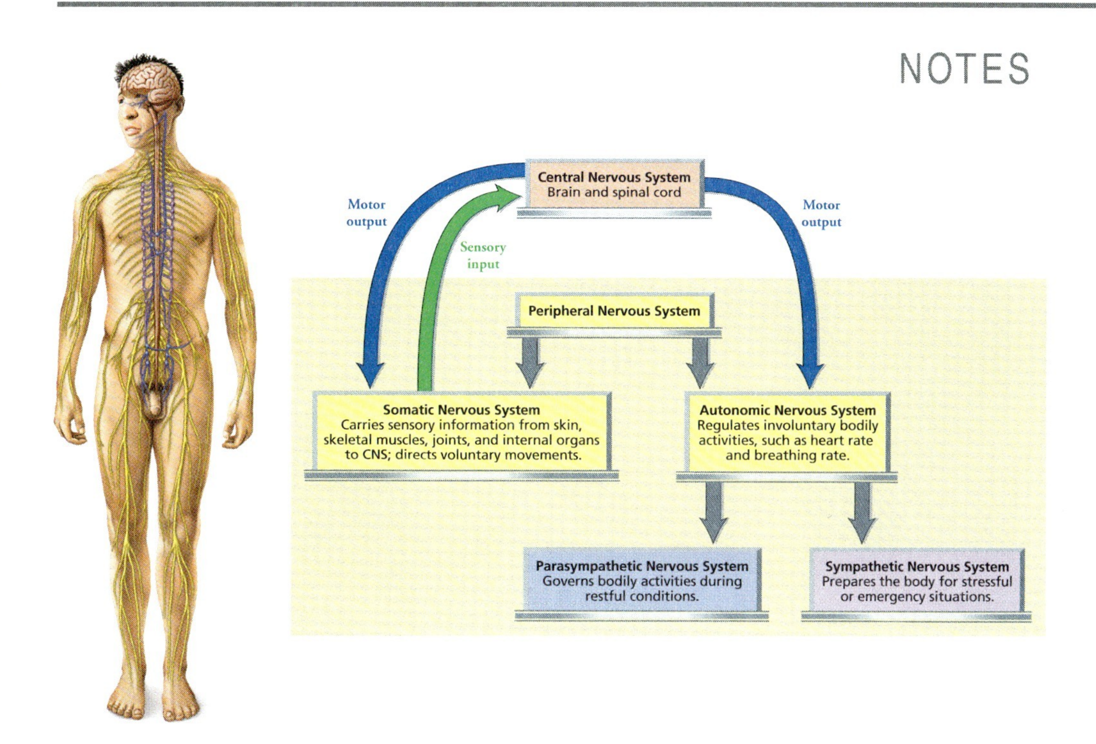

Fig. 8.1

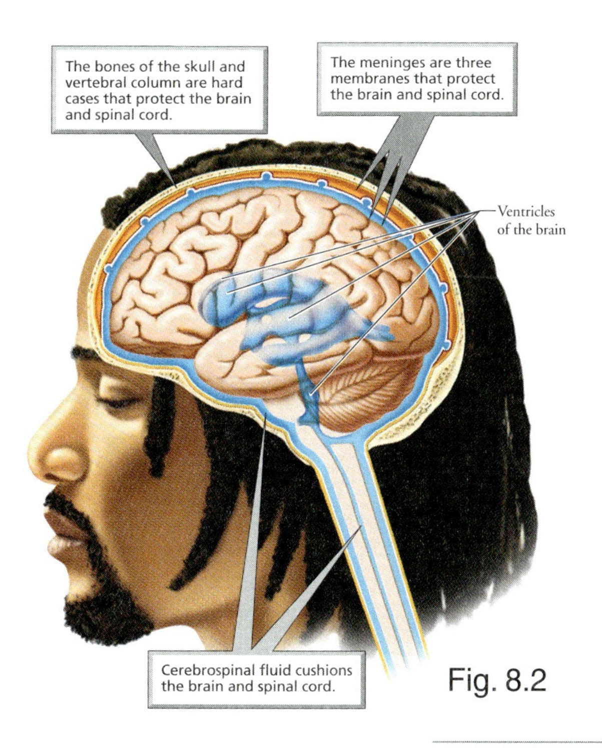

Fig. 8.2

NOTES

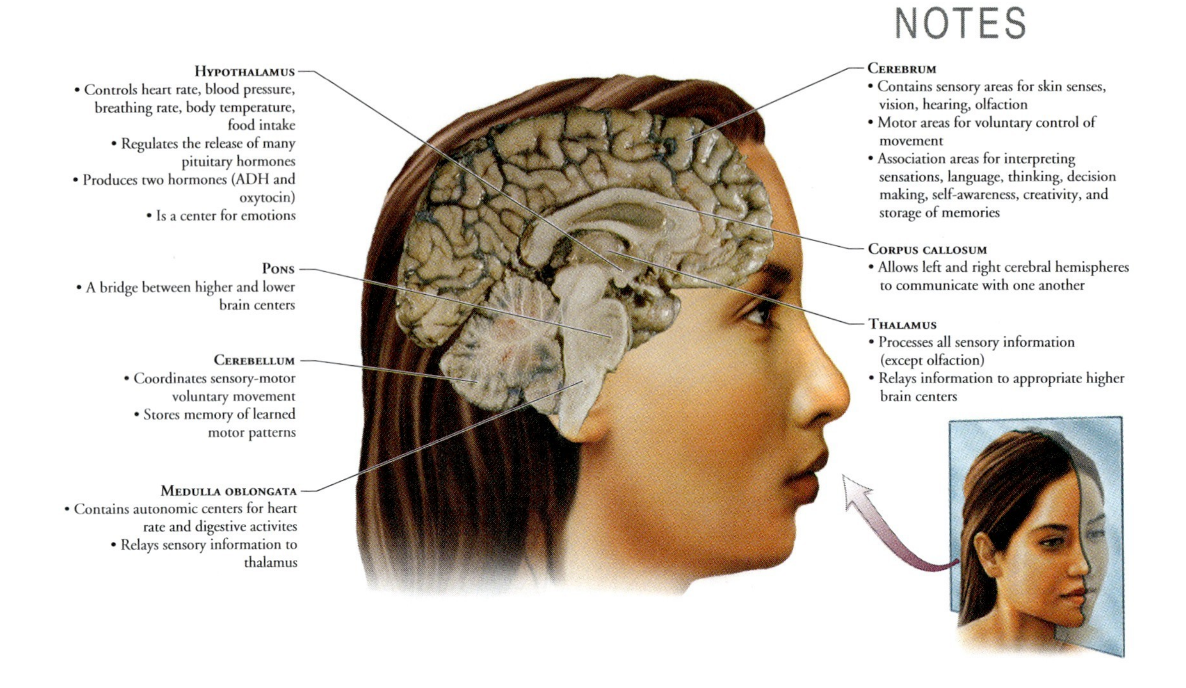

Fig. 8.3

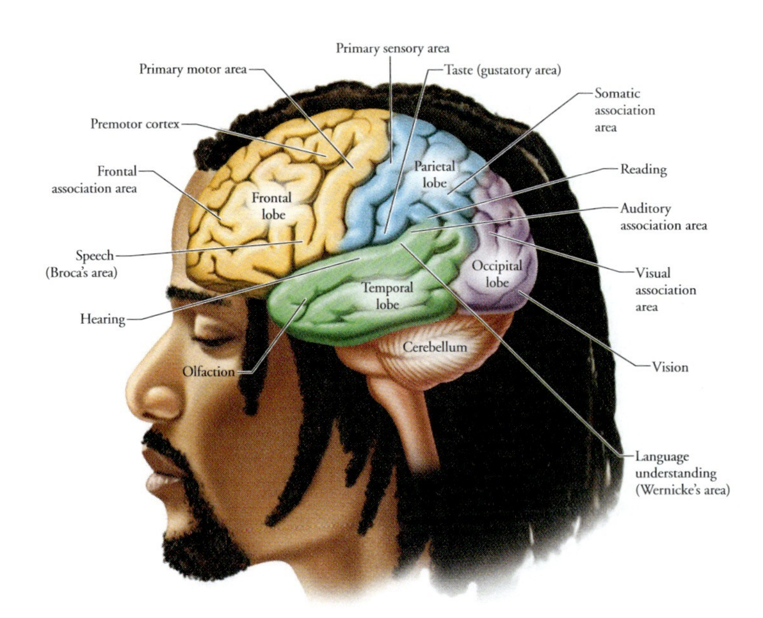

Fig. 8.6

NOTES

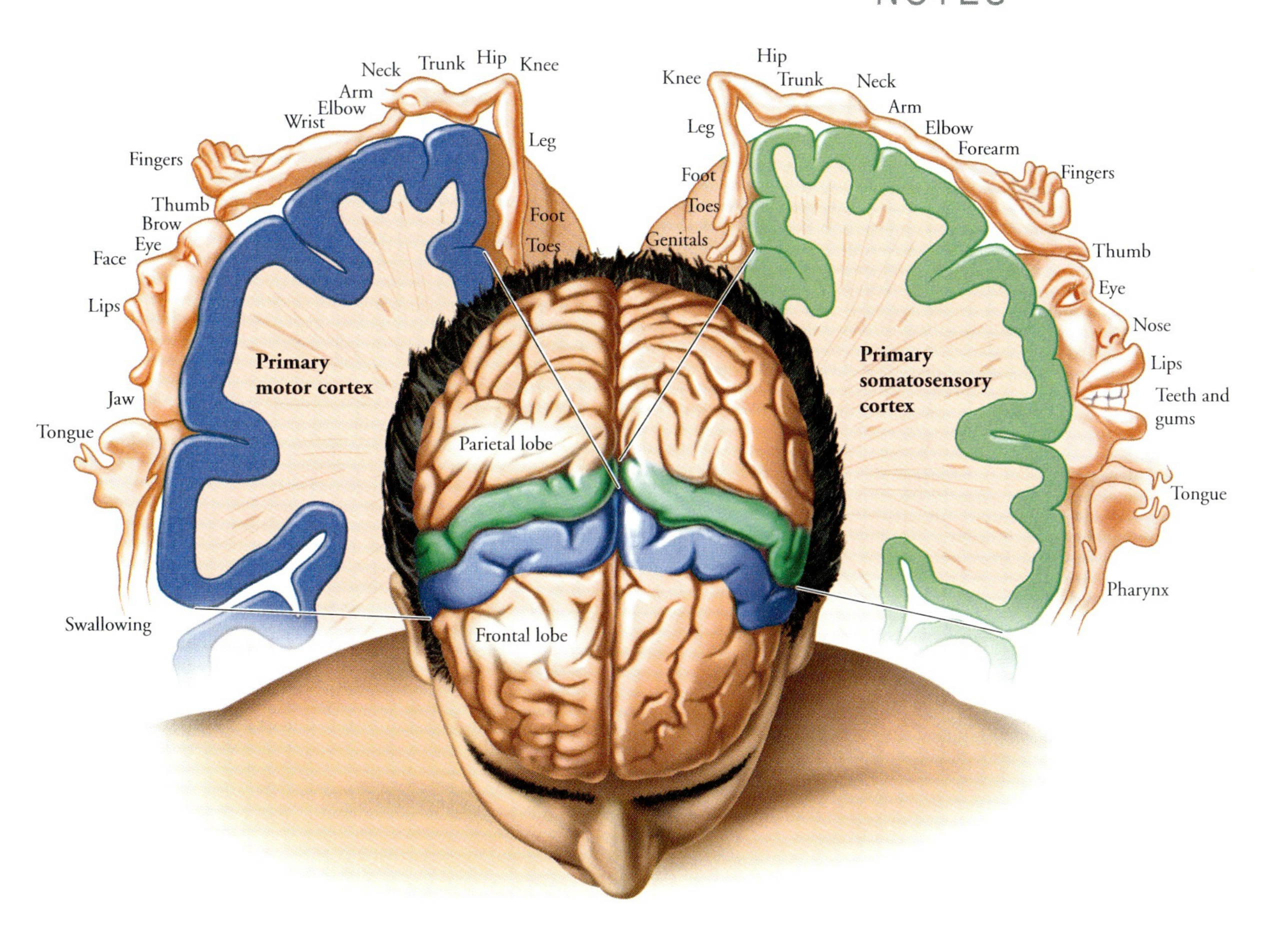
Neck
Trunk
Hip
Knee
Arm
Elbow
Wrist
Leg
Fingers
Thumb
Brow
Eye
Face
Foot
Toes
Lips
Primary motor cortex
Jaw
Tongue
Swallowing
Knee
Hip
Trunk
Neck
Arm
Elbow
Forearm
Leg
Foot
Toes
Genitals
Fingers
Thumb
Eye
Nose
Primary somatosensory cortex
Lips
Teeth and gums
Tongue
Pharynx
Parietal lobe
Frontal lobe

Fig. 8.7

NOTES

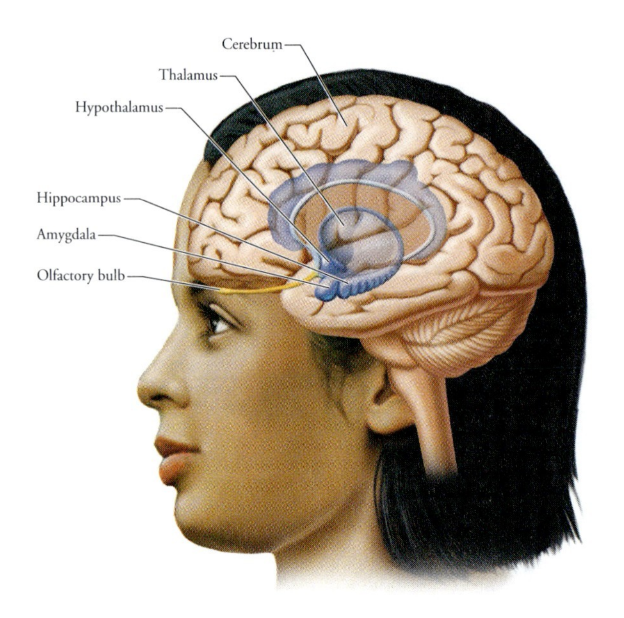

Fig. 8.8

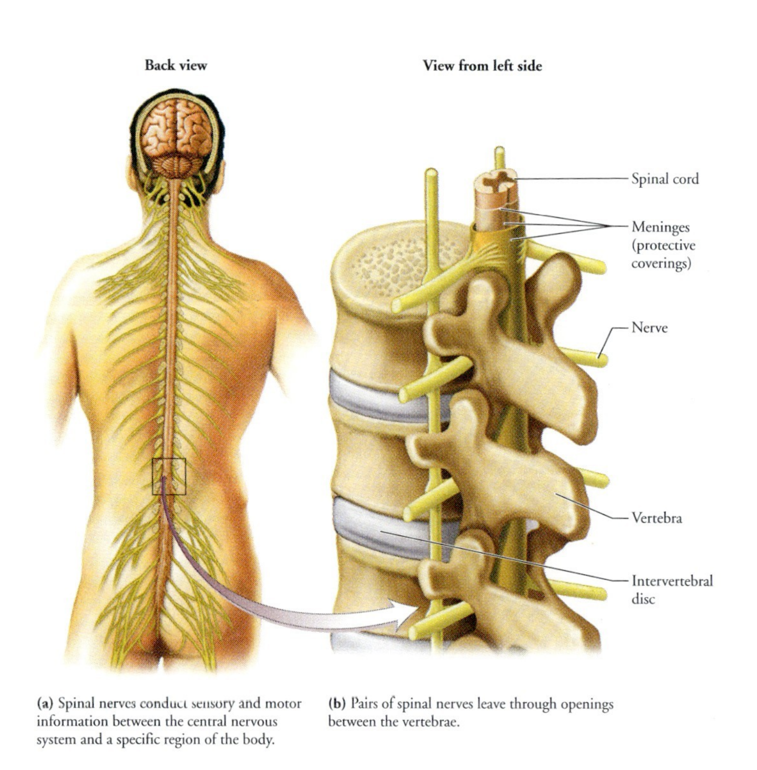

(a) Spinal nerves conduct sensory and motor information between the central nervous system and a specific region of the body.

(b) Pairs of spinal nerves leave through openings between the vertebrae.

Fig. 8.9

NOTES

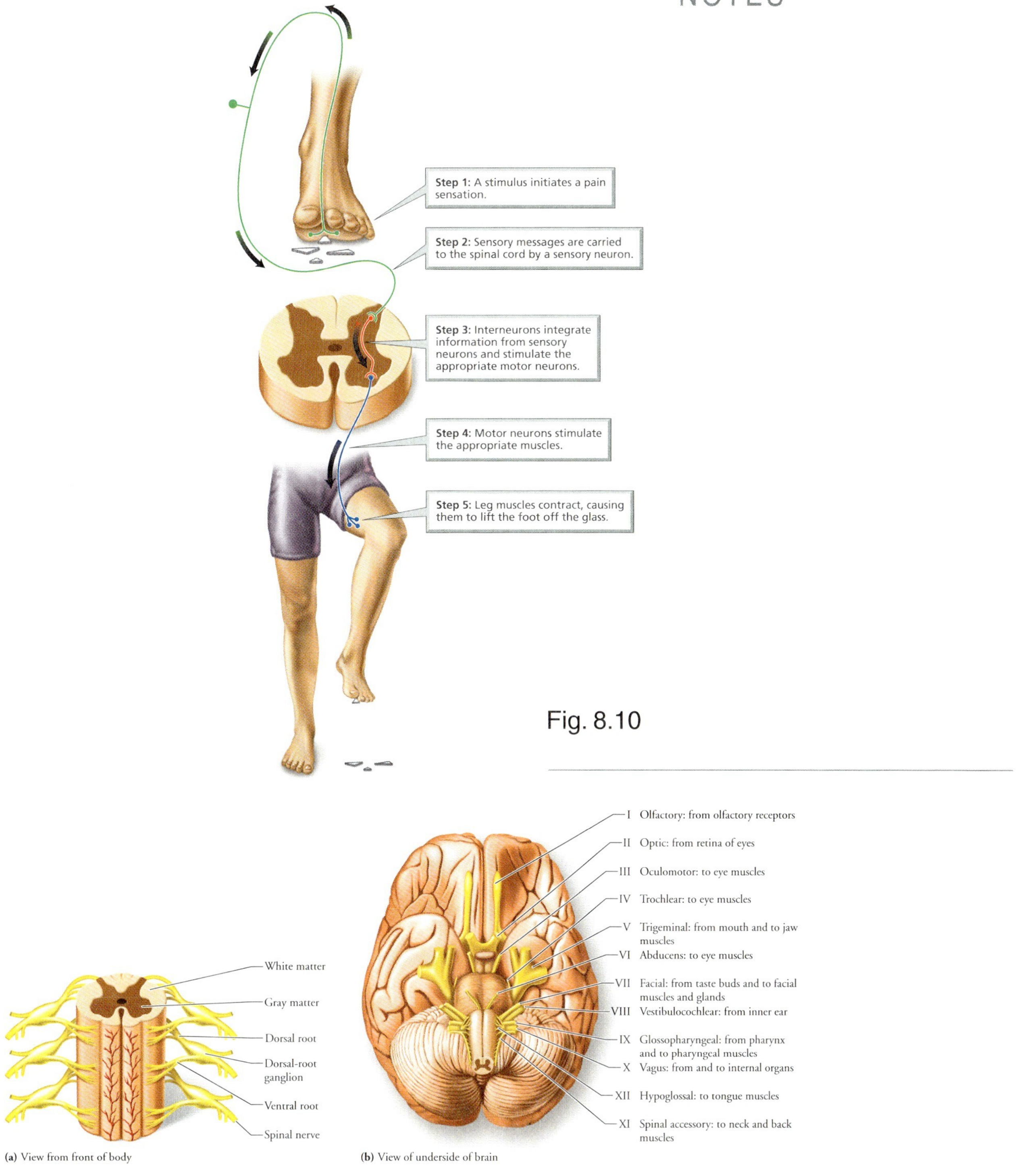

Fig. 8.10

(a) View from front of body

(b) View of underside of brain

Fig. 8.11

NOTES

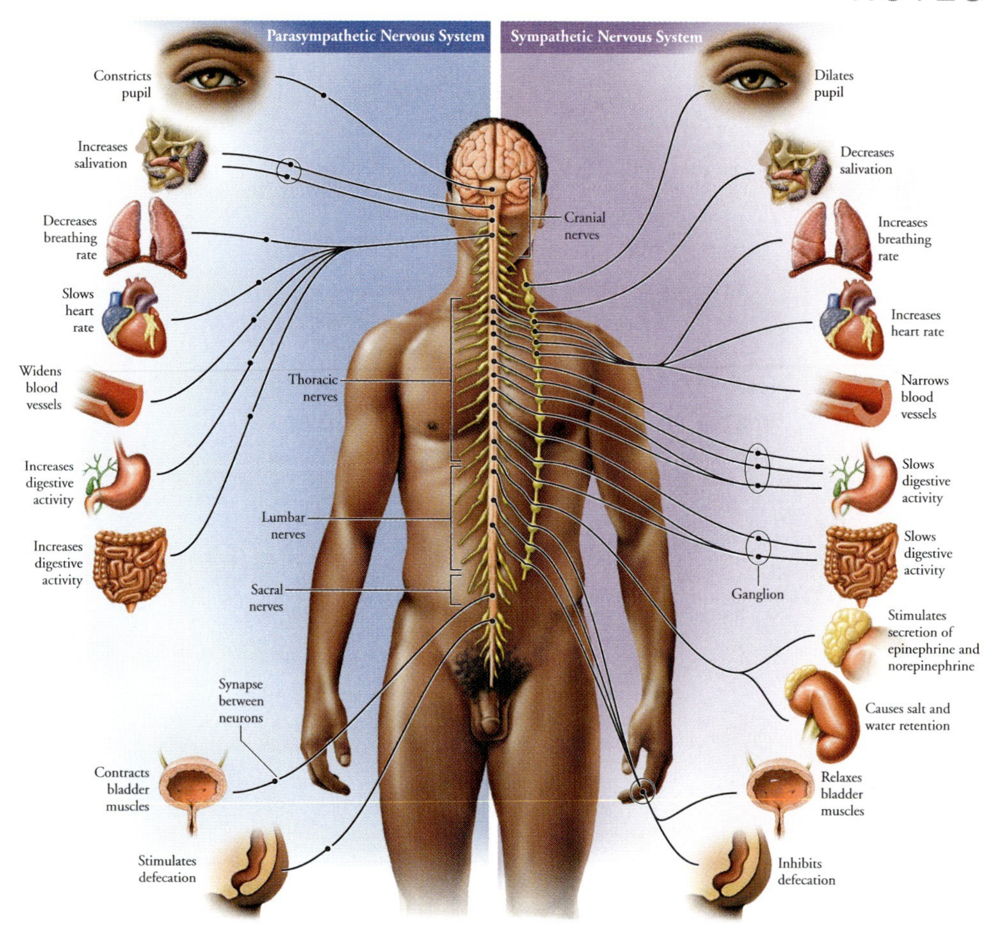

Fig. 8.12

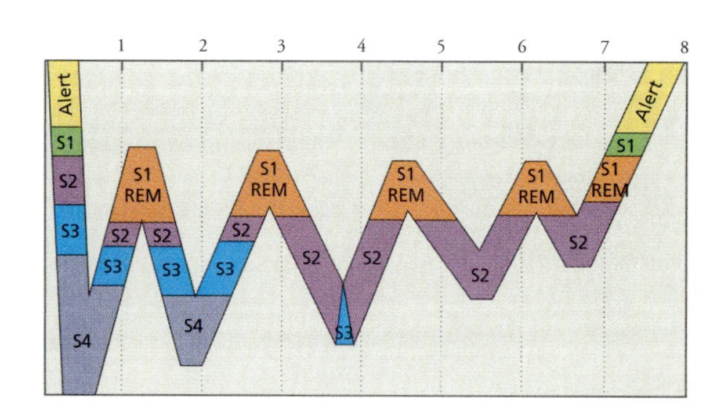

Fig. 8.A

NOTES

Neurotransmitter
Receptor
Cell membrane of receiving neuron
Produces effect

(a) The natural sequence of events: molecules of neurotransmitter released by one neuron diffuse across a gap and fit into receptors on the membrane of a receiving neuron, causing a response.

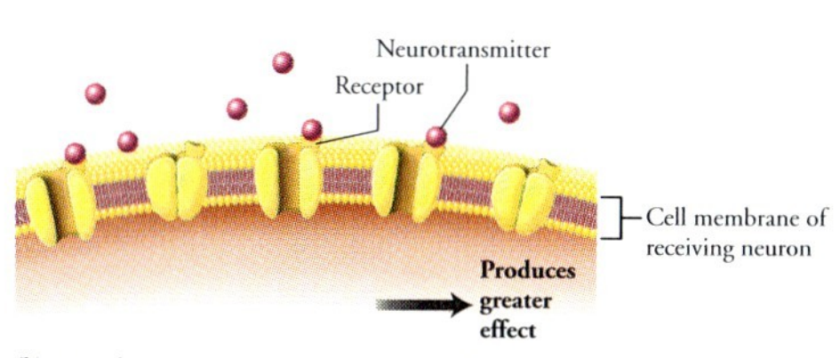

(b) A psychoactive drug may increase the number of neurotransmitter molecules released, increasing the response of the receiving neuron.

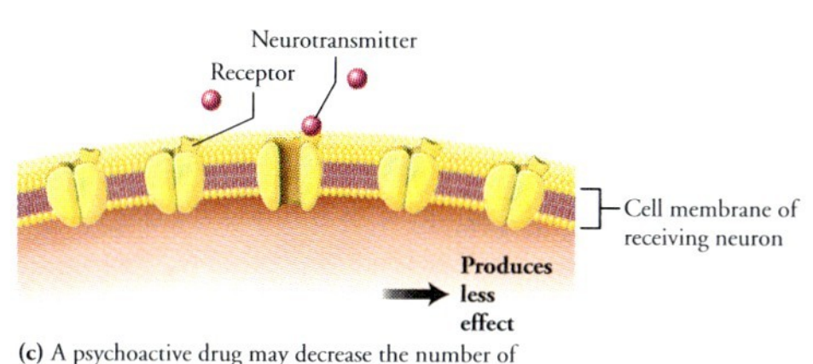

(c) A psychoactive drug may decrease the number of neurotransmitter molecules released, decreasing the response of the receiving neuron.

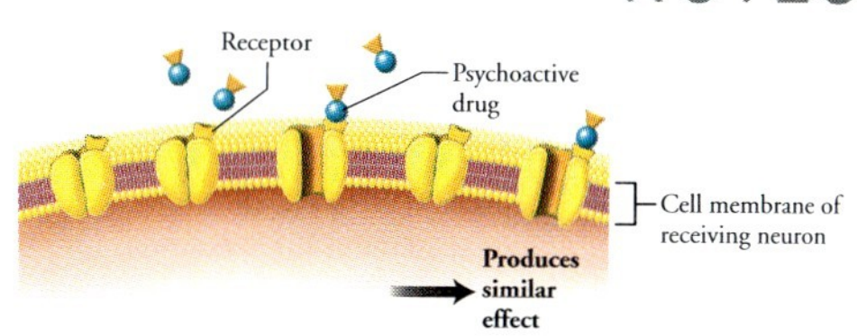

(d) A psychoactive drug may fit into the receptors for a neurotransmitter, causing a similar response by the receiving neuron.

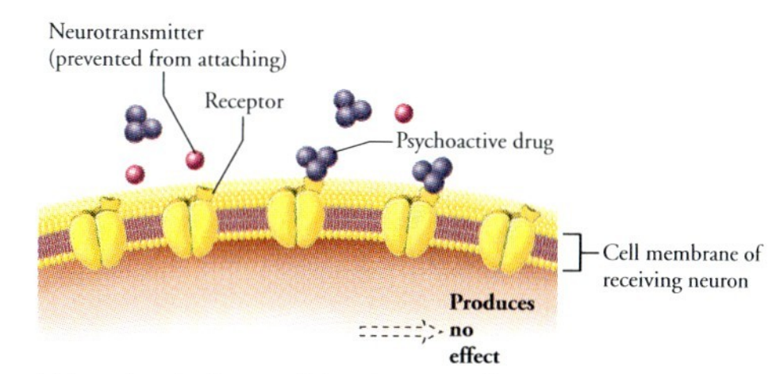

(e) A psychoactive drug may fit into the receptors for a neurotransmitter and prevent the neurotransmitter from entering the receptor, blocking the response by the receiving neuron.

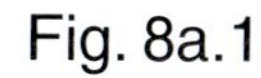
Fig. 8a.1

The Alcoholic Content of Selected Beverages

ALCOHOLIC BEVERAGE	PERCENT ALCOHOL (ETHANOL)
Light beer	4.5
Most beer	~5
Flavored malt beverages (coolers)	5
Ice beer	5.5–5.9
Dark beer (stout, porter, bock)	6–7
Malt liquor	8
French and German wines	8.5–10
American wine (most)	12–14
Sherry and port	18–21
Distilled liquor (vodka, gin, scotch, whiskey, rum, brandy, cognac)	Most 40% (80 proof); some 45% (90 proof) or 50% (100 proof)

Table 8a.1

NOTES

12 oz Most beer	5 oz Wine (12% alcohol)	1.5 oz Distilled spirits (80 proof gin, whisky, vodka, scotch)	8 oz Mixer Carbonated beverage	6 oz Mixer Fruit juice
1 bottle or can	1 glass	1 jigger	1 glass	1 glass
140–150 calories	110–200 calories	~100 calories	~70–120 calories	35–105 calories

Fig. 8a.2

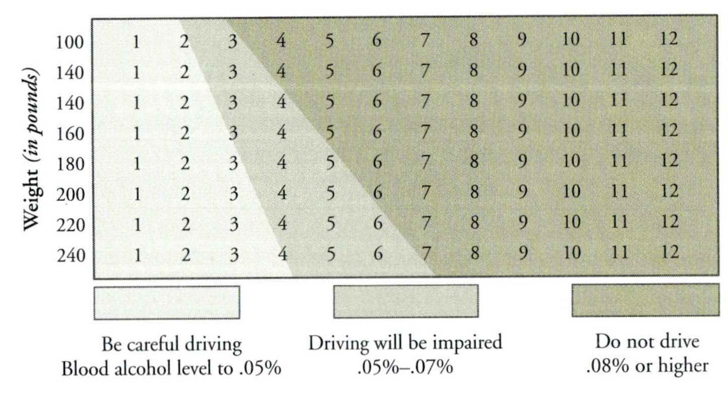

Fig. 8a.3

NOTES

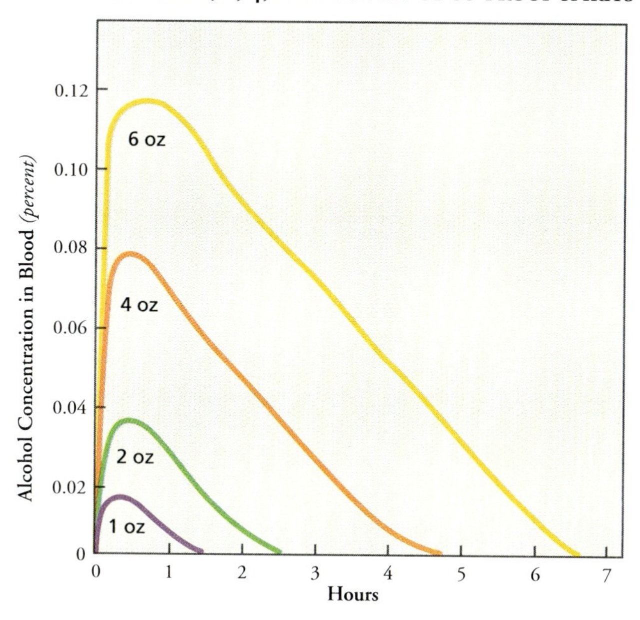

Source: Reproduced with permission from *The Encyclopedia Britannica*, © 2004 by Encyclopedia Britannica, Inc.

Fig. 8a.4

Caffeine Content in Selected Beverages and Foods

ITEM	CAFFEINE CONTENT (mg)
Coffee (5-oz cup)	50–150
Tea, major U.S. brands (5-oz cup)	20–90
Tea, imported brands (5-oz cup)	25–100
Cola soft drink (12 oz)	35–55
Dark chocolate (semisweet) (1 oz)	15–30
Milk chocolate (1 oz)	1–10

Table 8a.2

NOTES

NOTES

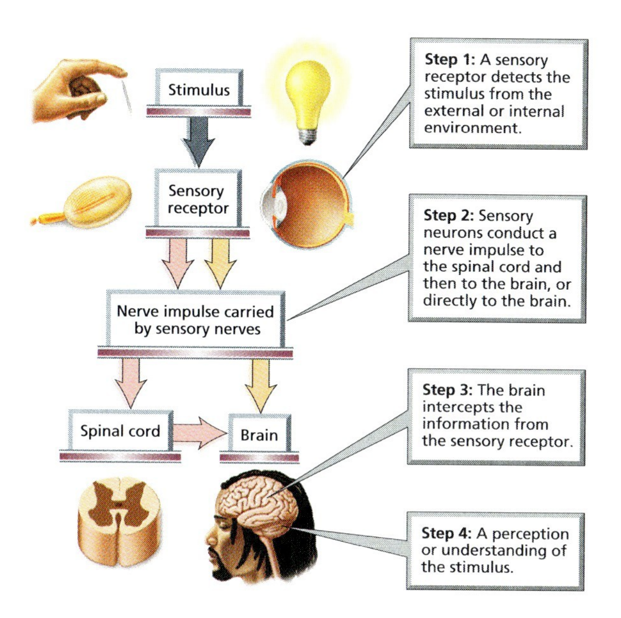

Fig. 9.1

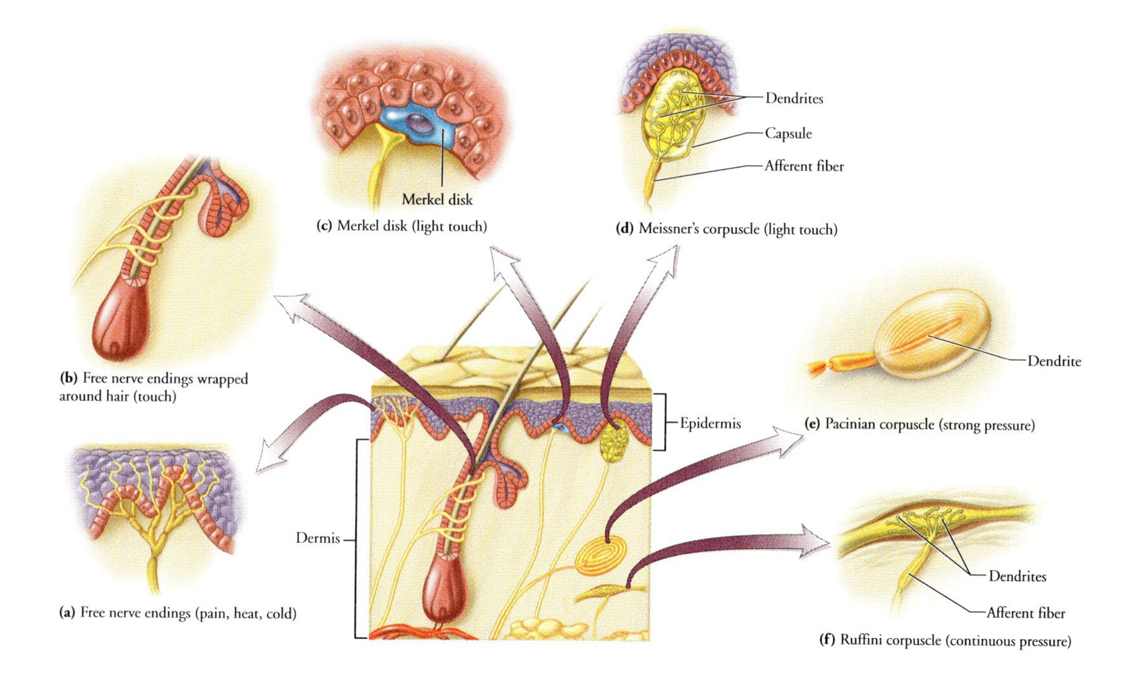

Fig. 9.2

NOTES

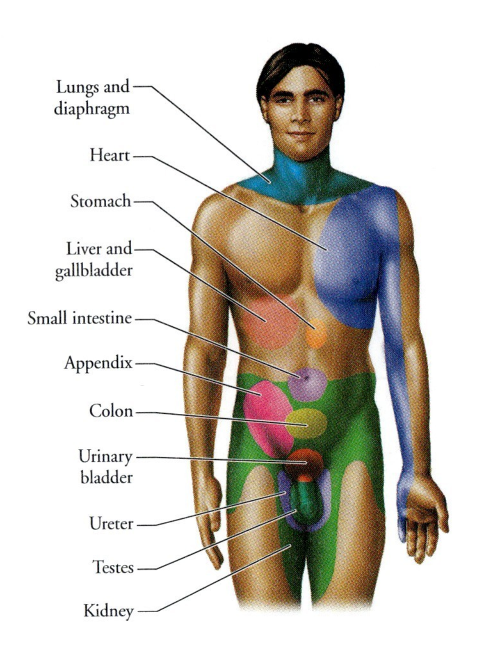

Fig. 9.3

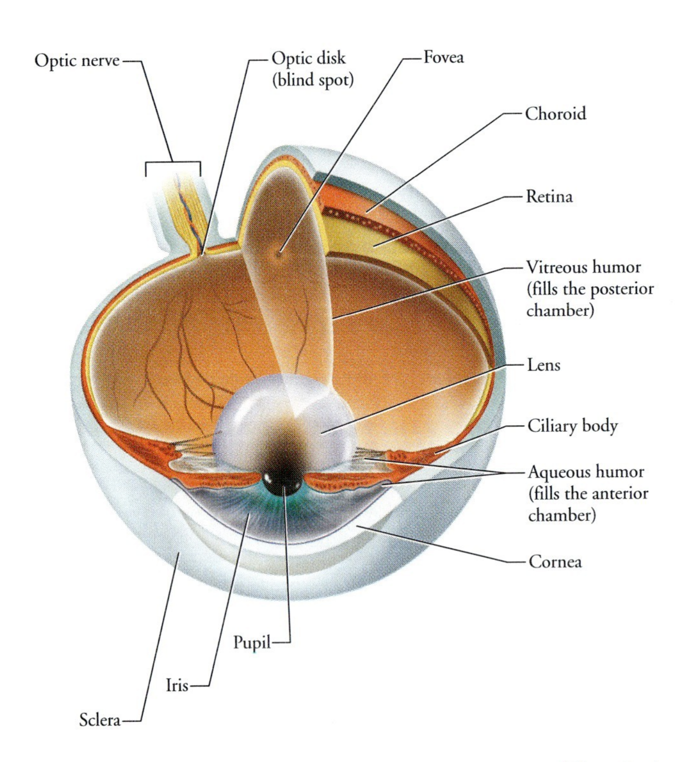

Fig. 9.4

NOTES

STRUCTURES OF THE EYE AND THEIR FUNCTIONS		
STRUCTURE	DESCRIPTION	FUNCTION
Outer layer		
Sclera	Outer layer of the eye	Protects the eyeball
Cornea	Transparent dome of tissue forming the outer layer at the front of the eye	Refracts light, focusing it on the retina
Middle layer		
Choroid	Pigmented layer containing blood vessels	Absorbs stray light; delivers nutrients and oxygen to tissues of eye
Ciliary body	Encircles lens; contains the ciliary muscles	Controls shape of lens; secretes aqueous humor
Iris	Colored part of the eye	Regulates the amount of light entering the eye through the pupil
Pupil	Opening at the center of the iris	Opening for incoming light
Inner layer		
Retina	Layer of tissue that contains the photoreceptors (rods and cones); also contains bipolar and ganglion cells involved in retinal processing	Receives light and generates neural messages
Rods	Photoreceptor	Responsible for black and white vision and vision in dim light
Cones	Photoreceptor	Responsible for color vision and visual acuity
Fovea	Small pit in the retina that has a high concentration of cones	Provides detailed color vision
Other structures of the eye		
Lens	Transparent, semispherical body of tissue behind the iris and pupil	Fine focusing of light onto retina
Aqueous humor	Clear fluid found between the cornea and the lens	Refracts light and helps maintain shape of the eyeball
Vitreous humor	Gelatinous substance found within the chamber behind the lens	Refracts light and helps maintain shape of the eyeball
Optic nerve	Group of axons from the eye to the brain	Transmits impulses from the retina to the brain

Table 9.1

NOTES

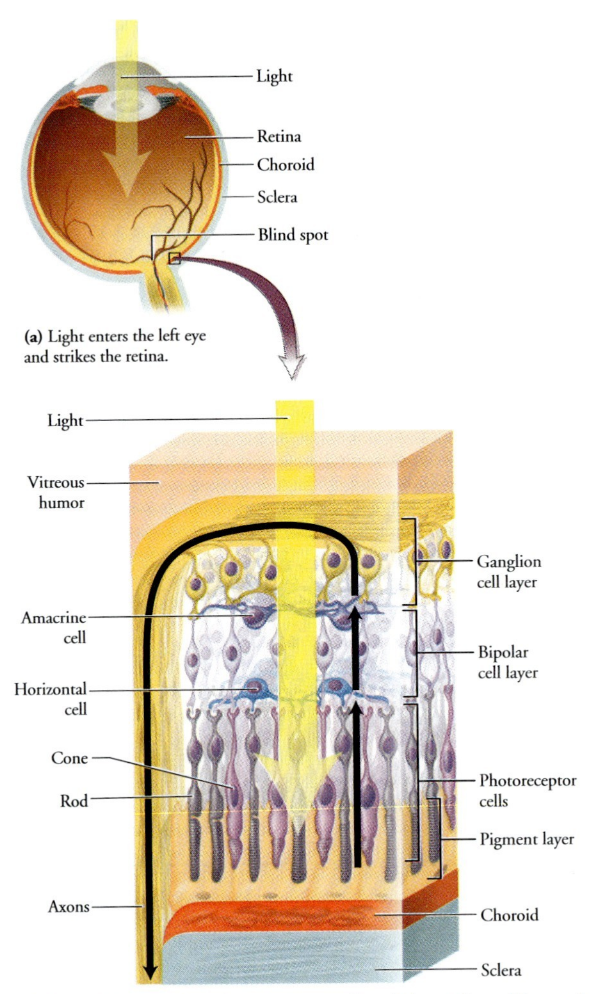

(a) Light enters the left eye and strikes the retina.

(b) When light is focused on the retina, it passes through the ganglion cell layer and bipolar cell layer before reaching the rods and cones. In response to light, the rods and cones generate electrical signals that are sent to bipolar cells and then to ganglion cells. These cells begin the processing of visual information.

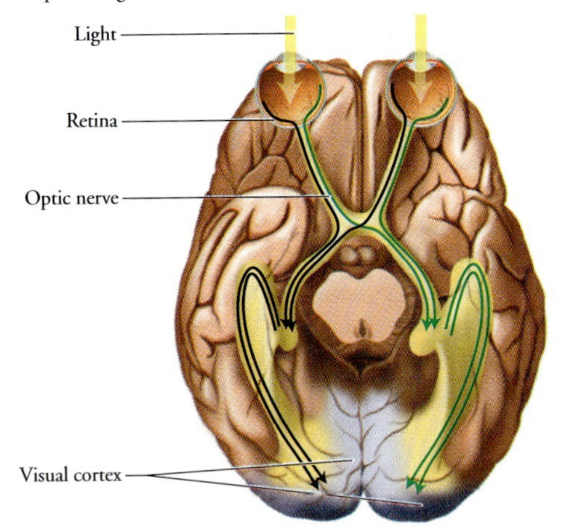

(c) The axons of the ganglion cells leave the eye at the blind spot, carrying nerve impulses to the brain (viewed from below) by means of the optic nerve.

Fig. 9.5

NOTES

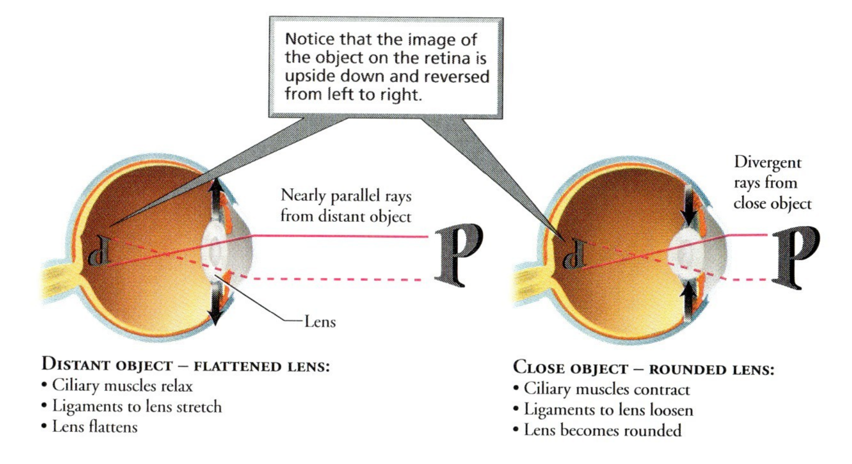

Fig. 9.6

PROCESSES IMPORTANT TO SHARP VISION

PROCESS	STRUCTURE(S) INVOLVED	RESULT
Refraction	Cornea	Produces most bending of light
Accommodation	Lens	Changes shape to further bend light and focus image on retina
Convergence	Extrinsic eye muscles	Directs eyes so that image falls on the fovea

Table 9.2

FOCUSING PROBLEMS

PROBLEM	DESCRIPTION	CAUSE	CORRECTION
Farsightedness	See distant objects more clearly than nearby objects	Eyeball too short or lens too thin; lens cannot become round enough	Convex lens; increases corneal curvature
Nearsightedness	See nearby objects more clearly than distant objects	Eyeball too long or lens too thick; lens cannot flatten enough	Concave lens; decreases corneal curvature
Astigmatism	Visual image is distorted	Irregularities in curvature of cornea or lens	Lenses that correct for the asymmetrical bending of light

Table 9.3

NOTES

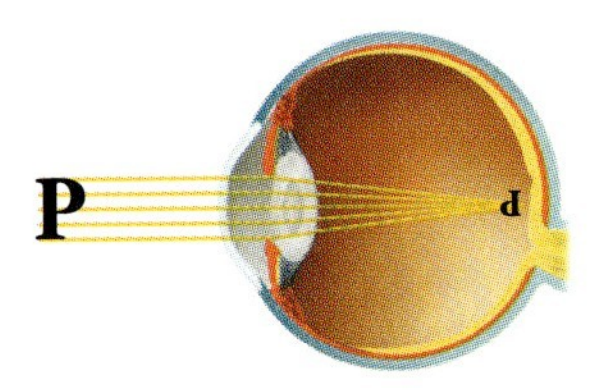

(a) Normal eye
- Close and distant object seen clearly
- Image focuses on retina

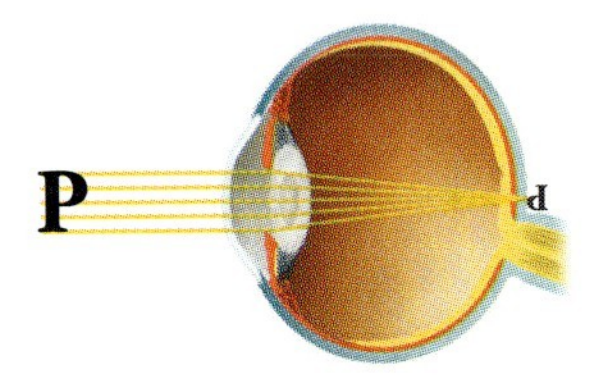

(b) Farsightedness
- Distant objects seen clearly
- Close objects out of focus
- Short eyeball causes image to focus behind retina

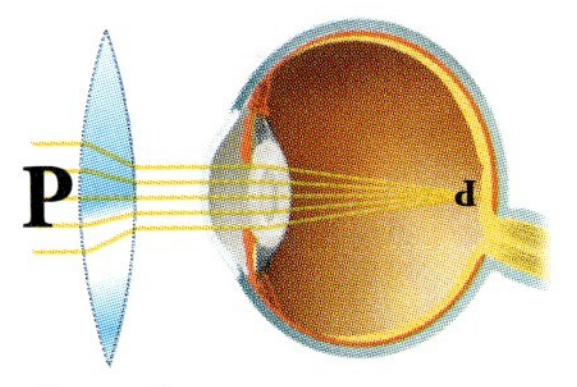

Convex lens
- Causes light rays to converge so that the image focuses on the retina

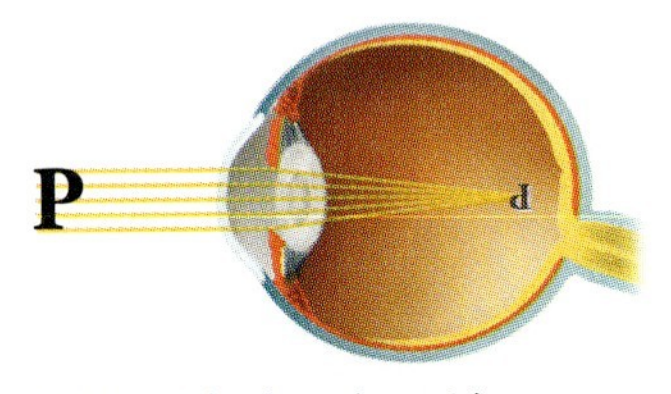

(c) Nearsightedness (myopia)
- Close objects seen clearly
- Distant objects out of focus
- Long eyeball causes image to focus in front of retina

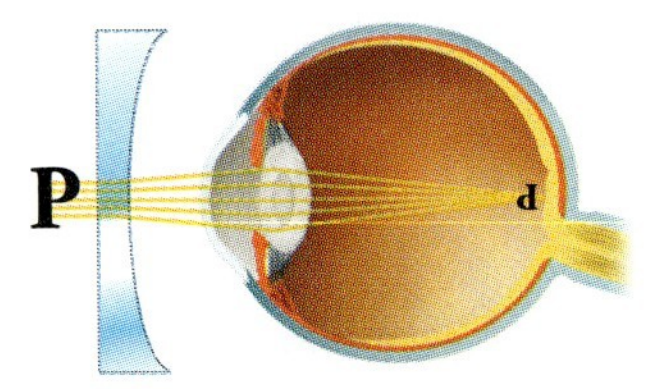

Concave lens
- Causes light rays to diverge so that the image focuses on the retina

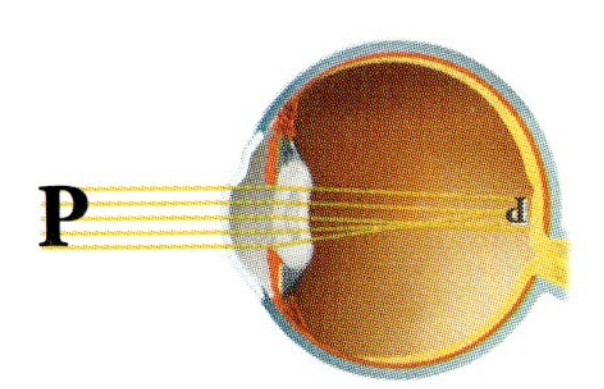

(d) Astigmatism
- Image blurred
- Irregular curvature of cornea or lens causes light rays to focus unevenly

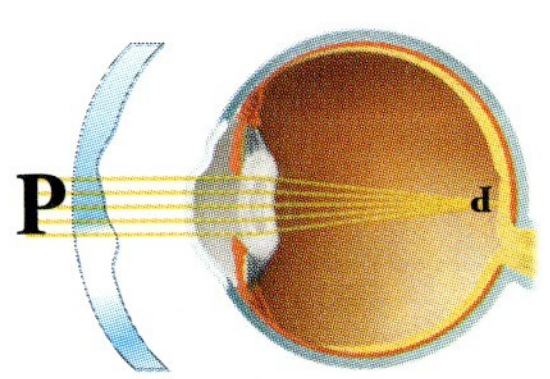

Uneven lens
- Focuses entire image on retina

Fig. 9.7

NOTES

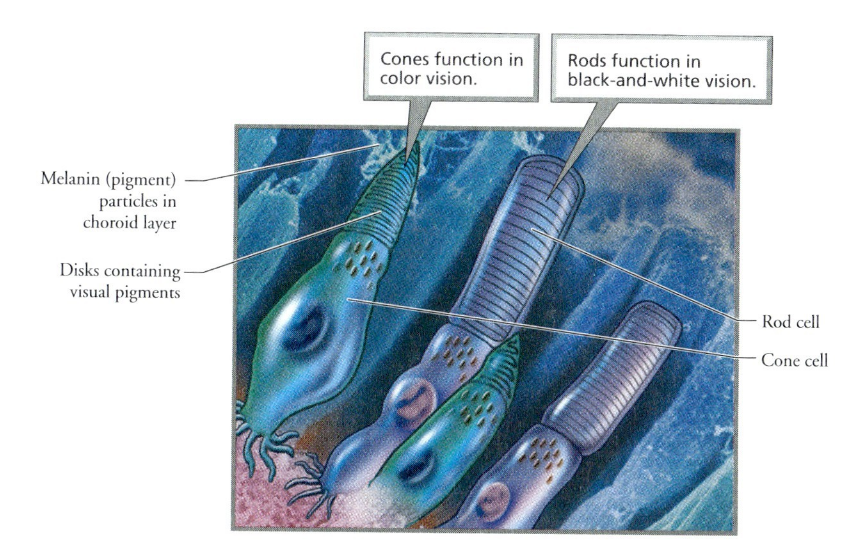

Fig. 9.8

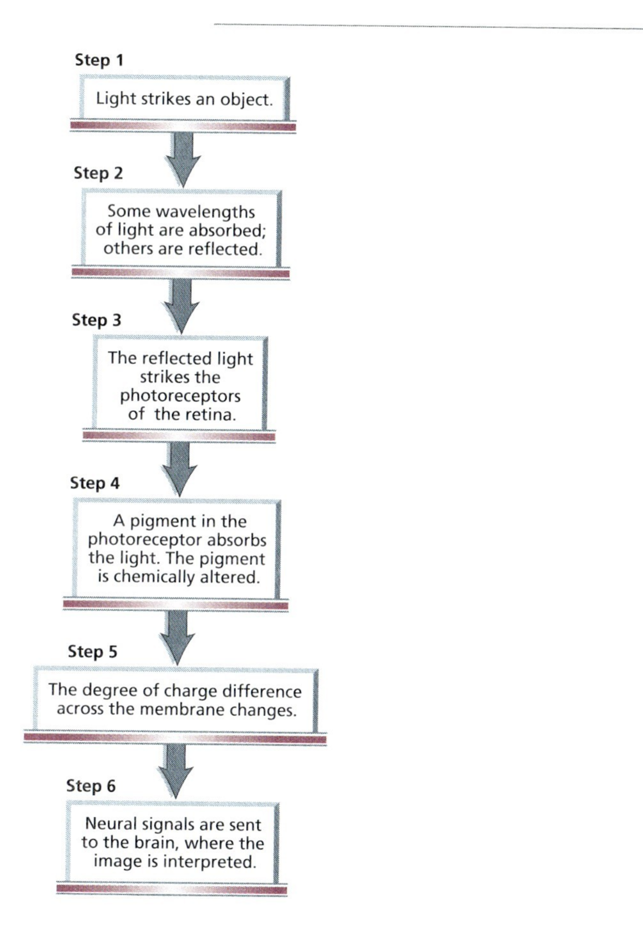

Fig. 9.9

NOTES

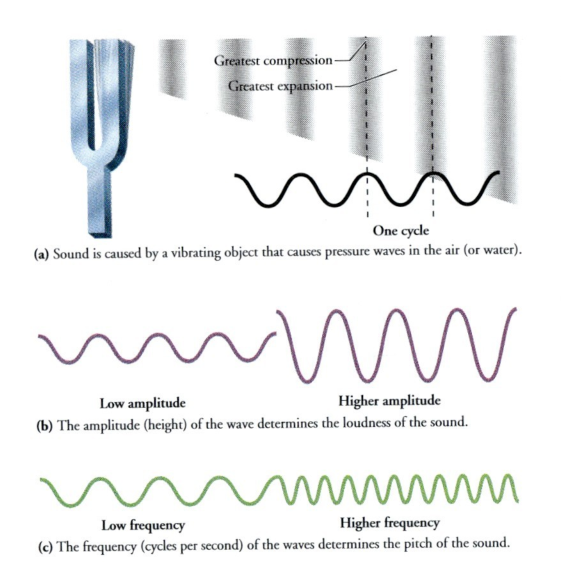

(a) Sound is caused by a vibrating object that causes pressure waves in the air (or water).

(b) The amplitude (height) of the wave determines the loudness of the sound.

(c) The frequency (cycles per second) of the waves determines the pitch of the sound.

Fig. 9.10

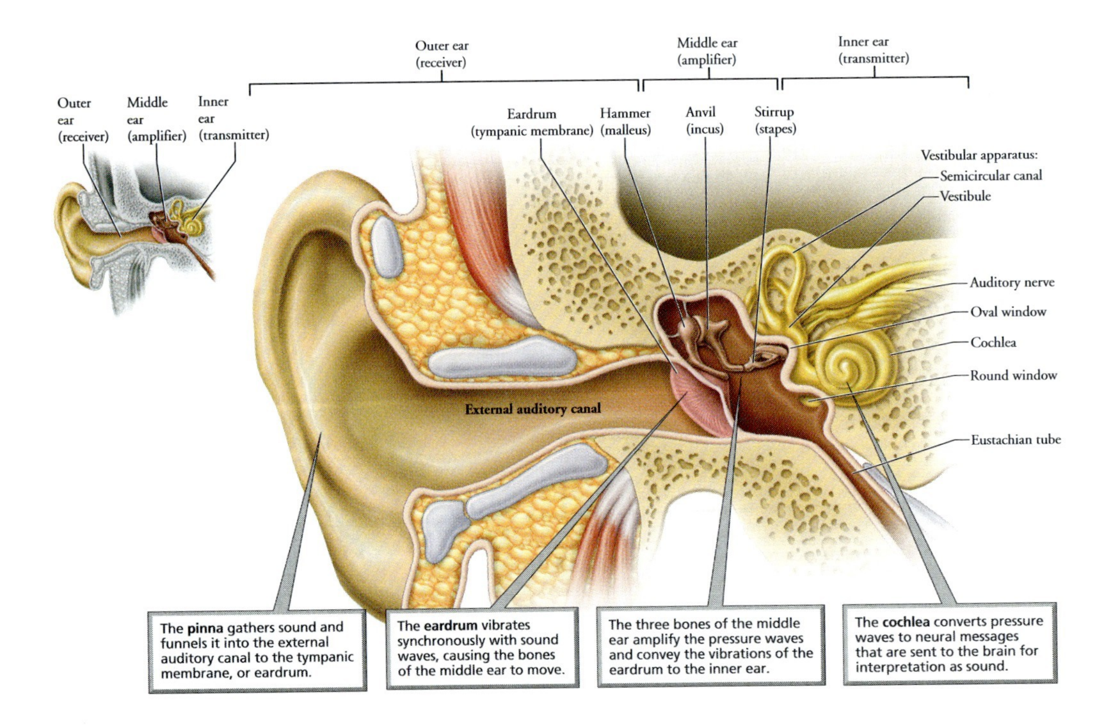

Fig. 9.11

NOTES

STRUCTURES OF THE EAR AND THEIR FUNCTIONS		
STRUCTURE	**DESCRIPTION**	**FUNCTION**
Outer ear		
Pinna	Fleshy, funnel-shaped part of the ear protruding from the side of the head	Collects and directs sound waves
External auditory canal	Canal between pinna and tympanic membrane	Directs sound to the middle ear
Middle ear		
Eardrum (tympanic membrane)	Membrane spanning the end of the external auditory canal	Vibrates in response to sound waves
Hammer (malleus), anvil (incus), and stirrup (stapes)	Three tiny bones of the middle ear	Amplify the vibrations of the tympanic membrane and transmit vibrations to inner ear
Eustachian tube	A tube that connects the middle ear with the throat	Allows equalization of pressure in middle ear with external air pressure
Inner ear		
Cochlea	Fluid-filled, bony, snail-shaped chamber	Houses organ of Corti and has openings called oval window and round window
Organ of Corti	Contains hair cells	The organ of hearing
Oval window	Membrane between the middle and inner ear that the stapes presses against	Transmits the movements of the stapes to the fluid in the inner ear
Round window	Membrane at the end of the lower canal in cochlea	Relieves pressure created by the movements of the oval window
Vestibular apparatus	Fluid-filled chambers and canals	Monitors position and movement of the head
Vestibule (utricle and saccule)	Two fluid-filled chambers	Maintain static equilibrium (body and head stationary, information on position of head)
Semicircular canals	Three fluid-filled chambers oriented at right angles to one another	Maintain dynamic equilibrium (body or head moving)

Table 9.4

NOTES

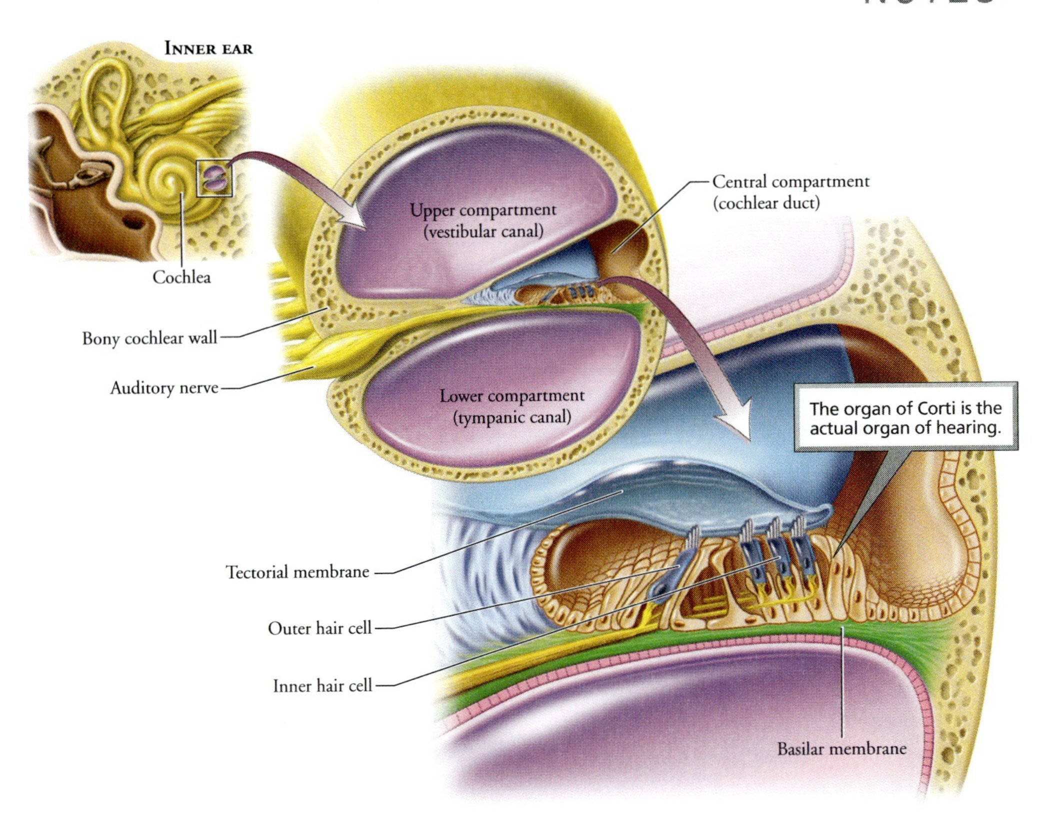

Fig. 9.12

NOTES

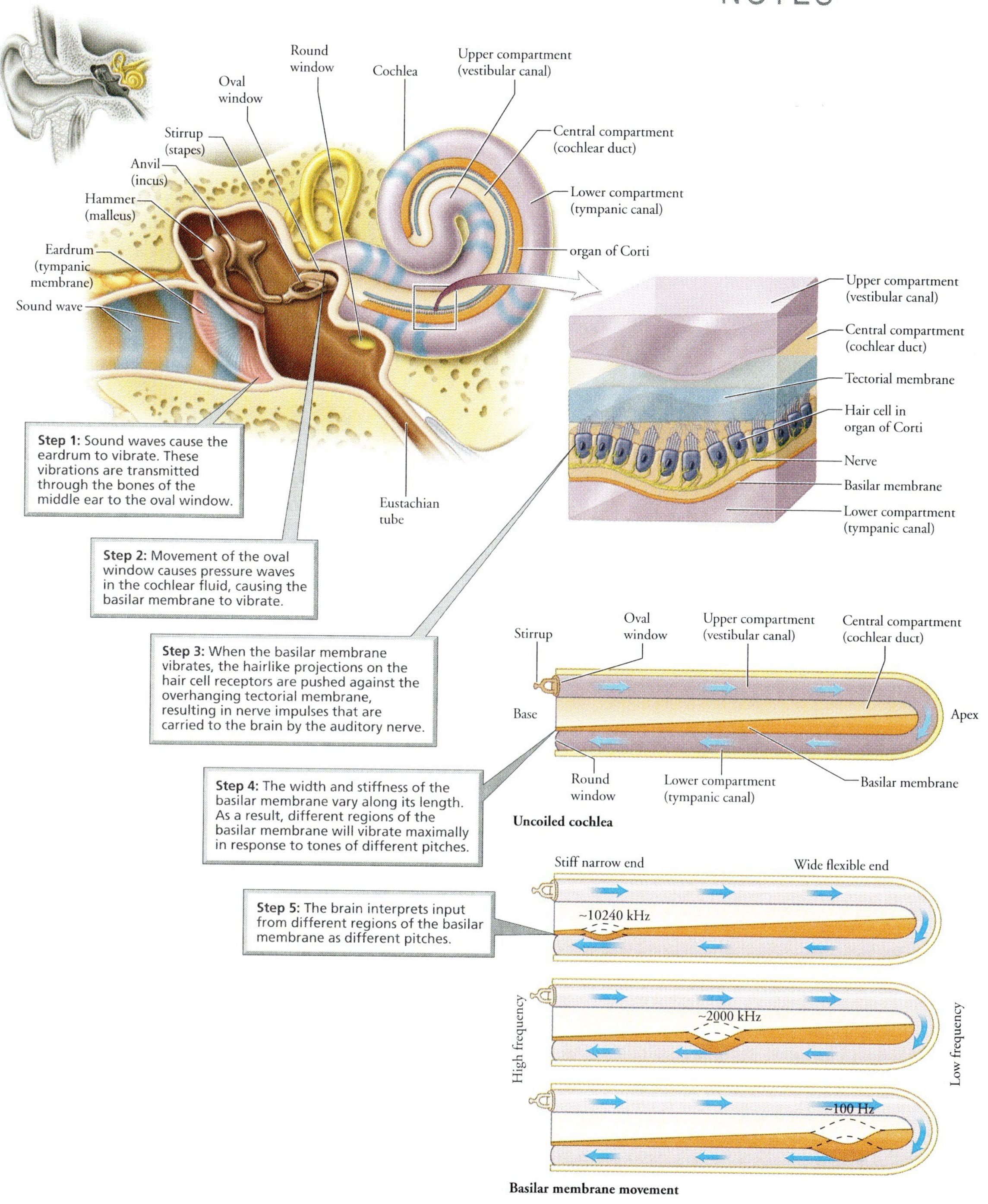

Fig. 9.13

NOTES

EFFECTS OF NOISE POLLUTION		
SOUND SOURCE	LOUDNESS (dB)	EFFECT FROM PROLONGED EXPOSURE
Jet plane at takeoff	150	Eardrum rupture
Deck of aircraft carrier	140	Very painful; traumatic injury
Rock-and-roll band (at maximum volume)	130	Irreversible damage
Jet plane at 152 m (500 ft)	110	Loss of hearing
Subway, lawn mower	100	
Electric blender	90	Annoying
Washing machine, freight train at 15 m (50 ft)	80	
Traffic noise	70	Intrusive
Normal conversation	65	
Chirping bird	60	
Quiet neighborhood (daytime)	50	
Soft background music	40	Quiet
Library	30	
Whisper	20	Very quiet
Breathing, rustling leaves	10	
	0	Threshold of hearing

Table 9.A

NOTES

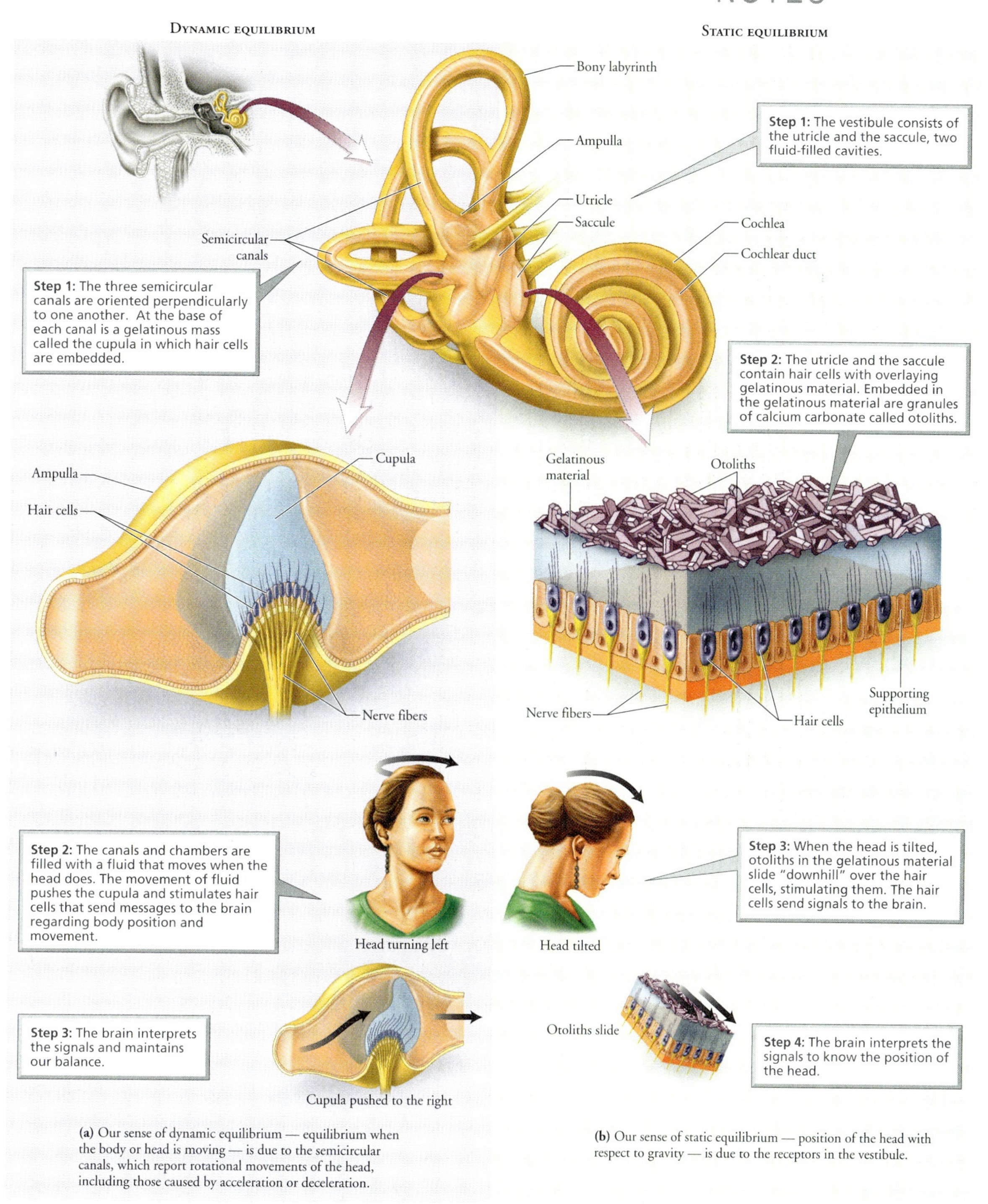

(a) Our sense of dynamic equilibrium — equilibrium when the body or head is moving — is due to the semicircular canals, which report rotational movements of the head, including those caused by acceleration or deceleration.

(b) Our sense of static equilibrium — position of the head with respect to gravity — is due to the receptors in the vestibule.

Fig. 9.14

NOTES

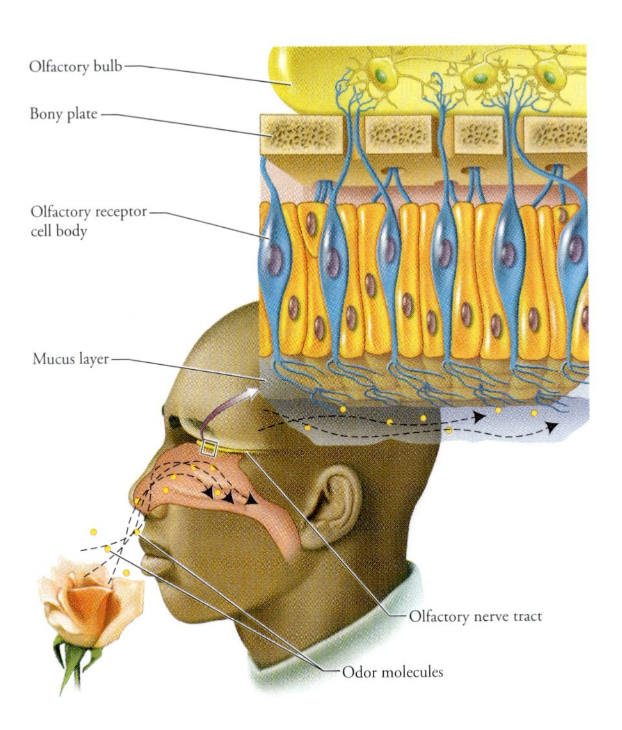

Fig. 9.15

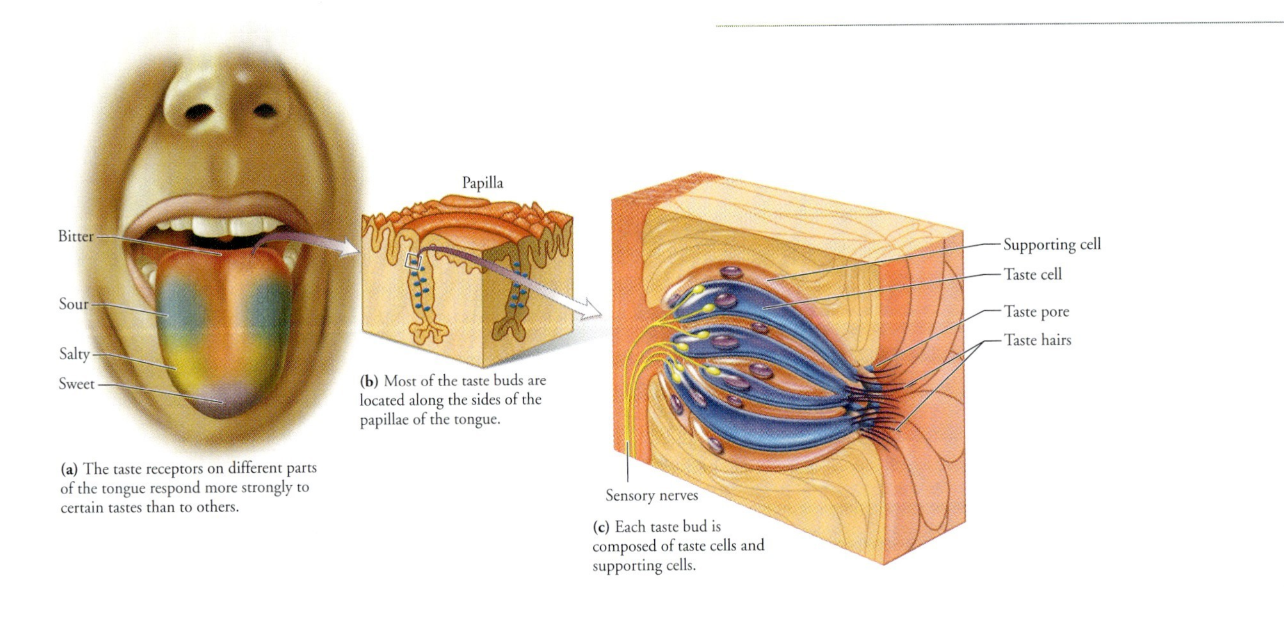

(a) The taste receptors on different parts of the tongue respond more strongly to certain tastes than to others.

(b) Most of the taste buds are located along the sides of the papillae of the tongue.

(c) Each taste bud is composed of taste cells and supporting cells.

Fig. 9.16

NOTES

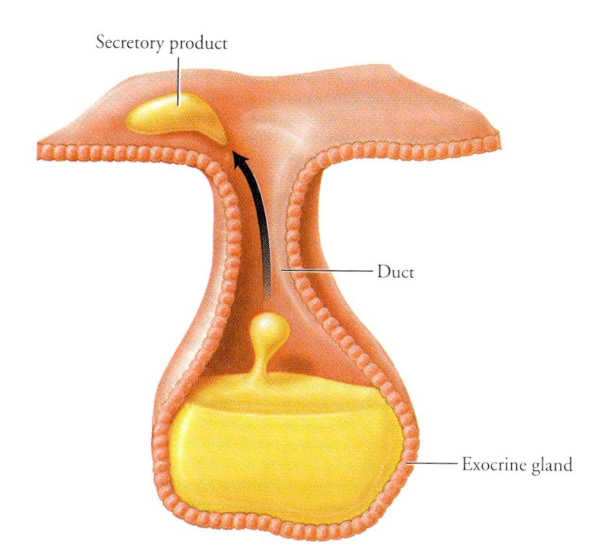

(a) Exocrine glands secrete their products into ducts that open onto the surface of the body, into the spaces within organs, or into cavities within the body.

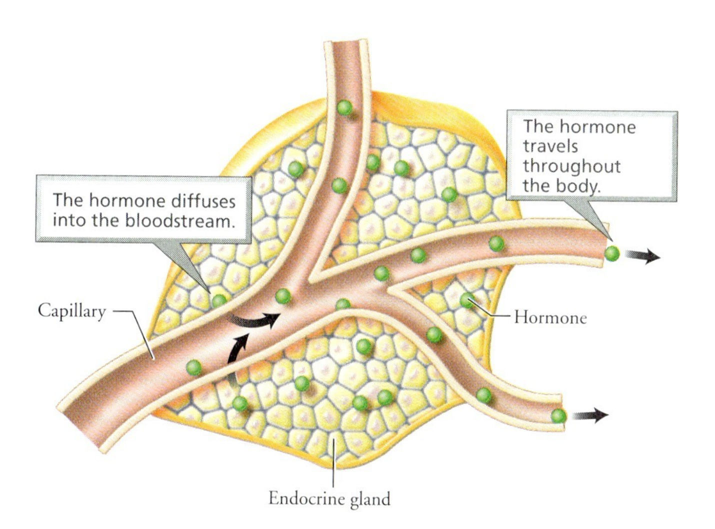

(b) Endocrine glands release their products, called hormones, into the fluid just outside cells. The hormones then diffuse into the bloodstream to be transported throughout the body.

Fig. 10.1

NOTES

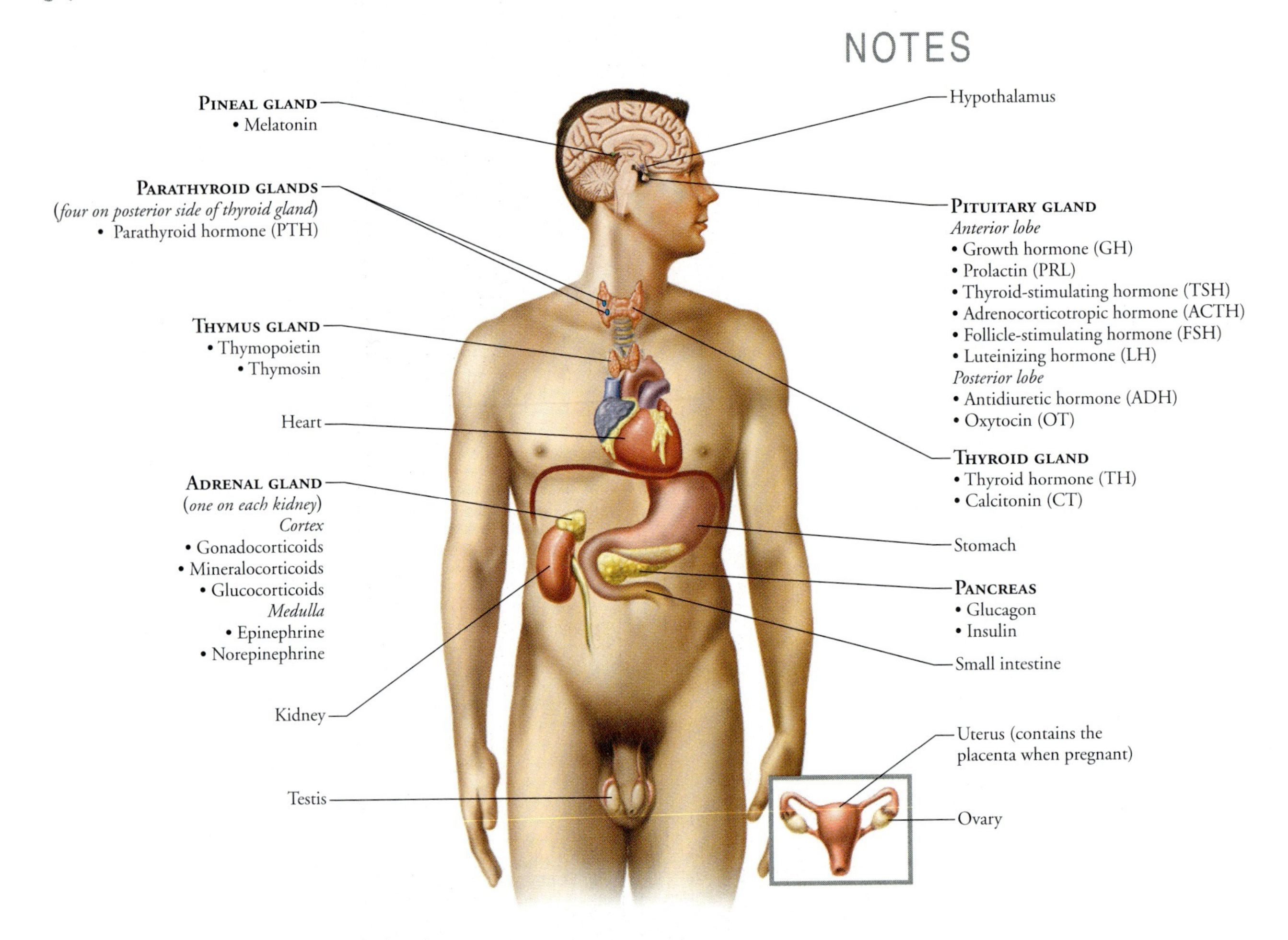

Fig. 10.2

NOTES

SOME ENDOCRINE GLANDS AND THEIR HORMONES

GLAND OR ORGAN	HORMONE	FUNCTION
Anterior lobe of pituitary	Growth hormone (GH)	Stimulates growth, particularly of muscle, bone, and cartilage Stimulates breakdown of fat
	Prolactin (PRL)	Stimulates breasts to produce milk
	Thyroid-stimulating hormone (TSH)	Stimulates synthesis and release of hormones from thyroid gland
	Adrenocorticotropic hormone (ACTH)	Stimulates synthesis and release of glucocorticoid hormones from adrenal glands
	Follicle-stimulating hormone (FSH)	Stimulates gamete production in males and females Stimulates secretion of estrogen by the ovaries
	Luteinizing hormone (LH)	Causes ovulation and stimulates ovaries to secrete estrogen and progesterone Stimulates cells of testes to develop and secrete testosterone
Posterior lobe of pituitary	Antidiuretic hormone (ADH)	Promotes water reabsorption by the kidneys
	Oxytocin (OT)	Stimulates milk ejection from the breasts Stimulates uterine contractions during childbirth
Thyroid	Thyroid hormone (TH)	Regulates metabolism and heat production Promotes normal development and functioning of nervous, muscular, skeletal, and reproductive systems
	Calcitonin (CT)	Decreases blood levels of calcium (stimulates absorption of calcium by bone)
Parathyroid	Parathyroid hormone (PTH)	Increases blood levels of calcium (stimulates breakdown of bone and rate at which calcium is removed from urine and absorbed from gastrointestinal tract)
Adrenal cortex	Gonadocorticoids (androgens, estrogens)	Amounts secreted by adults are so low that effects are probably insignificant
	Mineralocorticoids (aldosterone)	Increase sodium reabsorption by kidneys Increase potassium excretion by kidneys
	Glucocorticoids (cortisol, corticosterone, cortisone)	Stimulate glucose synthesis Inhibit the inflammatory response
Adrenal medulla	Epinephrine	Fight-or-flight response Response to stress
	Norepinephrine	Fight-or-flight response Response to stress
Pancreas	Glucagon	Increases blood glucose level (prompts liver to increase conversion of glycogen to glucose and formation of glucose from fatty and amino acids)
	Insulin	Decreases blood glucose level (stimulates transport of glucose into cells, inhibits breakdown of glycogen to glucose, prevents conversion of fatty and amino acids to glucose)
Thymus	Thymopoietin, thymosin	Promote maturation of white blood cells
Pineal	Melatonin	May influence daily rhythms, fertility, and aging

Table 10.1

NOTES

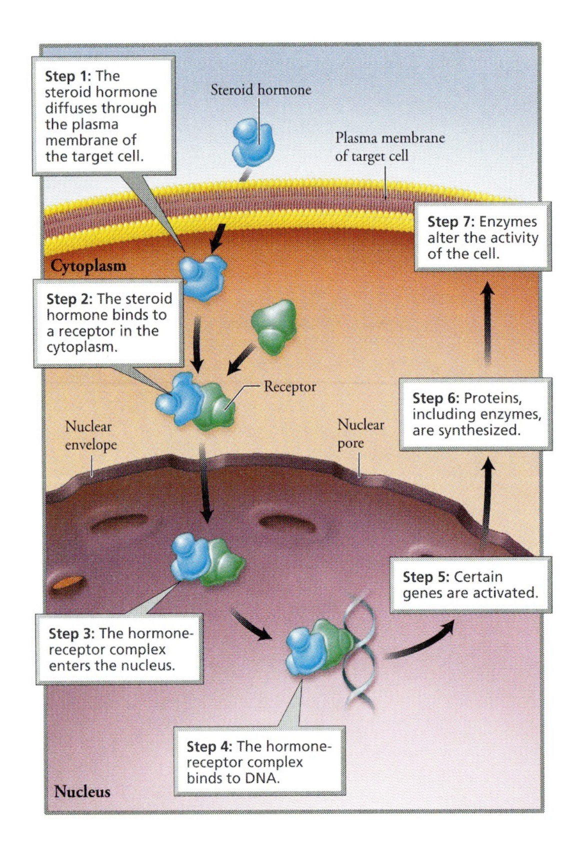

Fig. 10.3

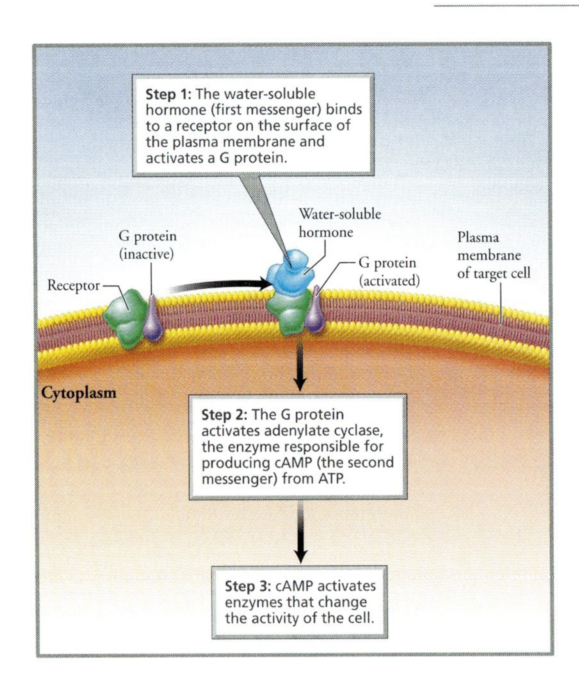

Fig. 10.4

NOTES

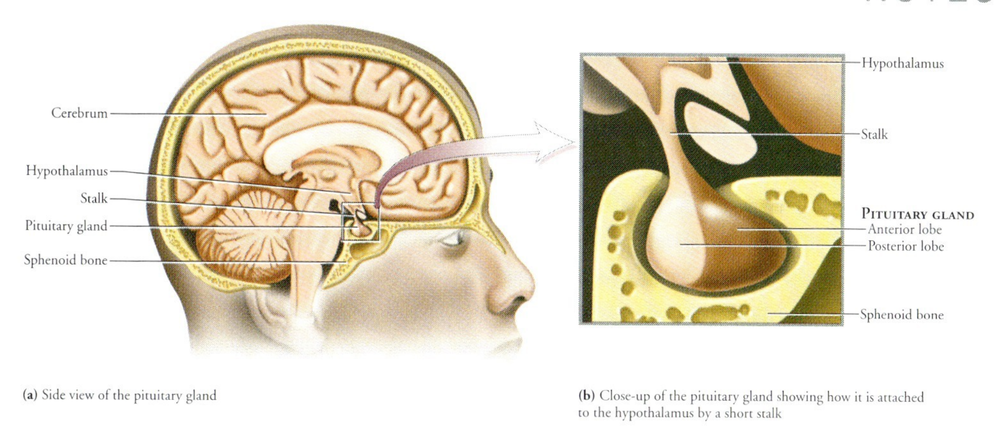

(a) Side view of the pituitary gland

(b) Close-up of the pituitary gland showing how it is attached to the hypothalamus by a short stalk

Fig. 10.5

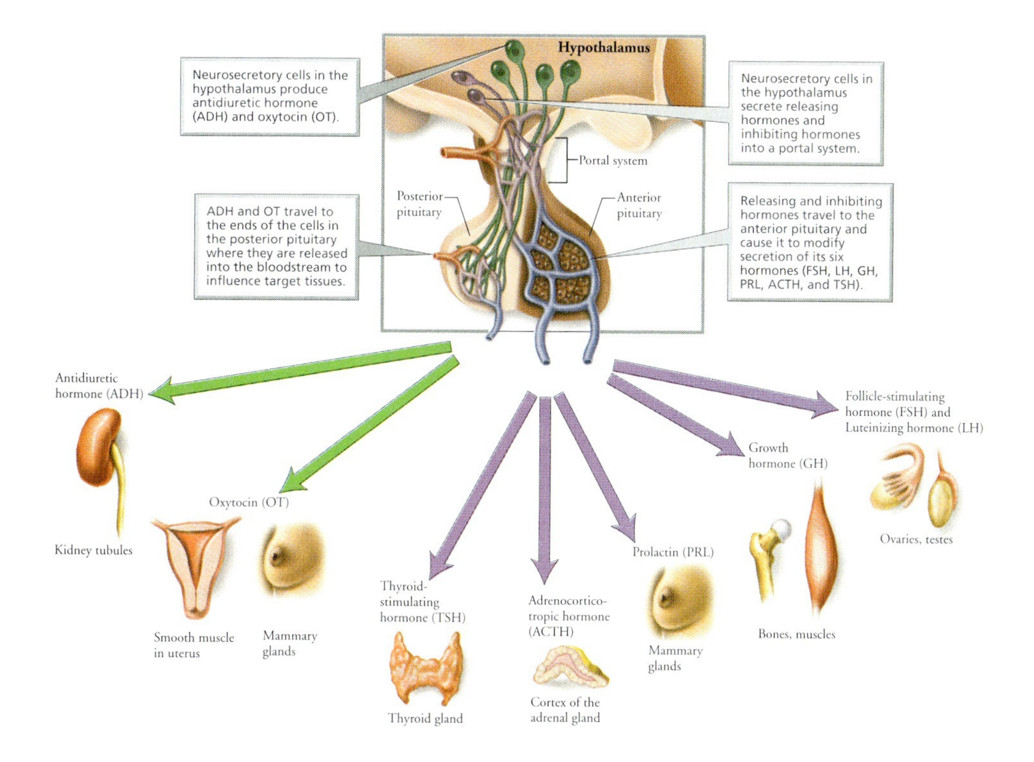

Fig.10.6

NOTES

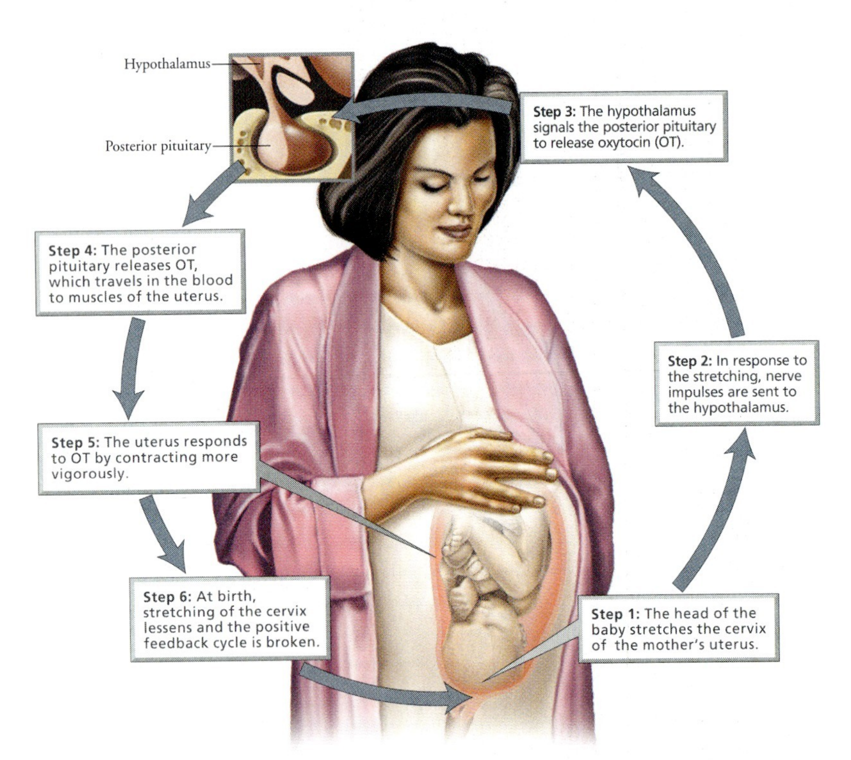

Fig. 10.10

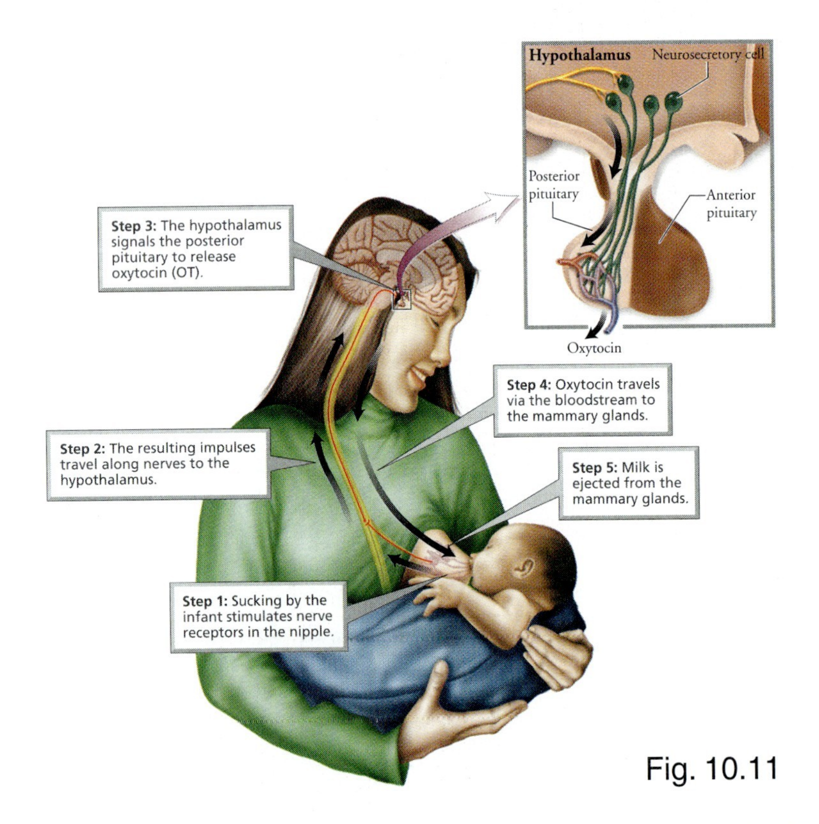

Fig. 10.11

NOTES

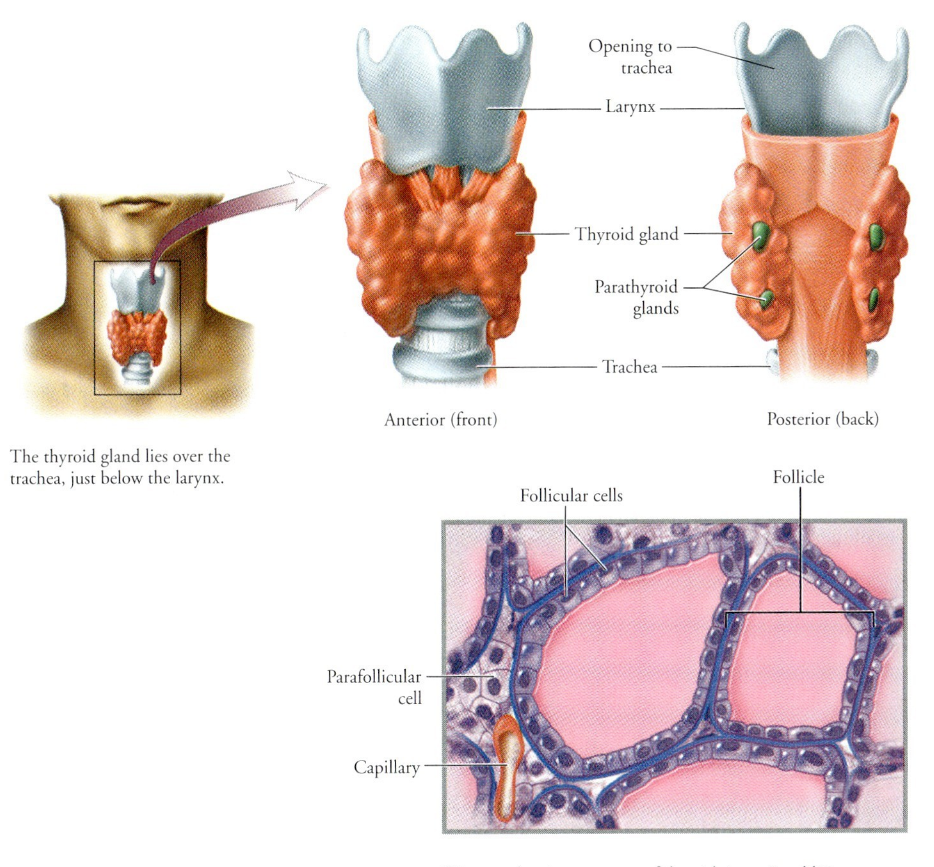

The thyroid gland lies over the trachea, just below the larynx.

Diagram showing structure of thyroid tissue. In addition to the follicular cells, note the parafollicular cells that produce calcitonin.

Fig. 10.12

NOTES

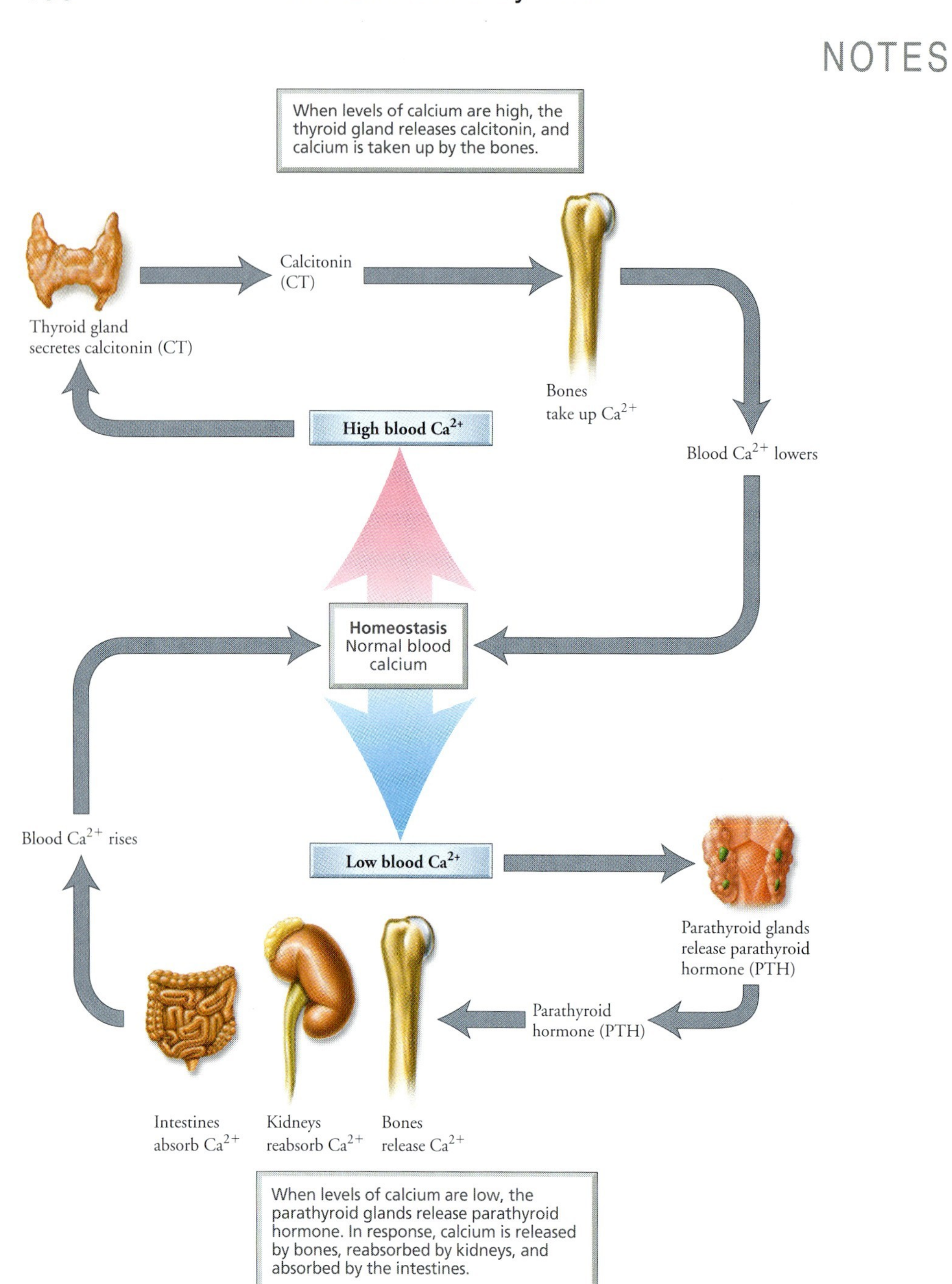

Fig. 10.16

NOTES

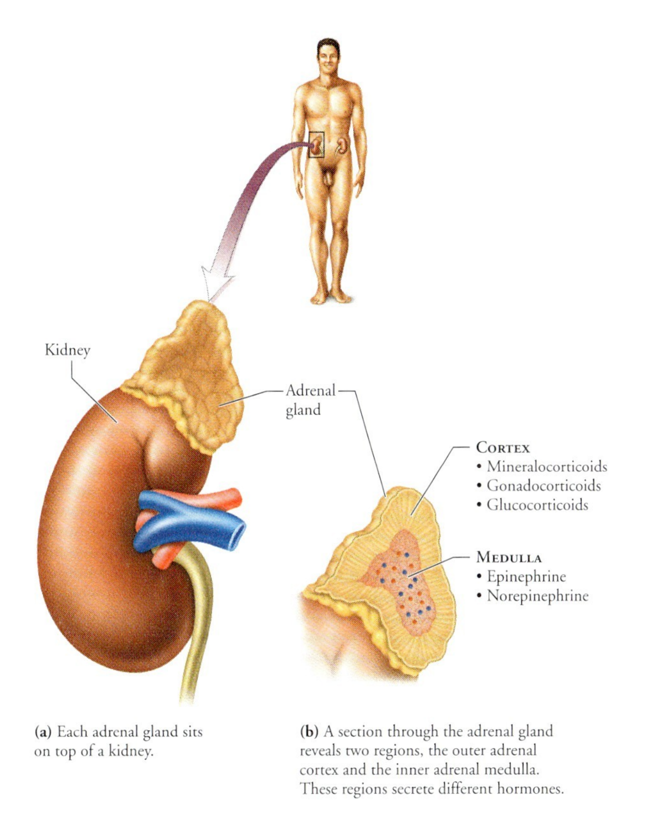

(a) Each adrenal gland sits on top of a kidney.

(b) A section through the adrenal gland reveals two regions, the outer adrenal cortex and the inner adrenal medulla. These regions secrete different hormones.

Fig. 10.17

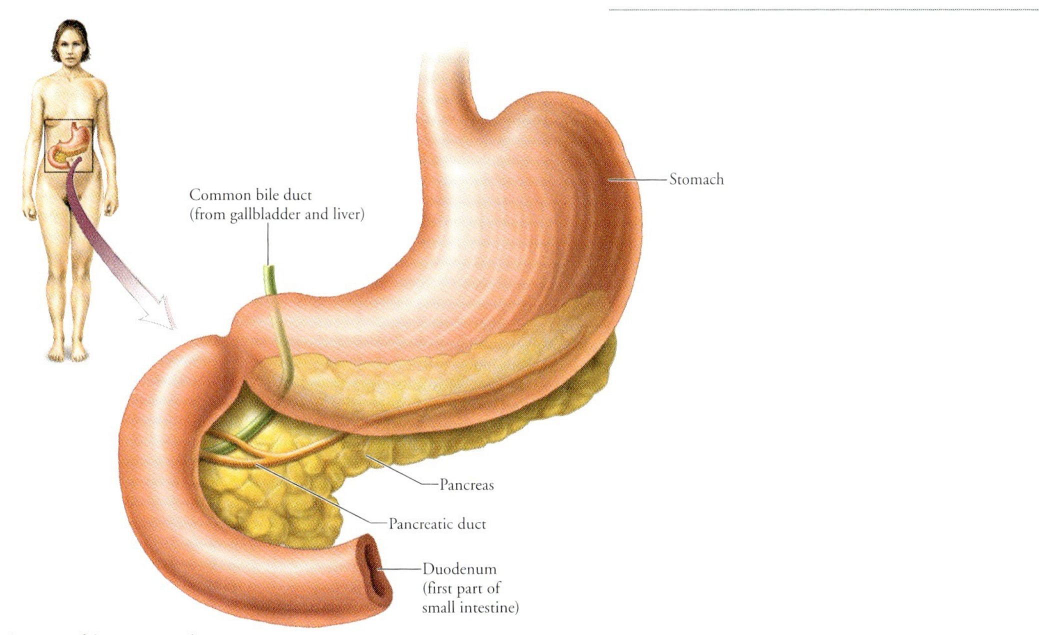

Structure of the pancreas and associated ducts. Exocrine cells of the pancreas secrete digestive enzymes into the pancreatic duct, which unites with the common bile duct before entering the small intestine.

Fig. 10.19

NOTES

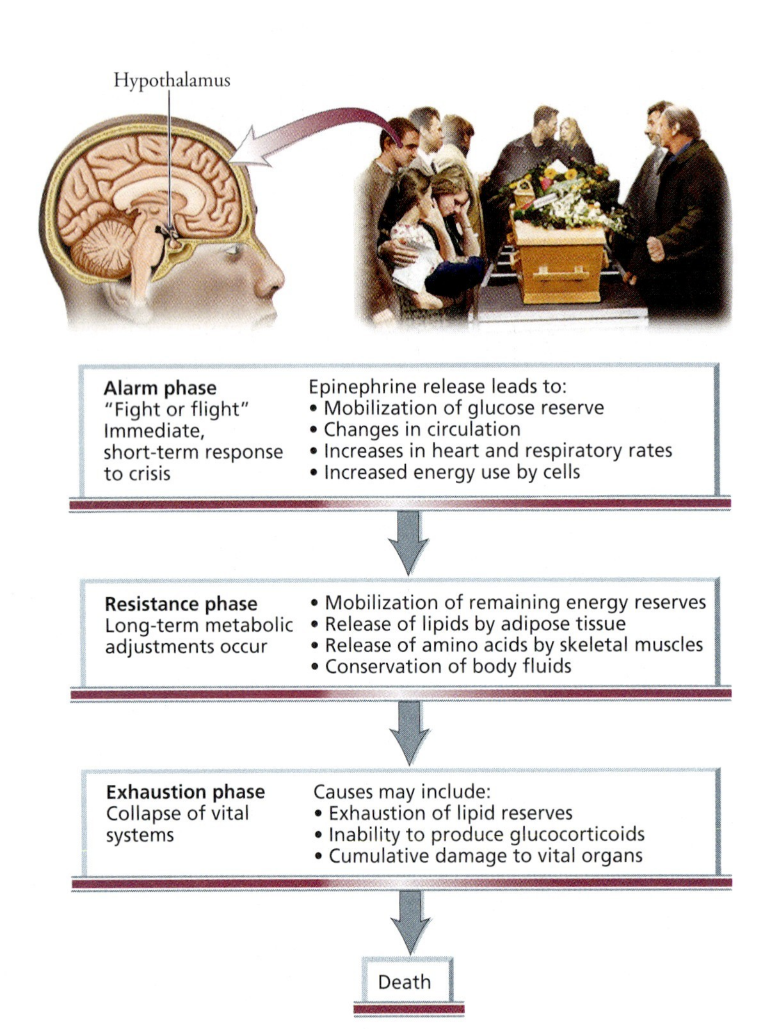

Fig. 10.A

NOTES

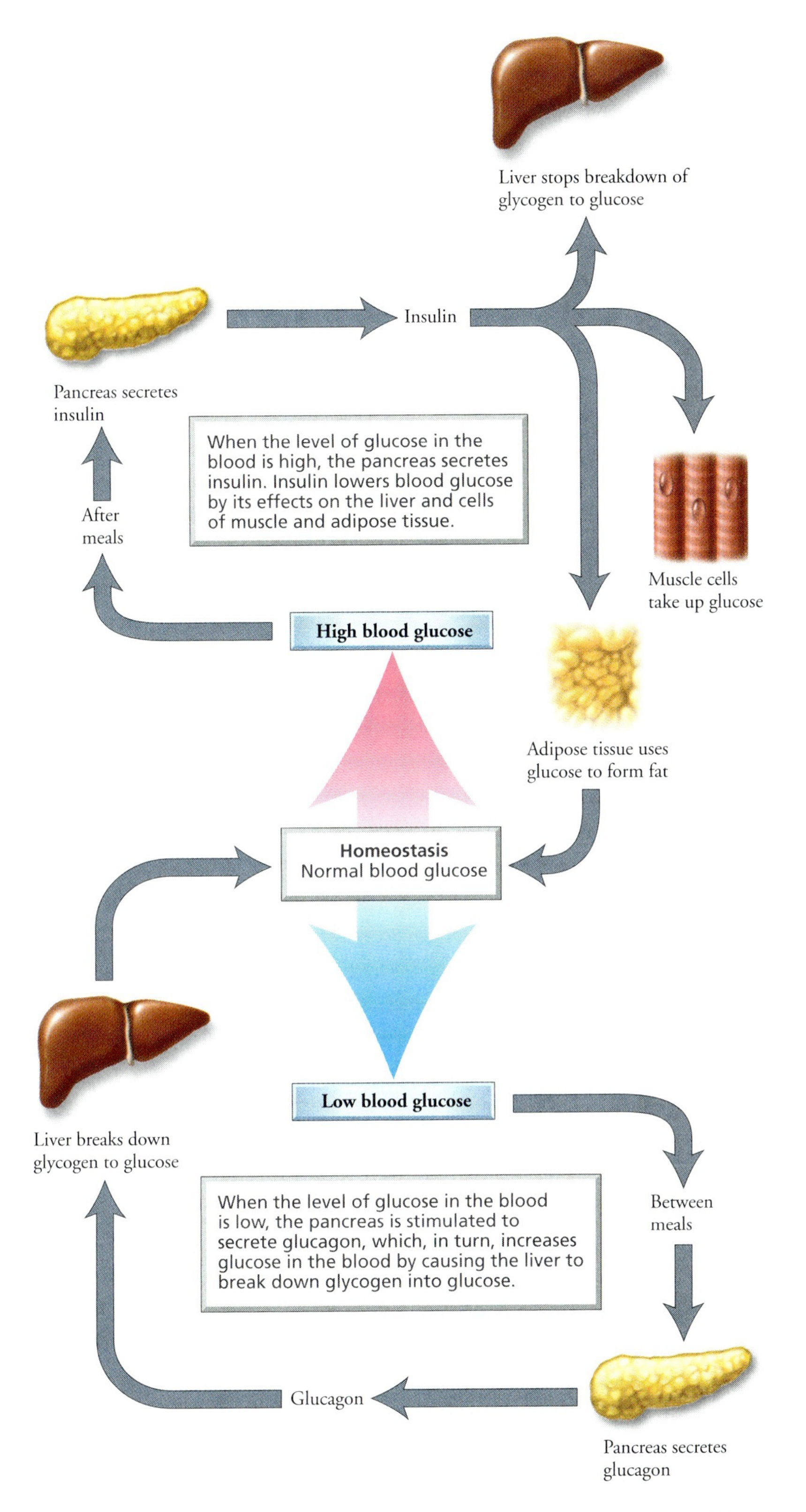

Fig. 10.20

NOTES

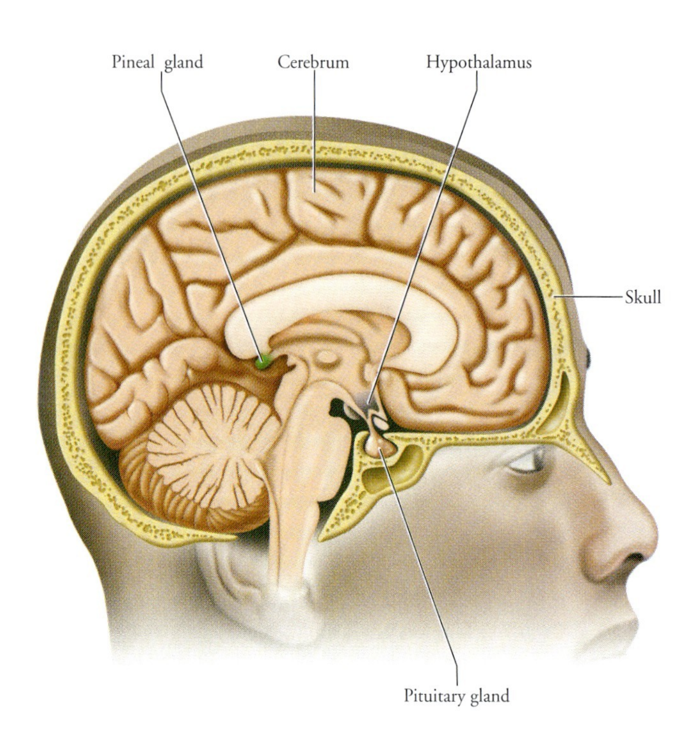

Fig. 10.21

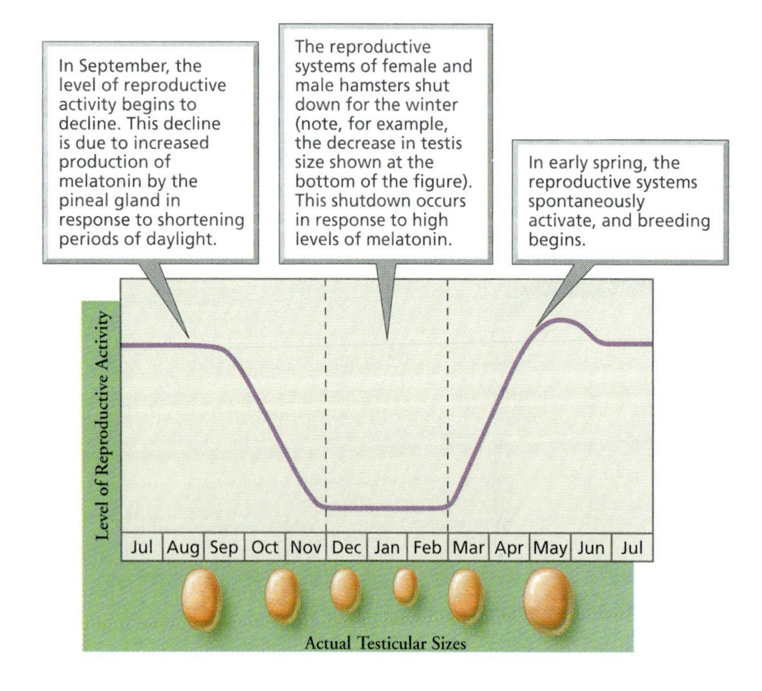

Fig. 10.C

NOTES

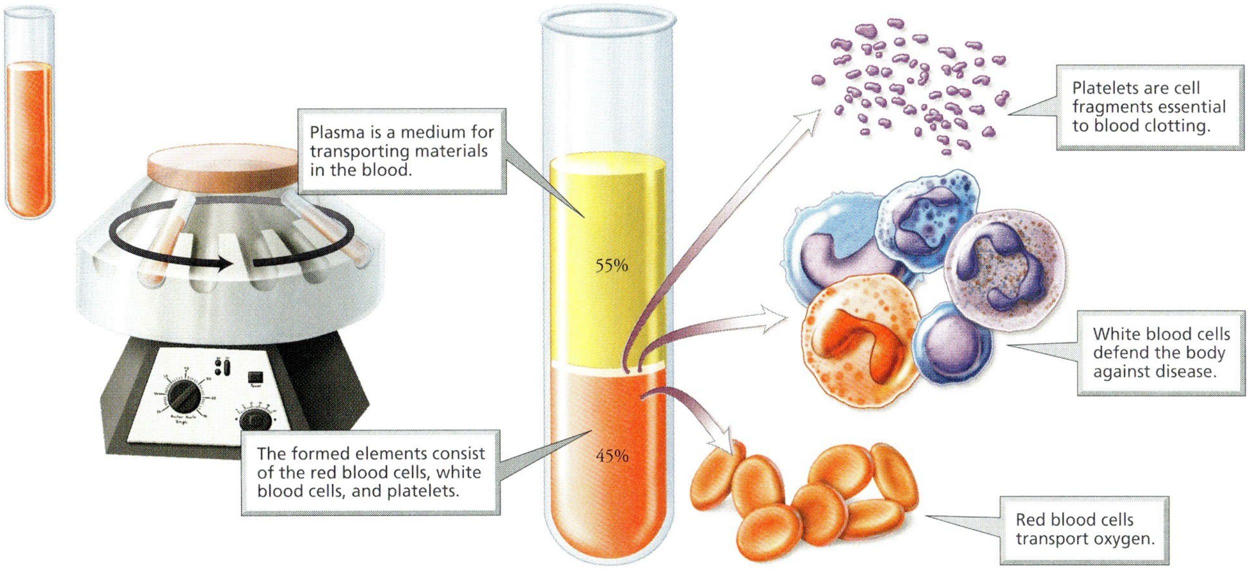

Plasma is a medium for transporting materials in the blood.

55%

45%

The formed elements consist of the red blood cells, white blood cells, and platelets.

Platelets are cell fragments essential to blood clotting.

White blood cells defend the body against disease.

Red blood cells transport oxygen.

Fig 11.1

NOTES

The Formed Elements of Blood

Type of Formed Element	Description	No. of Cells/mm^3	Life Span	Cell Function
Platelets	Fragments of a megakaryocyte; small, purple-stained granules in cytoplasm	250,000–500,000	5–10 days	Play role in blood clotting
Leukocytes (white blood cells; WBCs)				
Granulocytes				
Neutrophils	Multilobed nucleus, clear-staining cytoplasm, inconspicuous granules	3000–7000	6–72 hours	Consume bacteria by phagocytosis
Eosinophils	Large granules in cytoplasm, pink-staining cytoplasm, bilobed nucleus	100–400	8–12 days	Consume antibody-antigen complex by phagocytosis; attack parasitic worms
Basophils	Large, purple cytoplasmic granules; bilobed nucleus	20–50	3–72 hours	Release histamine
Agranulocytes				
Monocytes	Gray-blue cytoplasm with no granules; U-shaped nucleus	100–700	Several months	Give rise to macrophages, which consume bacteria, dead cells, and cell parts by phagocytosis
Lymphocytes	Round nucleus that almost fills the cell	1500–3000	Many years	Attack damaged or diseased cells or produce antibodies
Erythrocytes (red blood cell; RBCs)	Biconcave disk, no nucleus	4–6 million	About 120 days	Transport oxygen and carbon dioxide

Table 11.1

NOTES

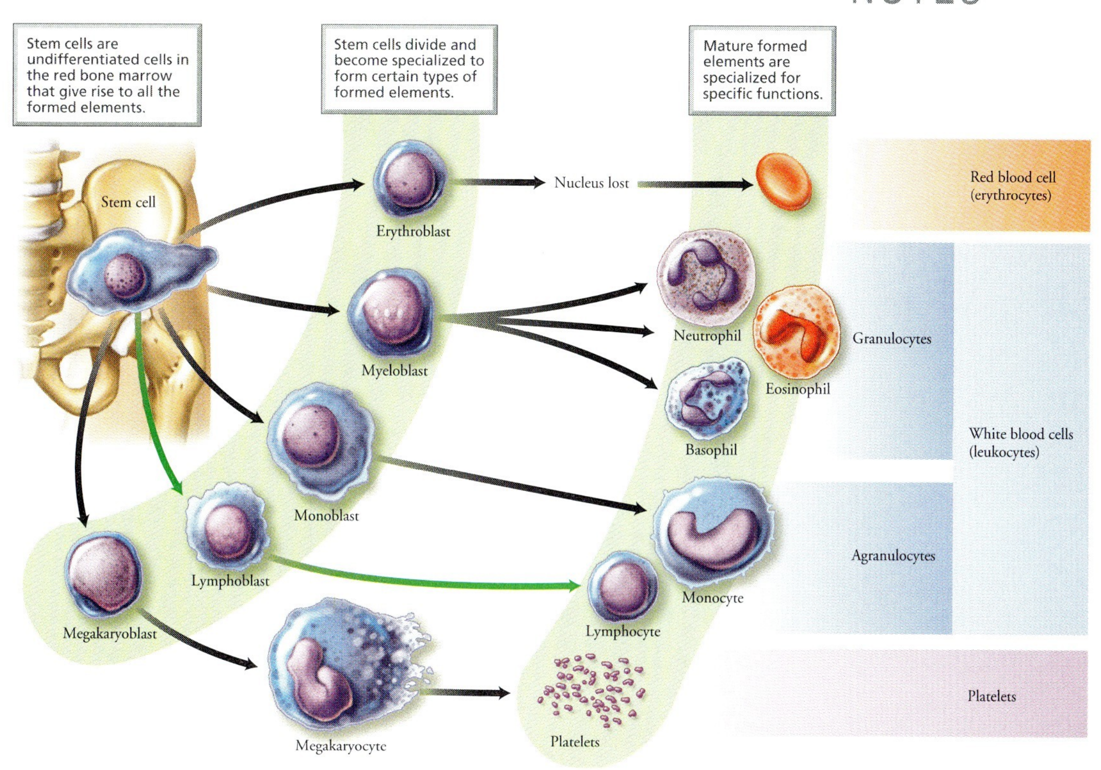

Fig. 11.2

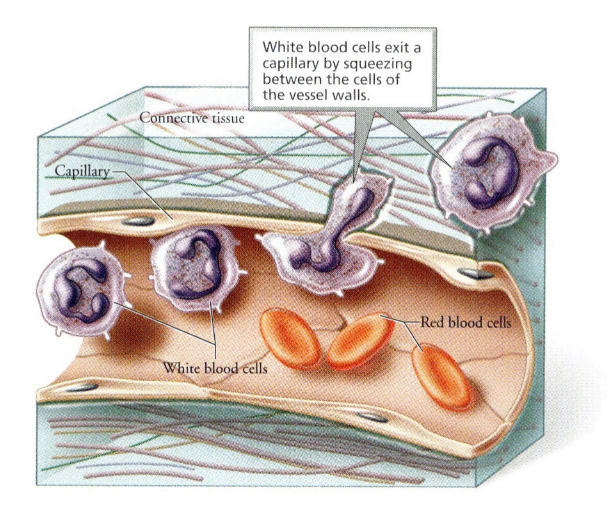

Fig. 11.3

NOTES

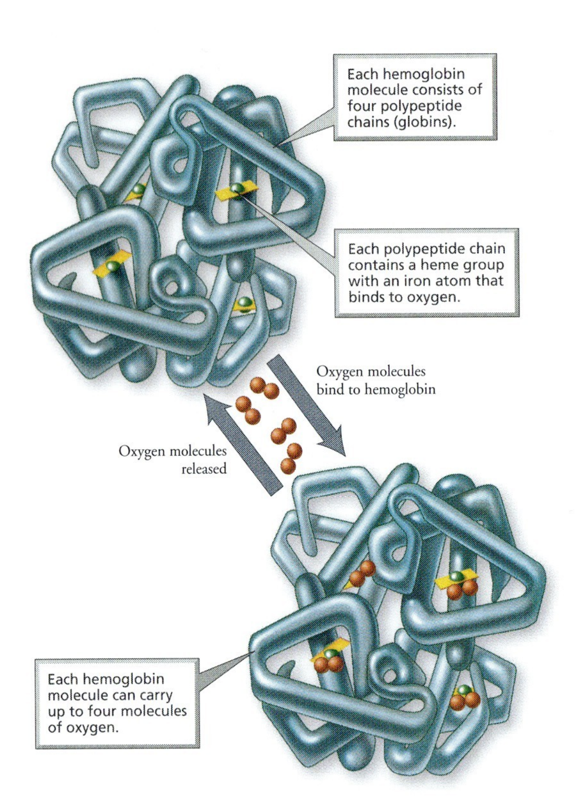

Fig. 11.5

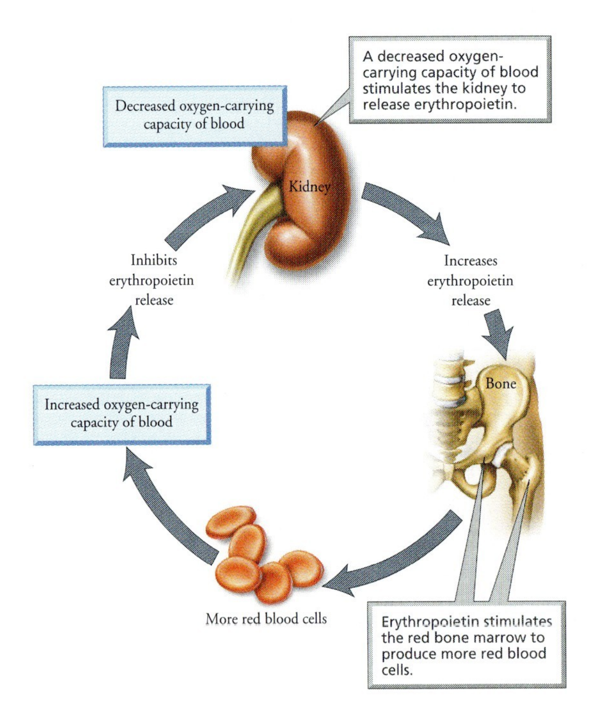

Fig. 11.6

NOTES

BLOOD TYPE	ANTIGENS ON RED BLOOD CELLS	ANTIBODIES IN PLASMA	BLOOD TYPES THAT CAN BE RECEIVED IN TRANSFUSIONS	INCIDENCE OF BLOOD TYPE IN U.S.
A	A	Anti-B	A, O	Caucasian, 40% African American, 27% Asian, 28% Native American, 8%
B	B	Anti-A	B, O	Caucasian, 10% African American, 20% Asian, 27% Native American, 1%
AB (universal recipient)	A and B	None	A, B, AB, O	Caucasian, 5% African American, 4% Asian, 5% Native American, 0%
O (universal donor)	None	Anti-A, anti-B	O	Caucasian, 45% African American, 49% Asian, 40% Native American, 91%

Table 11.2

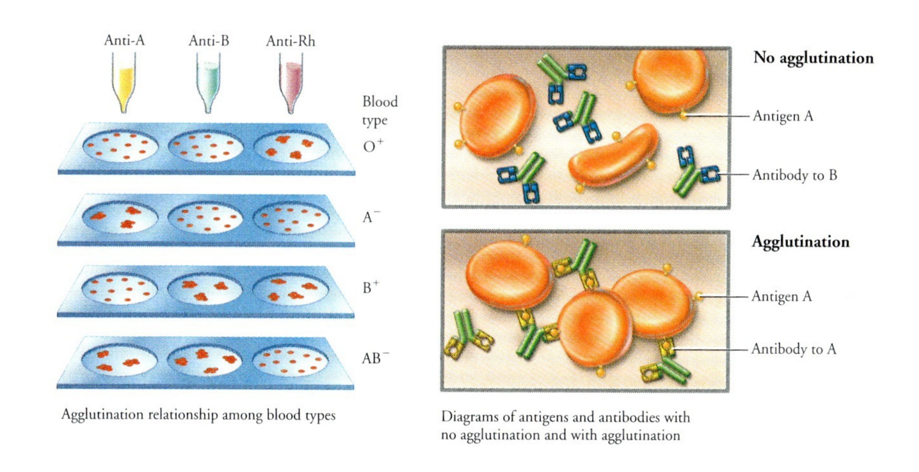

Agglutination relationship among blood types

Diagrams of antigens and antibodies with no agglutination and with agglutination

Fig. 11.7

NOTES

Rh+
Rh−

An Rh^+ male and an Rh^- female have a 50% chance of having an Rh+ baby.

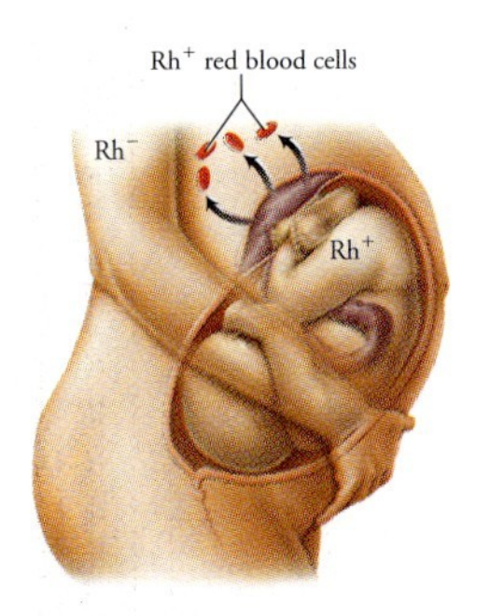
Rh^+ red blood cells
Rh−
Rh+

First pregnancy: At birth some of the Rh^+ blood of the fetus may enter the mother's circulation.

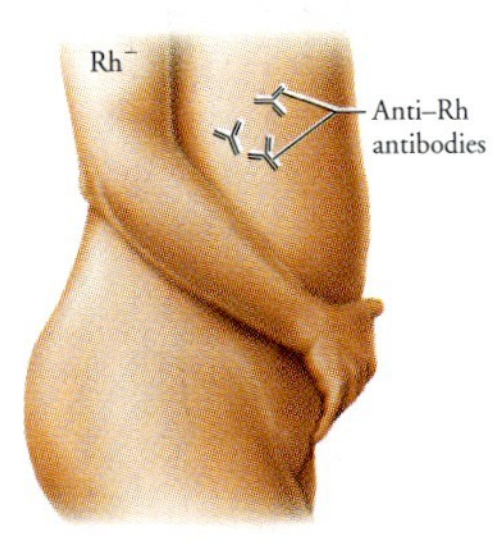
Rh−
Anti–Rh antibodies

The mother forms anti-Rh antibodies over the next few months.

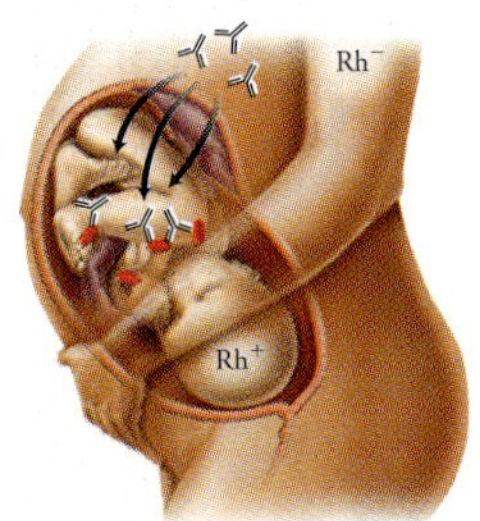
Rh−
Rh+

Second pregnancy with an Rh^+ fetus: Anti-Rh antibodies may pass into the fetus's blood, causing its blood cells to burst.

Fig. 11.8

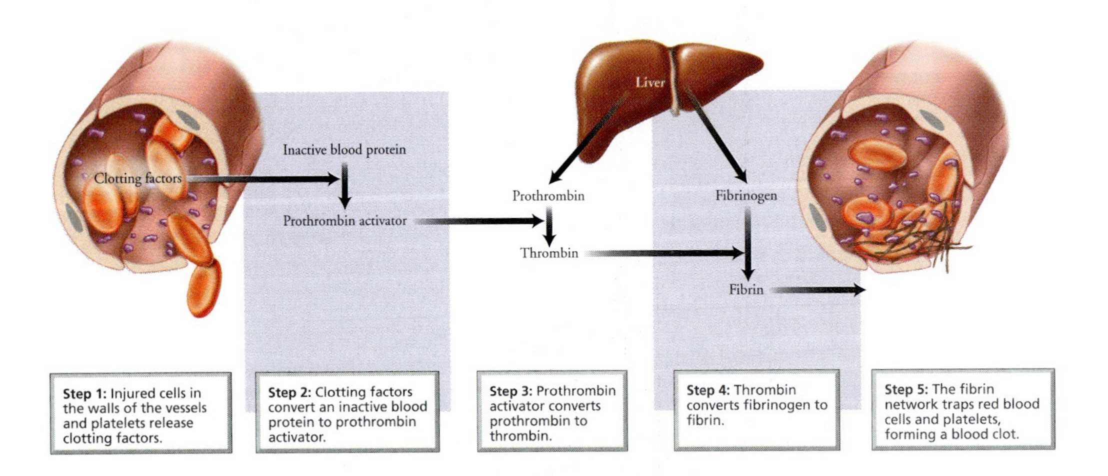
Clotting factors
Inactive blood protein
Prothrombin activator
Liver
Prothrombin
Thrombin
Fibrinogen
Fibrin
Step 1: Injured cells in the walls of the vessels and platelets release clotting factors.
Step 2: Clotting factors convert an inactive blood protein to prothrombin activator.
Step 3: Prothrombin activator converts prothrombin to thrombin.
Step 4: Thrombin converts fibrinogen to fibrin.
Step 5: The fibrin network traps red blood cells and platelets, forming a blood clot.

Fig. 11.9

NOTES

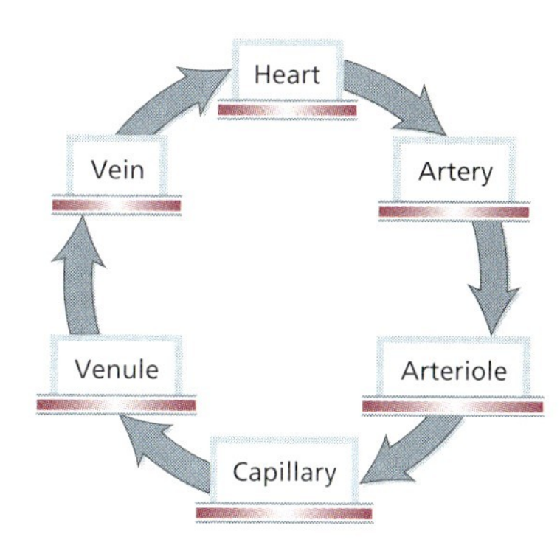

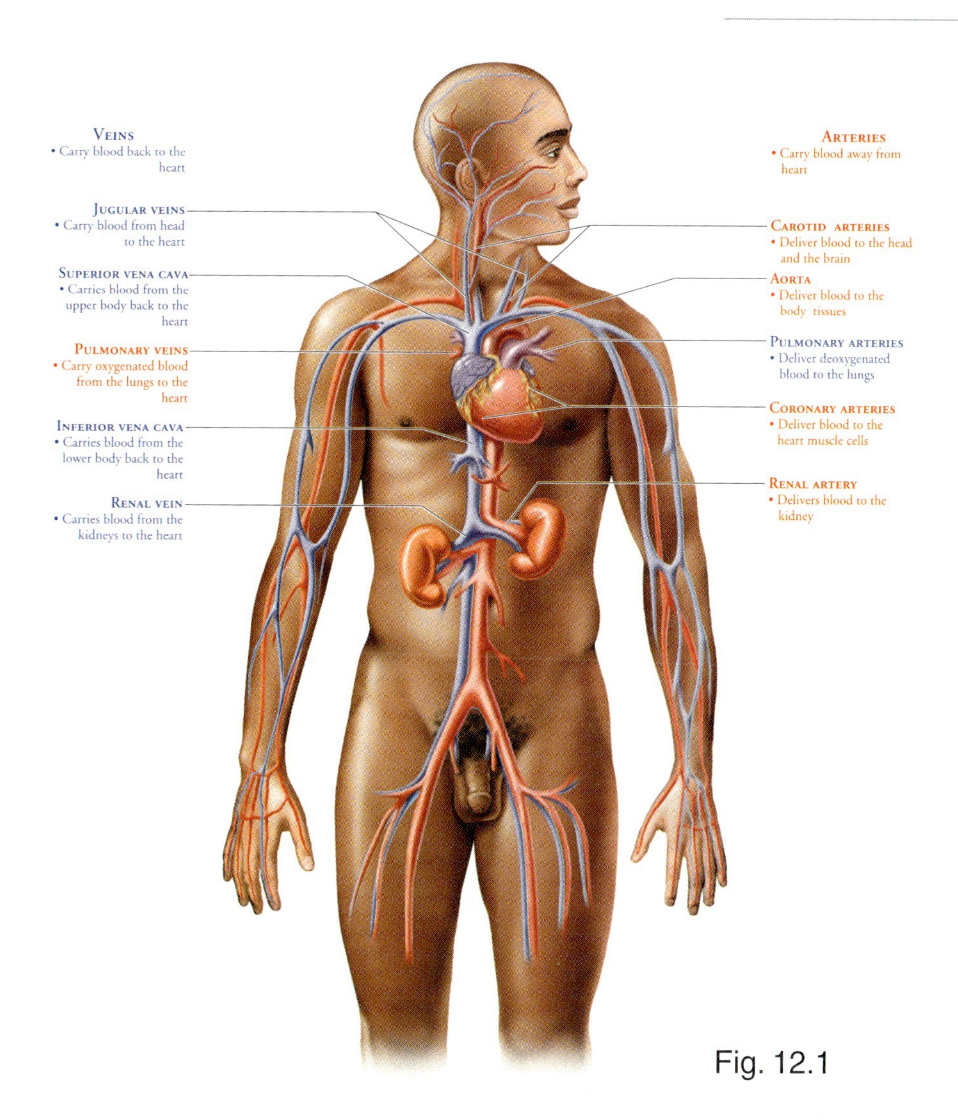

Fig. 12.1

NOTES

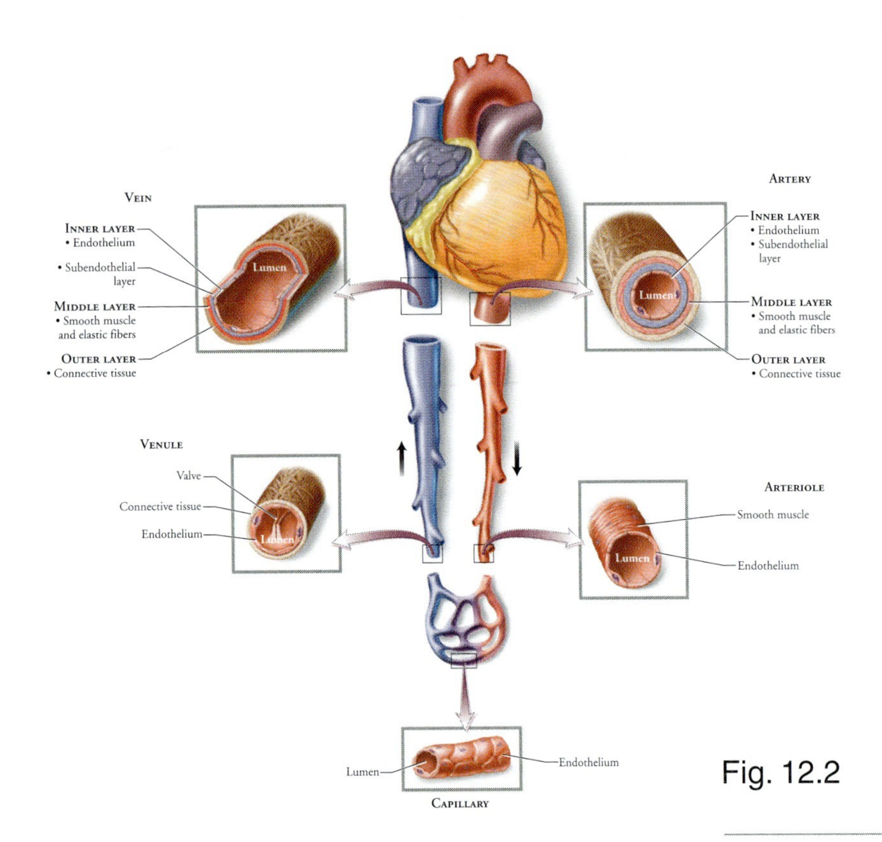

Fig. 12.2

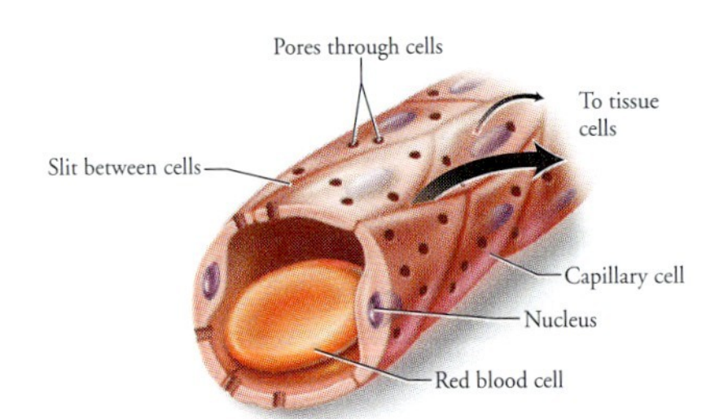

(a) Substances are exchanged between the blood and tissue fluid across the plasma membrane of the capillary.

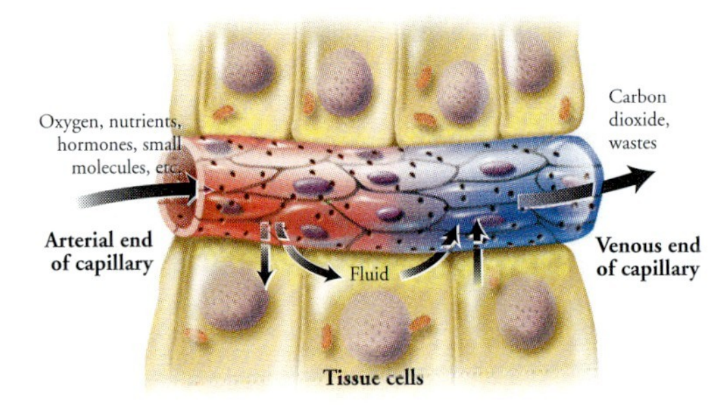

(b) At the arterial end of a capillary, blood pressure forces fluid out of the capillary to the fluid surrounding tissue cells. At the venous end, fluid is drawn back into the capillary by osmotic pressure.

Fig. 12.3

NOTES

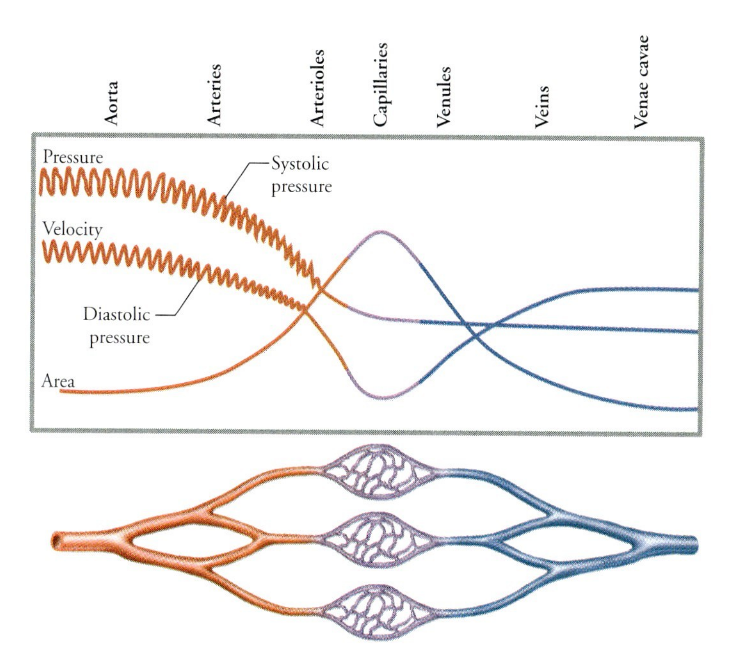

Fig. 12.4

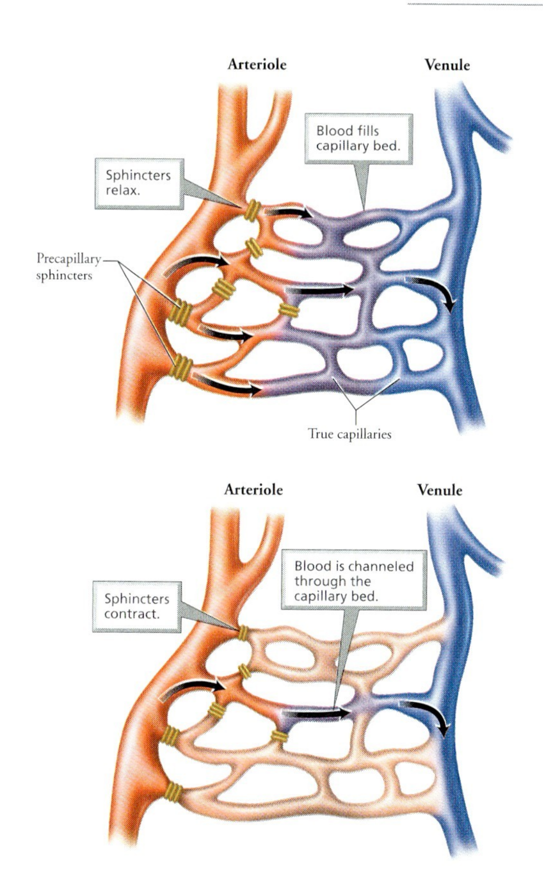

Fig. 12.5

NOTES

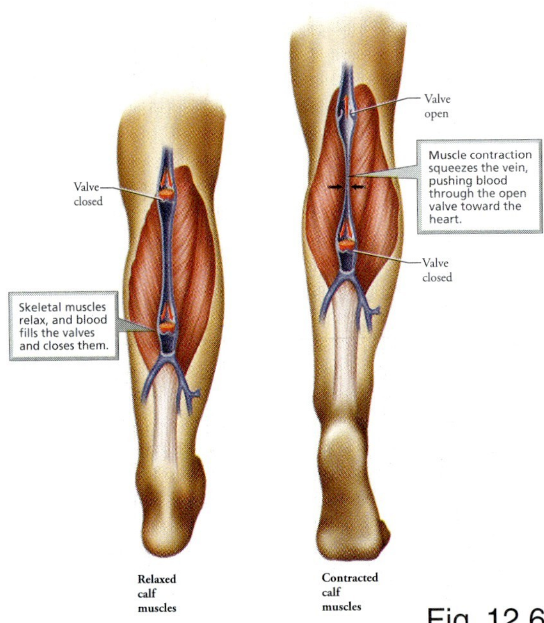

Fig. 12.6

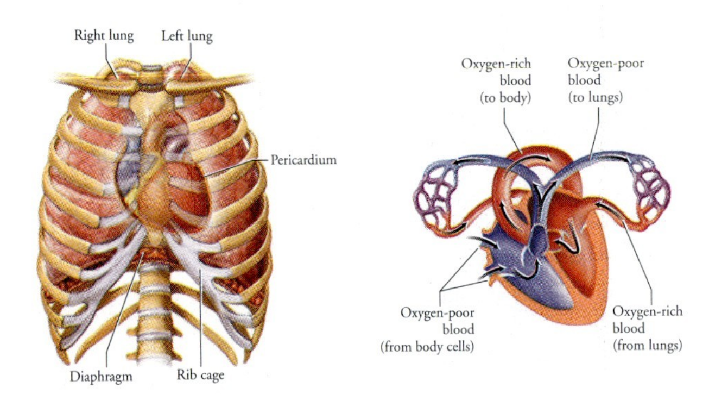

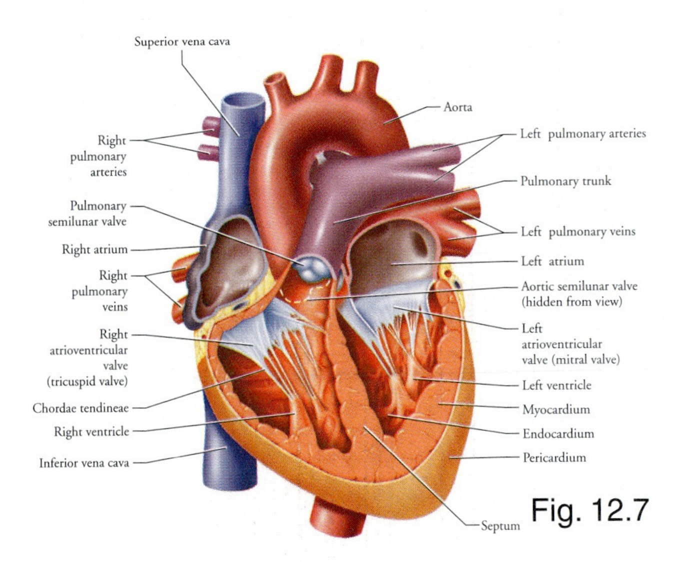

Fig. 12.7

NOTES

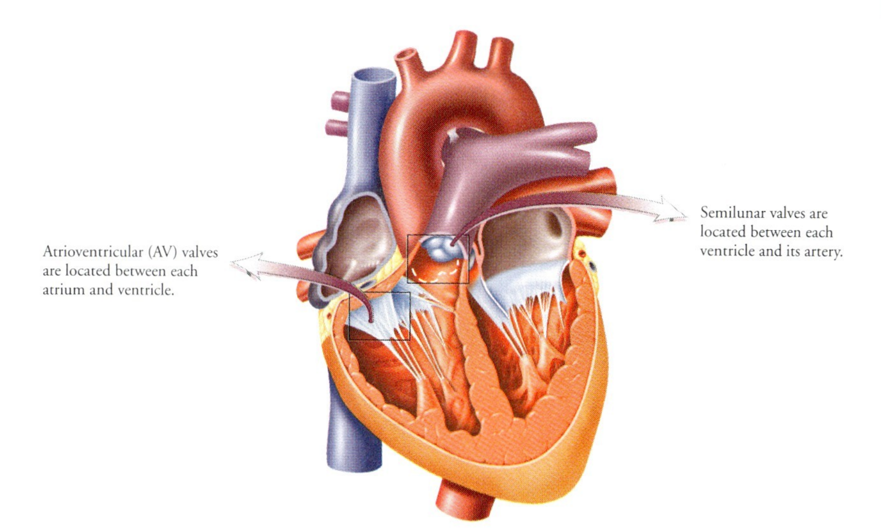

Fig 12.8

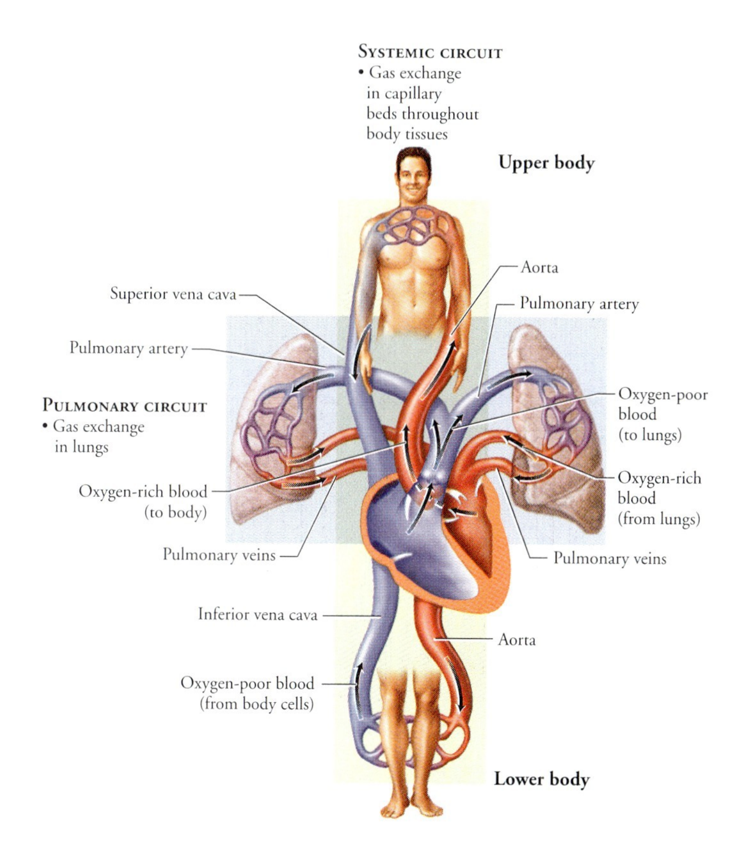

Fig. 12.10

NOTES

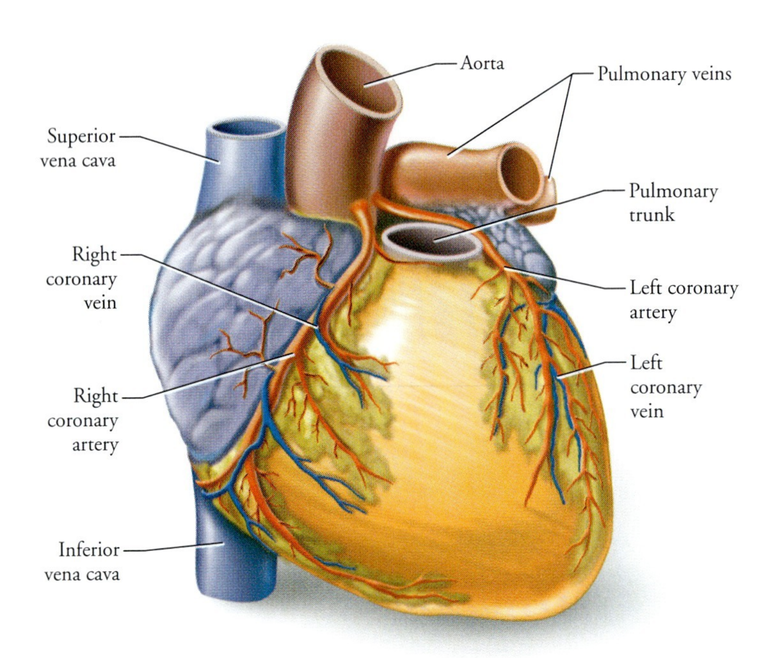

Fig. 12.11

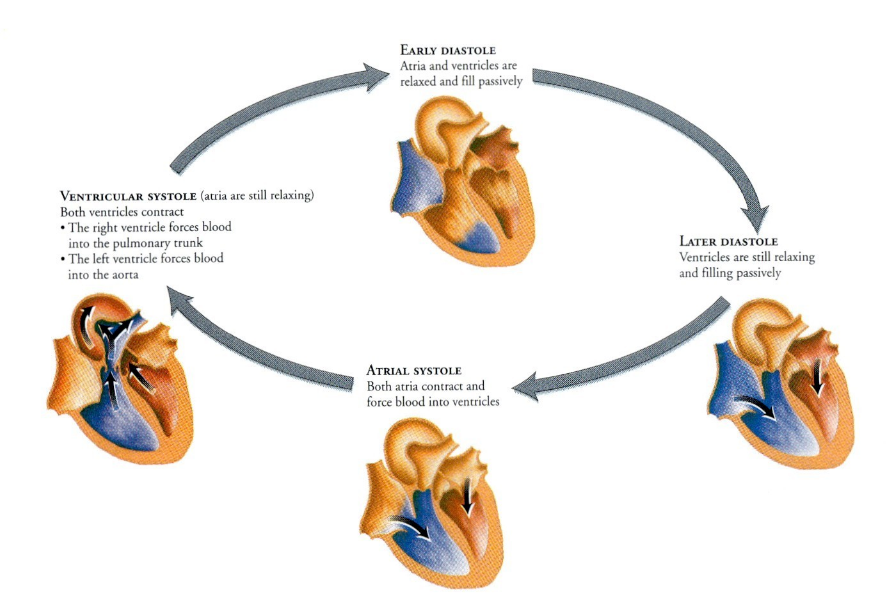

Fig. 12.12

NOTES

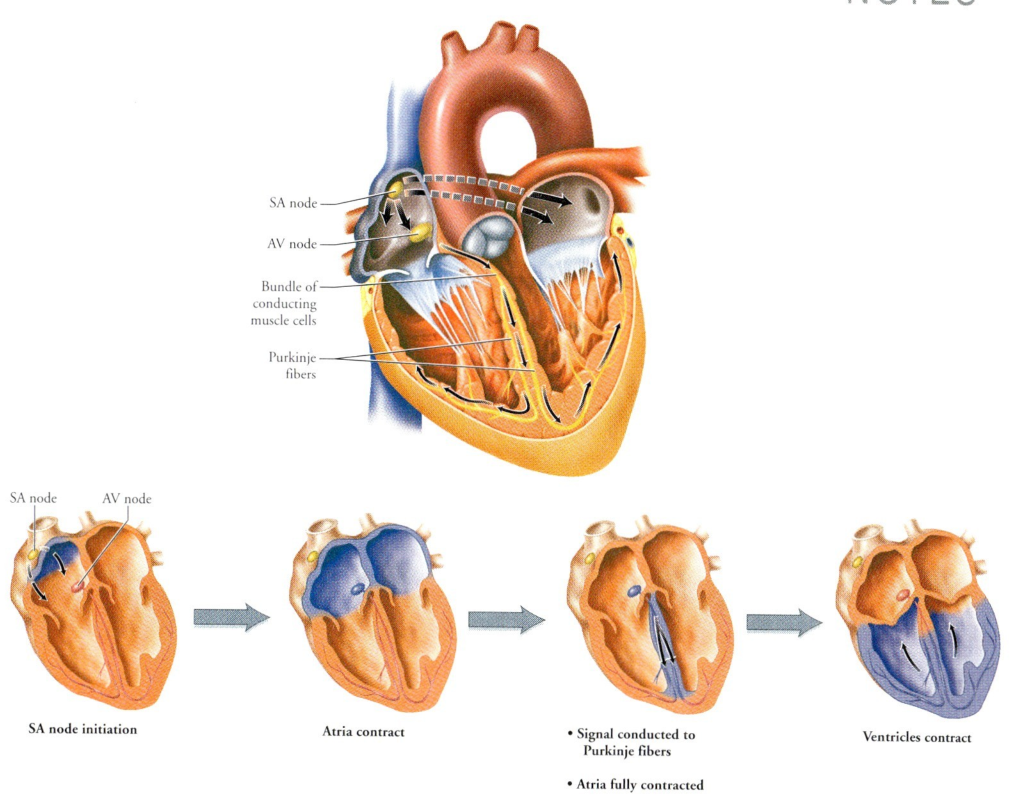

Fig. 12.13

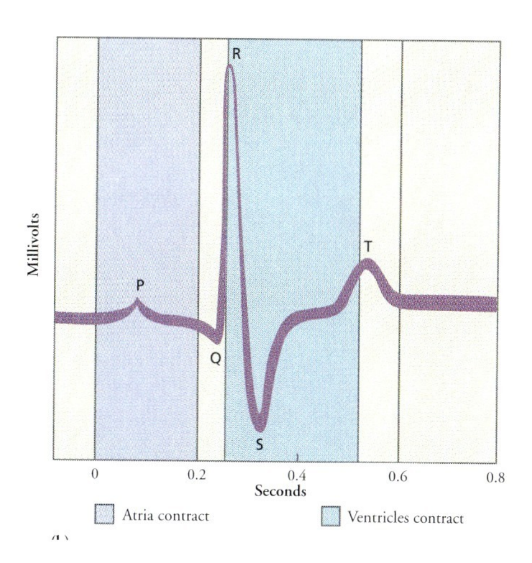

Fig. 12.14

NOTES

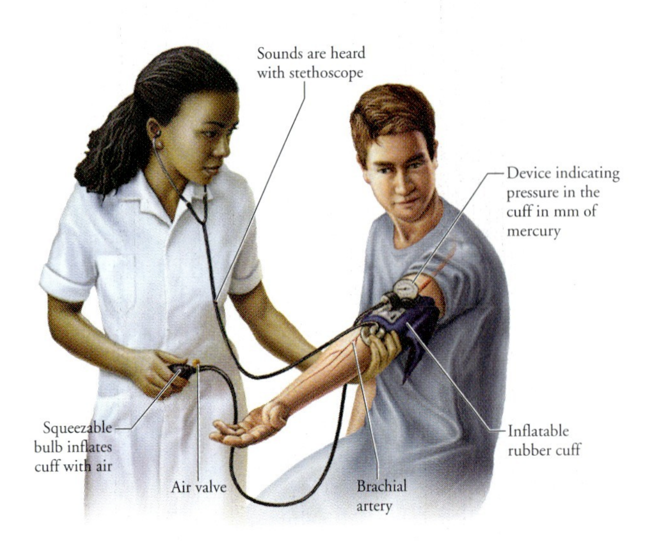

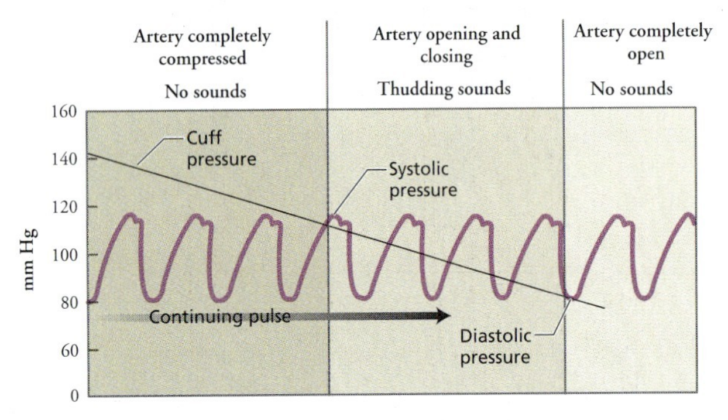

Fig. 12.15

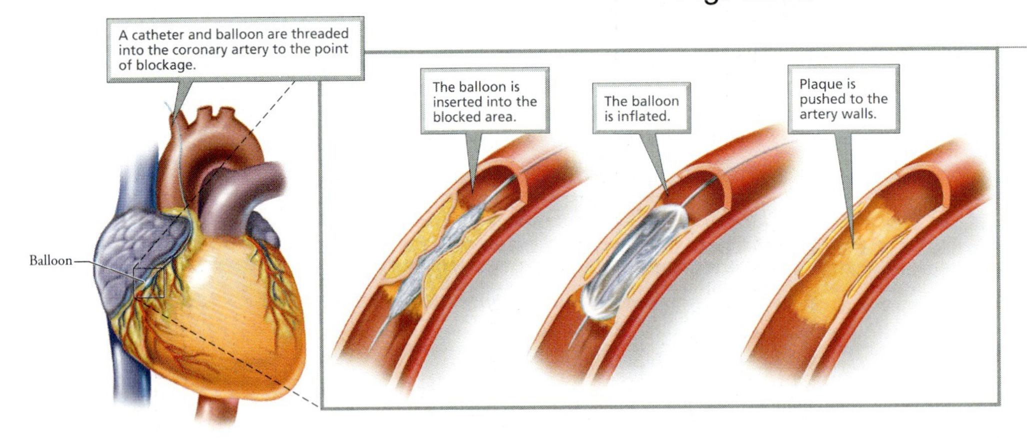

Fig. 12.17

NOTES

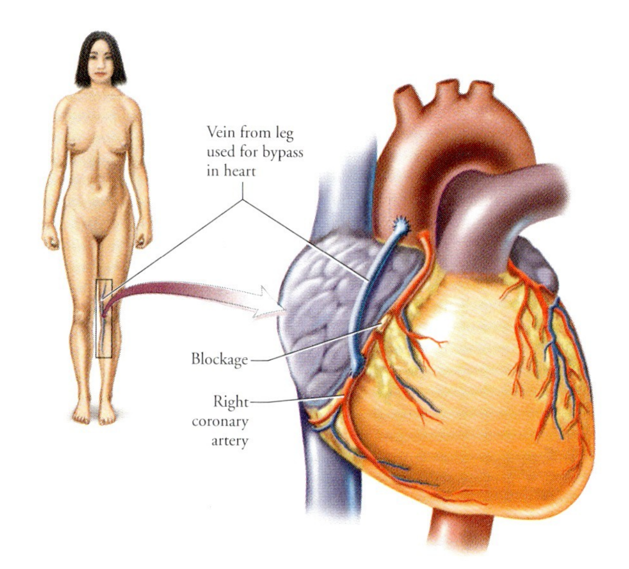

Fig. 12.18

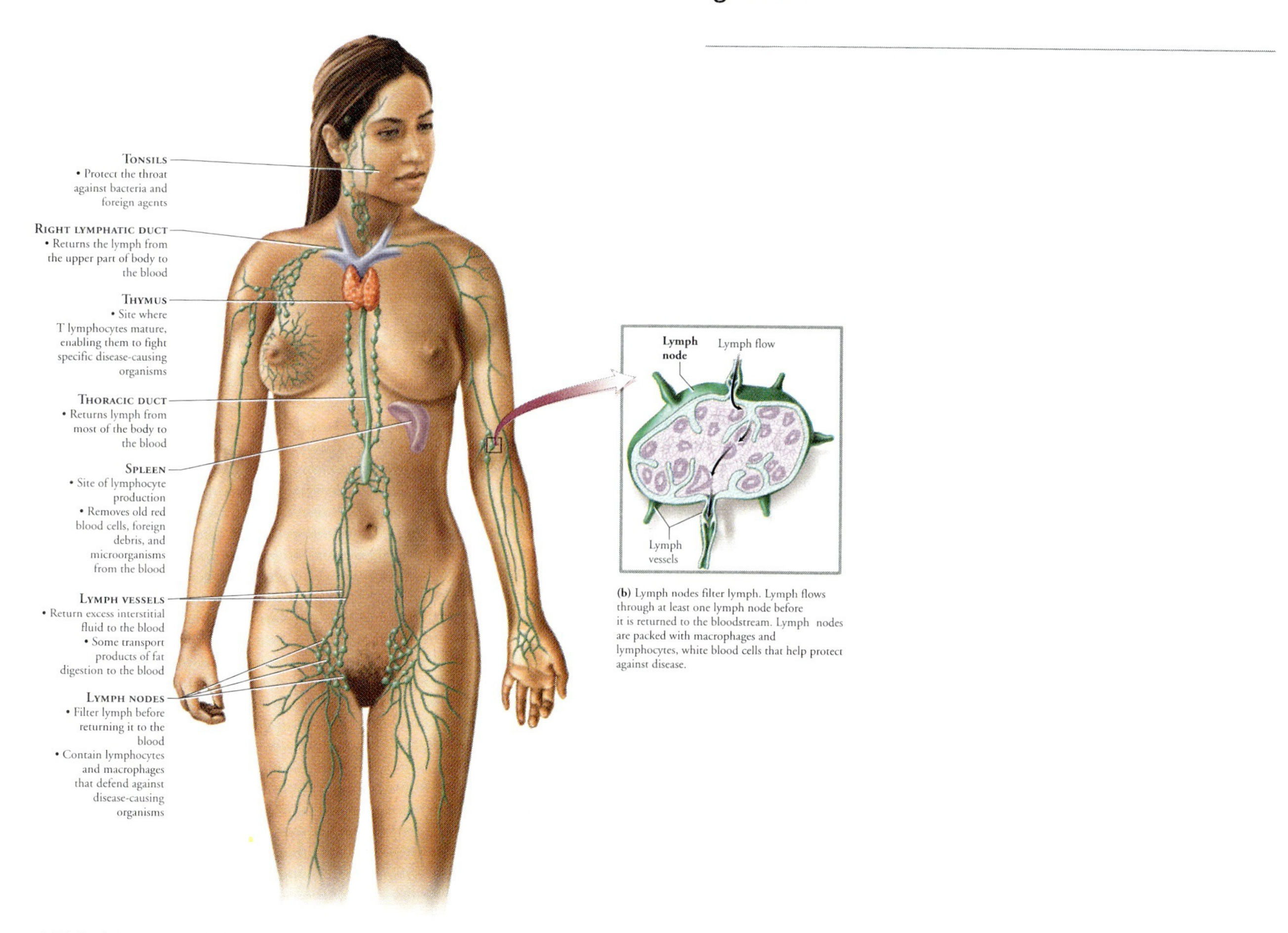

(a) The lymphatic system returns the fluid to the bloodstream that previously left the capillaries to bathe the cells, protects against disease-causing organisms, and transports products of fat digestion from the small intestine to the bloodstream.

(b) Lymph nodes filter lymph. Lymph flows through at least one lymph node before it is returned to the bloodstream. Lymph nodes are packed with macrophages and lymphocytes, white blood cells that help protect against disease.

Fig. 12.20

NOTES

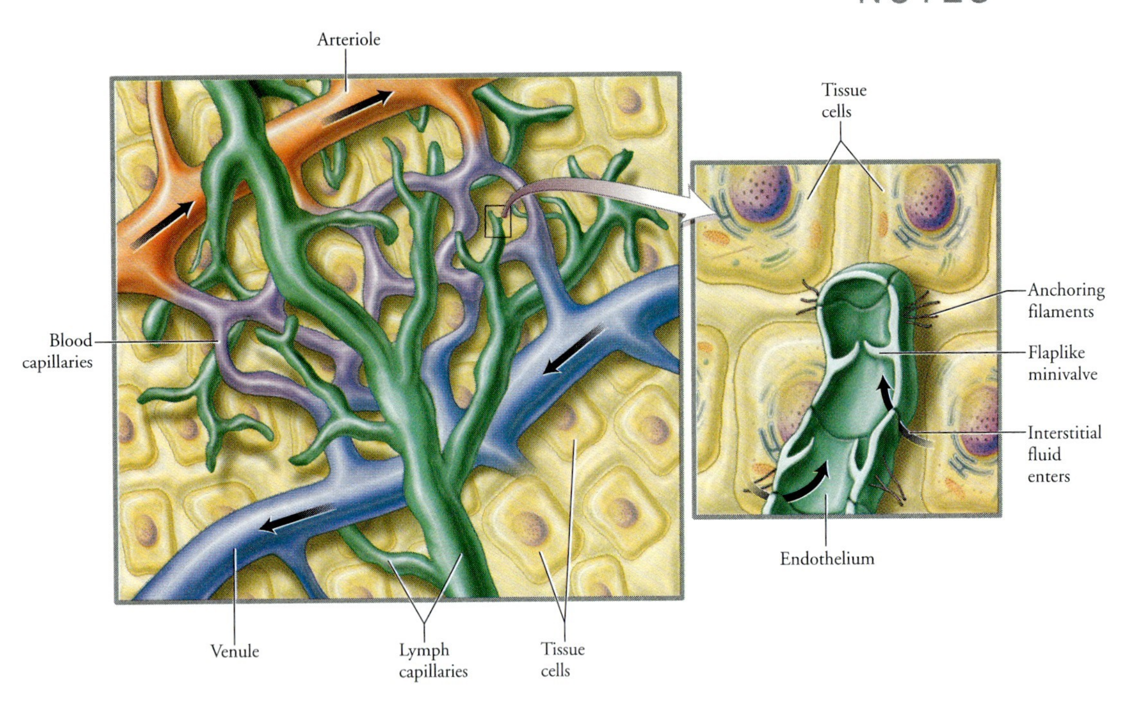

Fig. 12.22

NOTES

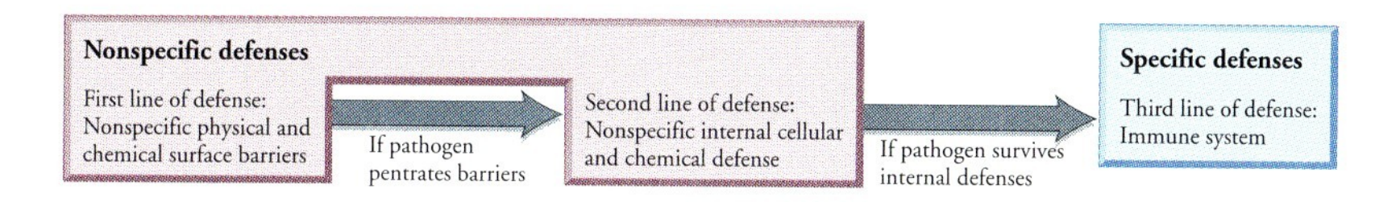

Fig. 13.2

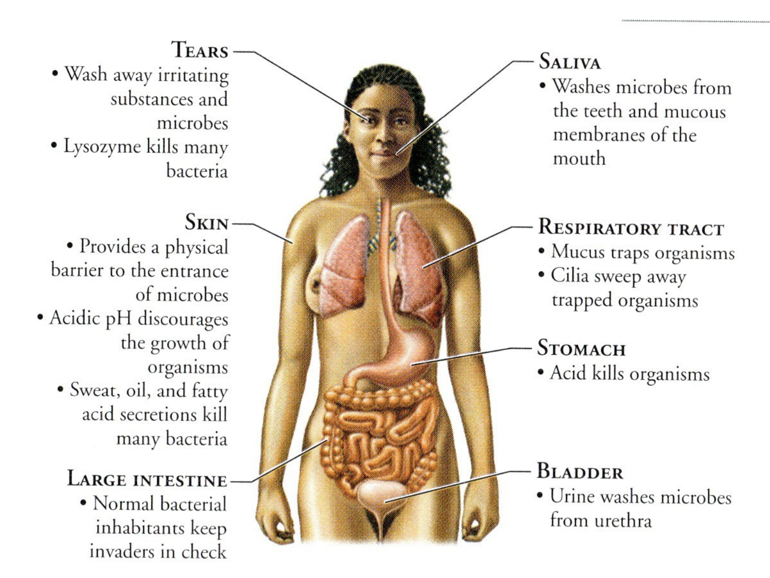

Fig. 13.3

THE SECOND LINE OF DEFENSE—NONSPECIFIC INTERNAL DEFENSES

DEFENSE	EXAMPLE	FUNCTION
Defensive cells	Phagocytic cells such as neutrophils and macrophages	Engulf invading organisms
	Eosinophils	Kill parasites
	Natural killer cells	Kill many invading organisms and cancer cells
Defensive proteins	Interferons	Slow the spread of viruses in the body
	Complement system	Stimulates histamine release; promotes phagocytosis; kills bacteria; enhances inflammation
Inflammation	Widening of blood vessels and increased capillary permeability lead to redness, heat, swelling, and pain	Brings in defensive cells and speeds healing
Fever	Abnormally high body temperature	Slows the growth of bacteria; speeds up body defenses

Table 13.1

NOTES

Fig. 13.4

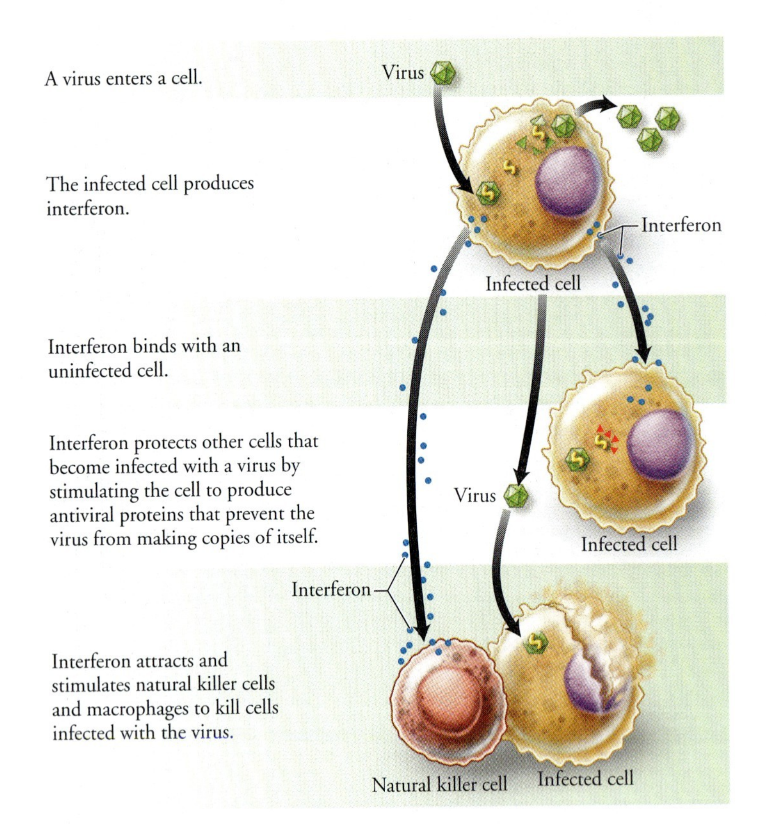
A virus enters a cell.
Virus
The infected cell produces interferon.
Interferon
Infected cell
Interferon binds with an uninfected cell.
Interferon protects other cells that become infected with a virus by stimulating the cell to produce antiviral proteins that prevent the virus from making copies of itself.
Virus
Infected cell
Interferon
Interferon attracts and stimulates natural killer cells and macrophages to kill cells infected with the virus.
Natural killer cell
Infected cell

Fig. 13.6

NOTES

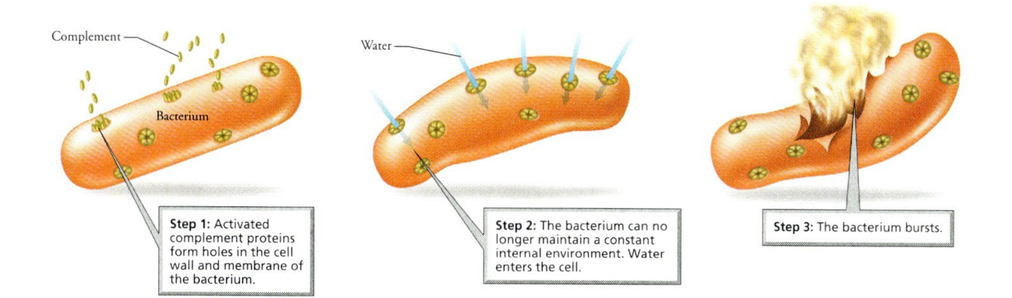

Fig. 13.7

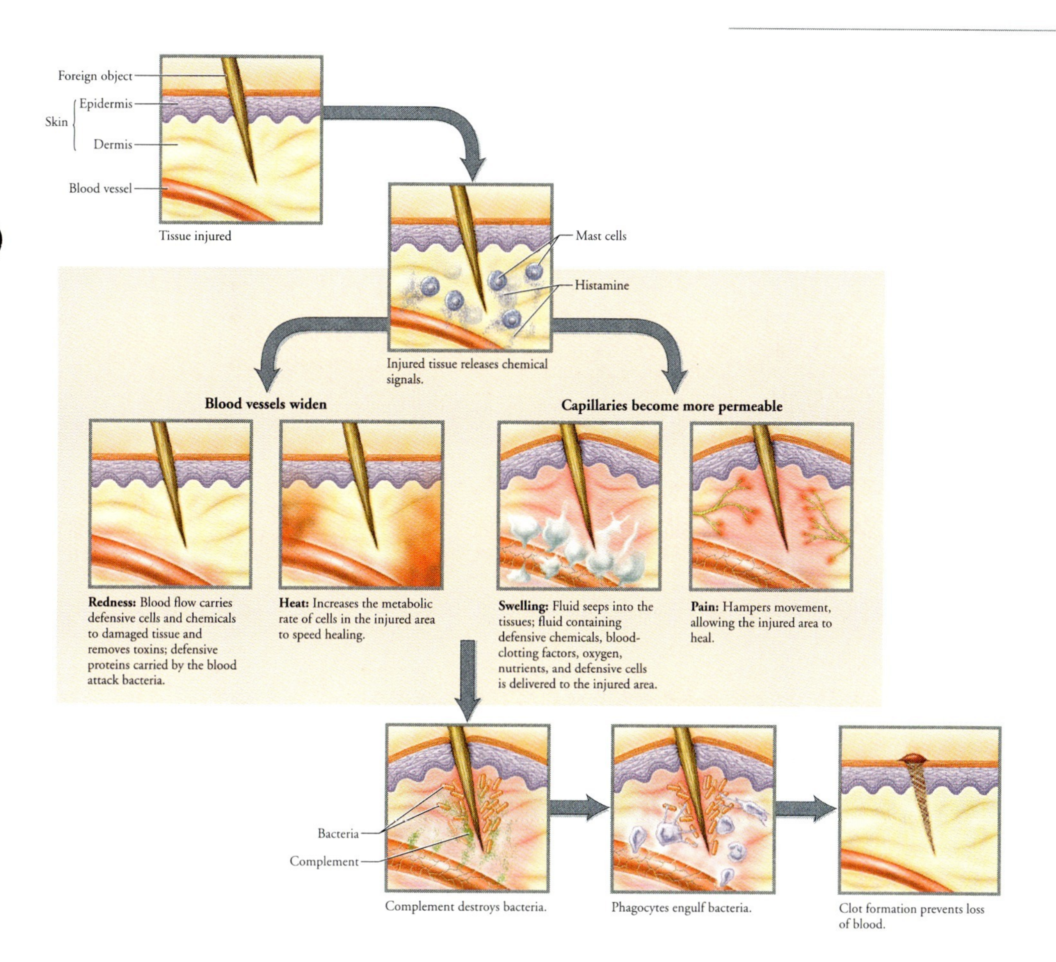

Fig. 13.8

NOTES

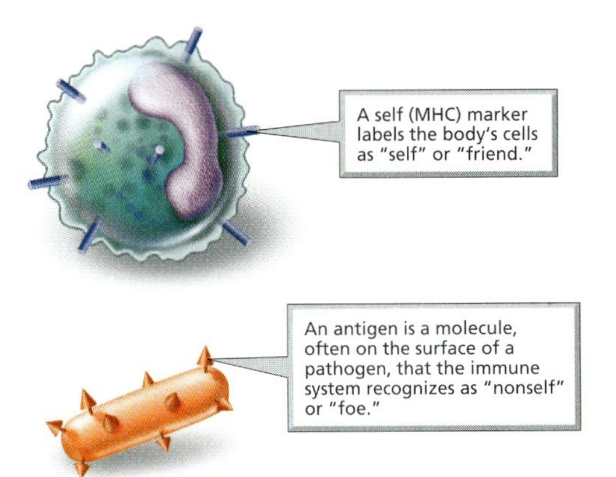

Fig. 13.10

CELLS INVOLVED IN THE IMMUNE RESPONSE	
CELL	**FUNCTIONS**
Macrophage	**An antigen-presenting cell** • Engulfs and digests antigens • Places a piece of consumed antigen on its plasma membrane • Presents the antigen to a helper T cell • Activates the helper T cell
T Cells	
Helper T cell	**The "on" switch for both lines of immune response** • After activation by macrophage, it divides, forming effector helper T cells and memory helper T cells • Helper T cells activate B cells and T cells
Cytotoxic T cell (effector T cell)	**Responsible for cell-mediated immune responses** • When activated by helper T cell, it divides to form effector cytotoxic T cells and memory cytotoxic T cells • Destroys cellular targets, such as virus-infected body cells, bacteria, fungi, parasites, and cancer cells
Suppressor T cell	**The "off" switch for immune responses** Suppresses the activity of B cells and T cells after the foreign cell or molecule has been successfully destroyed
B Cells	**Involved in antibody-mediated responses** When activated by helper T cell, it divides to form plasma cells and memory cells
Plasma cell	**Effector in antibody-mediated response** Secretes antibodies specific to the invader
Memory cells	**Responsible for memory of immune system** • Generated by B cells or any type of T cell during an immune response • Enable quick and efficient response on subsequent exposures of the antigen • May live for years

Table 13.2

NOTES

Steps in the Immune Response	
1. Threat	Foreign cell or molecule enters the body
2. Detection	Macrophage detects foreign cell or molecule and engulfs it
3. Alert	• Macrophage puts antigen from the pathogen on its surface and finds the helper T cell with correct receptors for that antigen • Macrophage presents antigen to the helper T cell • Macrophage alerts the helper T cell that there is an invader that "looks like" the antigen • Macrophage activates the helper T cell
4. Alarm	Helper T cell activates both lines of defense to fight that specific antigen
5. Build specific defense (clonal selection)	• Antibody-mediated defense—B cells are activated and divide to form plasma cells that secrete antibodies specific to the antigen • Cell-mediated defense—T cells divide to form cytotoxic T cells that attack cells with the specific antigen
6. Defense	• Antibody-mediated defense—antibodies specific to antigen eliminate the antigen • Cell-mediated defense—cytotoxic T cells cause cells with the antigen to burst
7. Continued surveillance	Memory cells formed when helper T cells, cytotoxic T cells, and B cells were activated remain to provide swift response if the antigen is detected again
8. Withdrawal of forces	Once the antigen has been destroyed, suppressor T cells shut down the immune response to that antigen

Table 13.3

NOTES

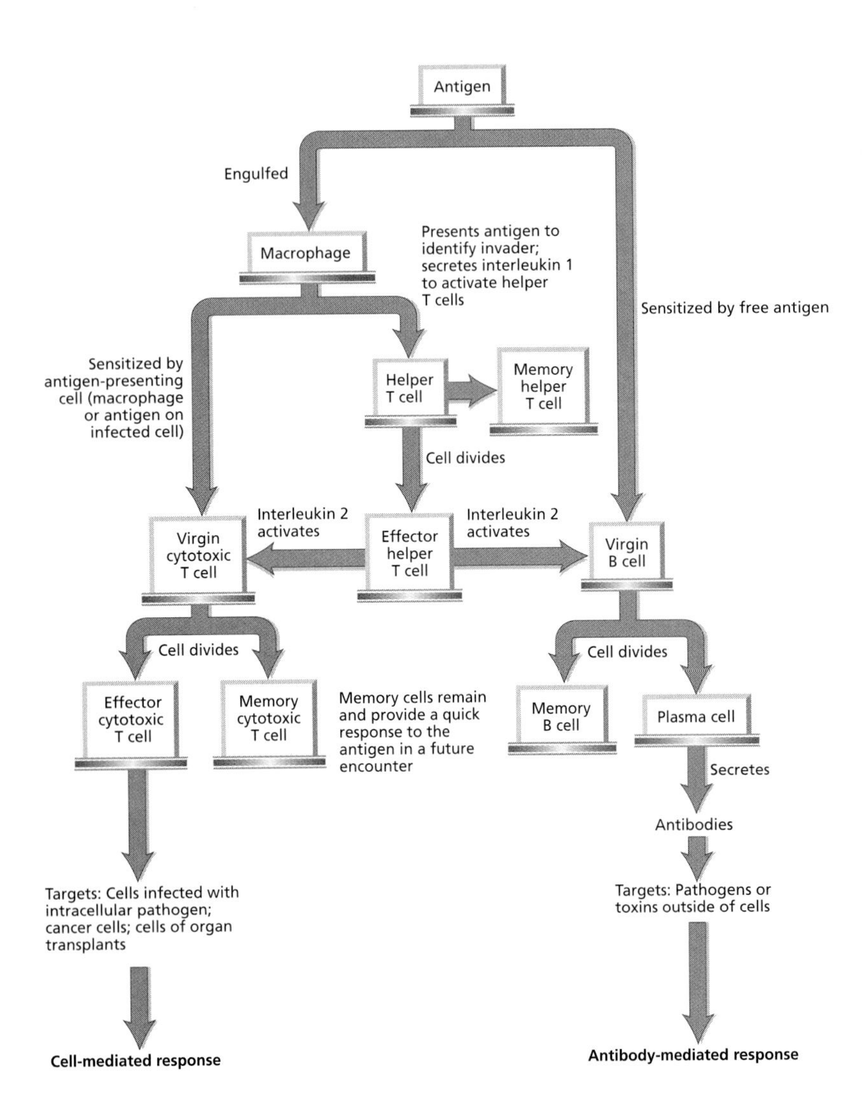

Fig. 13.11

NOTES

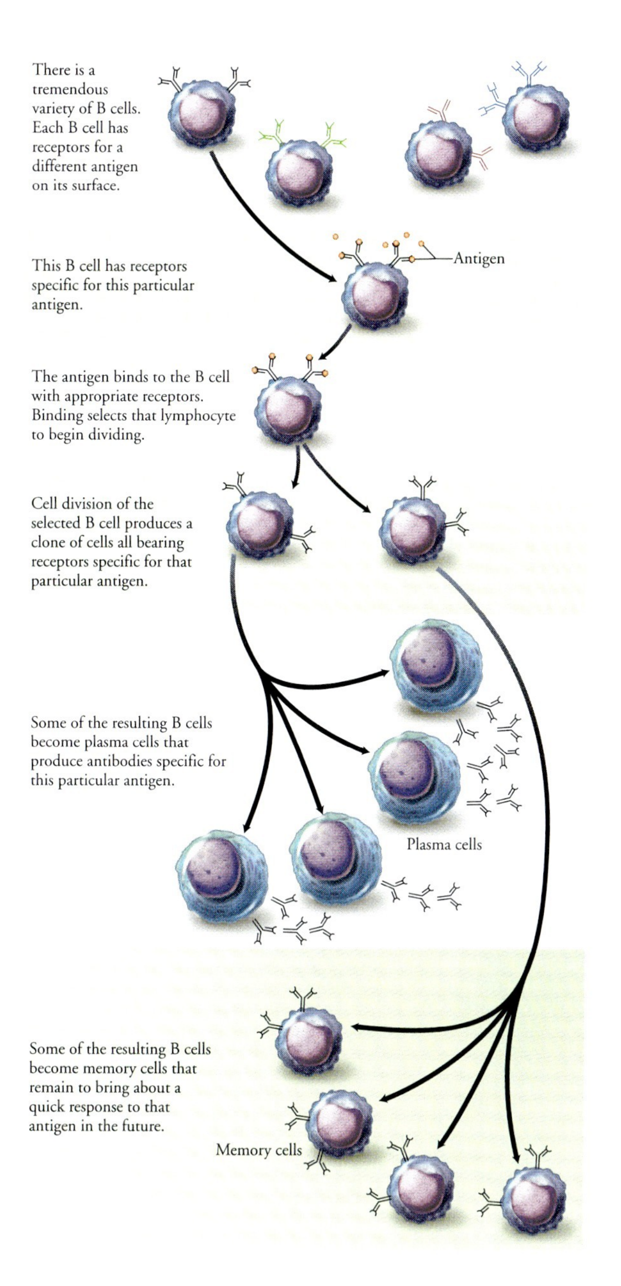

Fig. 13.12

NOTES

1. **Threat**
- An invader enters the body

2. **Detection**
- A macrophage encounters, engulfs, and digests the invader (e.g., a bacterium)
- The macrophage places a piece of the invader (antigen) on its surface with the "self" (MHC) marker

3. **Alert**
- The macrophage presents the antigen to a helper T cell
- The macrophage secretes interleukin 1, which activates the helper T cell
- A virgin, or memory B, cell is sensitized by binding to the antigen
- The B cell sensitized by an antigen binds to a helper T cell that has been activated by a macrophage

4. **Alarm**
- The helper T cell secretes interleukin 2, which stimulates the B cell to begin dividing
- Two populations of B cells are formed: memory cells and plasma cells

5. **Building specific defense**
- The B cell divides and forms plasma cells and memory cells

6. **Defense**
- Plasma cells secrete antibodies specific for that antigen
- Antibodies circulate in the blood, where they neutralize and agglutinate the target antigen and activate the complement system

7. **Continued surveillance**
- Memory B cells remain and mount a quick response if the invader is encountered again

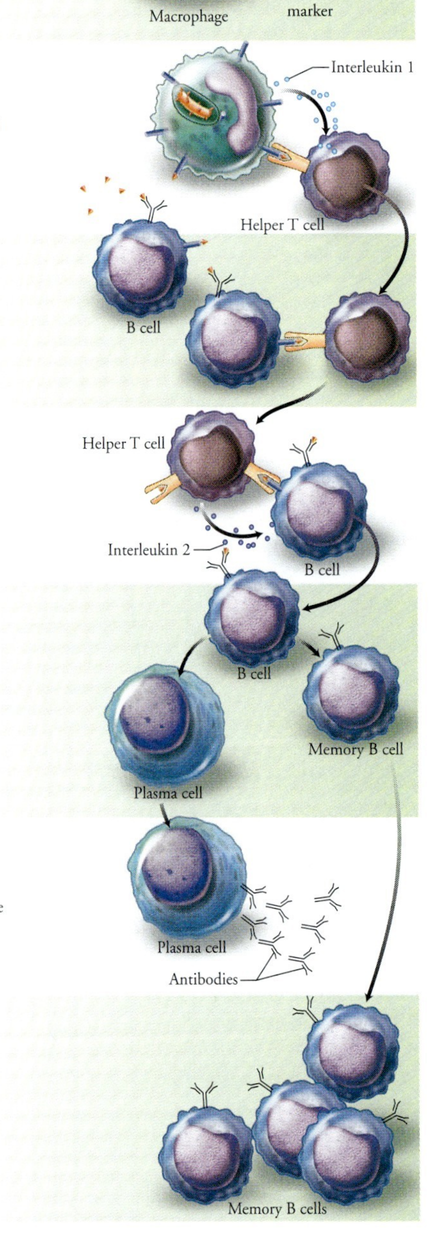

Fig. 13.13

NOTES

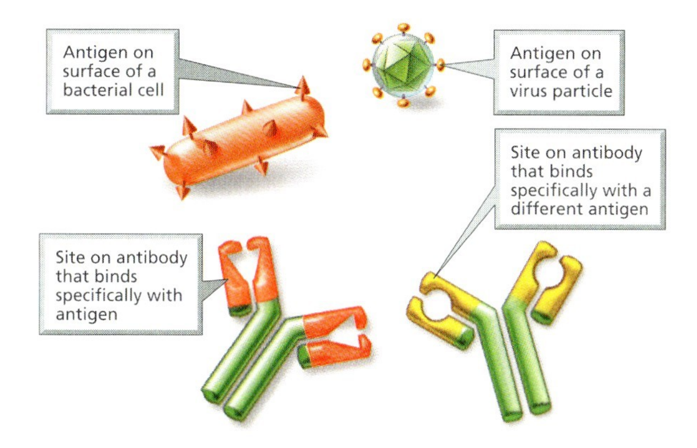

Fig. 13.14

CLASSES OF ANTIBODIES

CLASS	STRUCTURE	LOCATION	CHARACTERISTICS	PROTECTIVE FUNCTIONS
IgG	Monomer	Blood, lymph, and the intestines	Most abundant of all antibodies in body; involved in primary and secondary immune responses; can pass through placenta from mother to fetus and provides passive immune protection to fetus and newborn	Enhances phagocytosis; neutralizes toxins; triggers complement system
IgA	Dimer or monomer	Present in tears, saliva, and mucus as well in secretions of gastrointestinal system and excretory systems; present in breast milk	Levels decrease during stress, raising susceptibility to infection	Prevents pathogens from attaching to epithelial cells of surface lining
IgM	Pentamer	Attached to B cell where it acts as a receptor for antigens; free in blood and lymph	First Ig class released by plasma cell during primary response	Powerful agglutinating agent (10 antigen binding sites); activates complement
IgD	Monomer	Surface of many B cells; blood and lymph	Life span of about 3 days	Thought to be involved in recognition of antigen and in activating B cells
IgE	Monomer	Secreted by plasma cells in skin, mucous membranes of gastrointestinal and respiratory systems	Become bound to surface of mast cells and basophils	Involved in allergic reactions by triggering release of histamine and other chemicals from mast cells or basophils

Table 13.4

NOTES

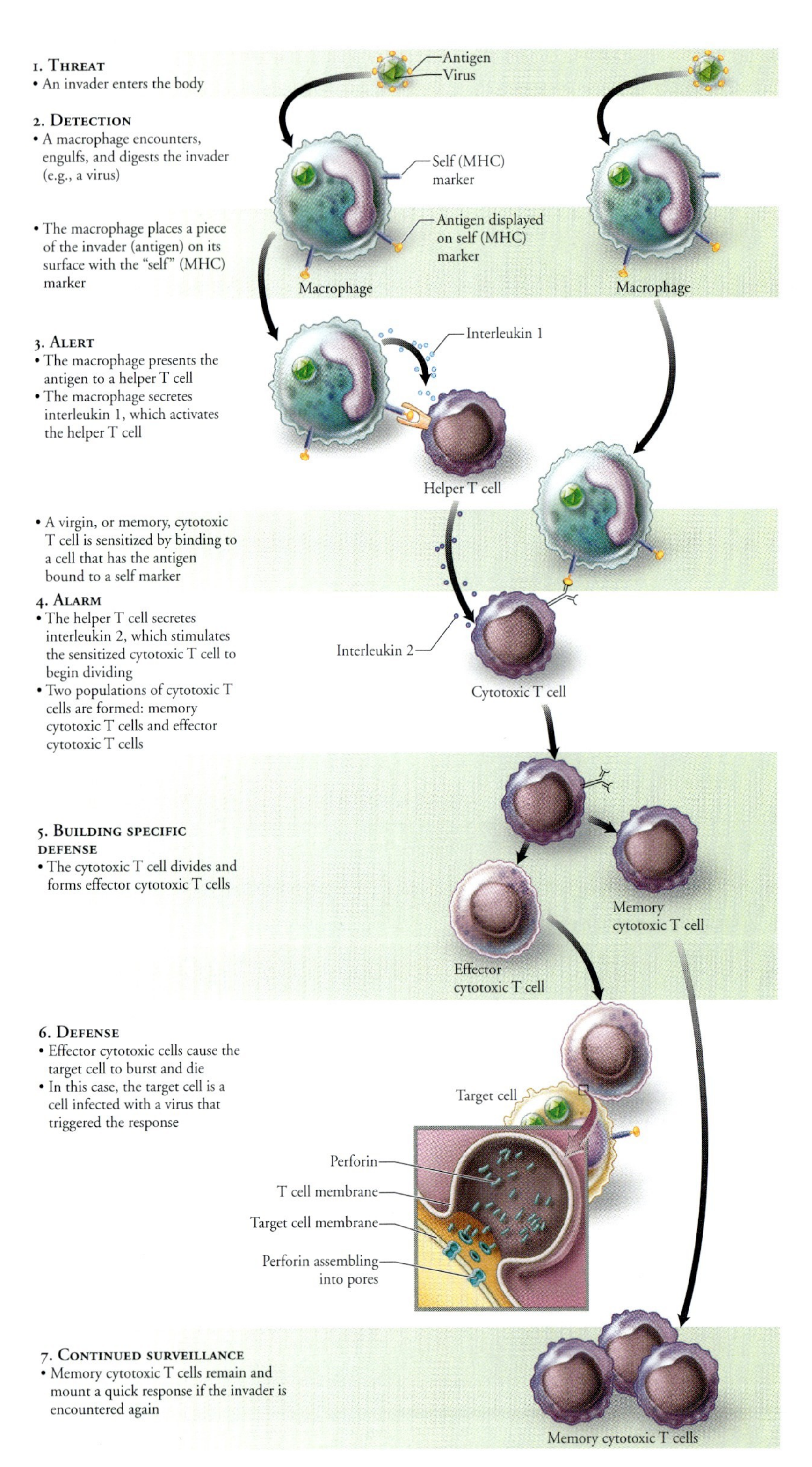
1. Threat
• An invader enters the body
Antigen
Virus
2. Detection
• A macrophage encounters, engulfs, and digests the invader (e.g., a virus)
Self (MHC) marker
• The macrophage places a piece of the invader (antigen) on its surface with the "self" (MHC) marker
Antigen displayed on self (MHC) marker
Macrophage
Macrophage
3. Alert
• The macrophage presents the antigen to a helper T cell
• The macrophage secretes interleukin 1, which activates the helper T cell
Interleukin 1
Helper T cell
• A virgin, or memory, cytotoxic T cell is sensitized by binding to a cell that has the antigen bound to a self marker
4. Alarm
• The helper T cell secretes interleukin 2, which stimulates the sensitized cytotoxic T cell to begin dividing
• Two populations of cytotoxic T cells are formed: memory cytotoxic T cells and effector cytotoxic T cells
Interleukin 2
Cytotoxic T cell
5. Building specific defense
• The cytotoxic T cell divides and forms effector cytotoxic T cells
Memory cytotoxic T cell
Effector cytotoxic T cell
6. Defense
• Effector cytotoxic cells cause the target cell to burst and die
• In this case, the target cell is a cell infected with a virus that triggered the response
Target cell
Perforin
T cell membrane
Target cell membrane
Perforin assembling into pores
7. Continued surveillance
• Memory cytotoxic T cells remain and mount a quick response if the invader is encountered again
Memory cytotoxic T cells

Fig. 13.15

NOTES

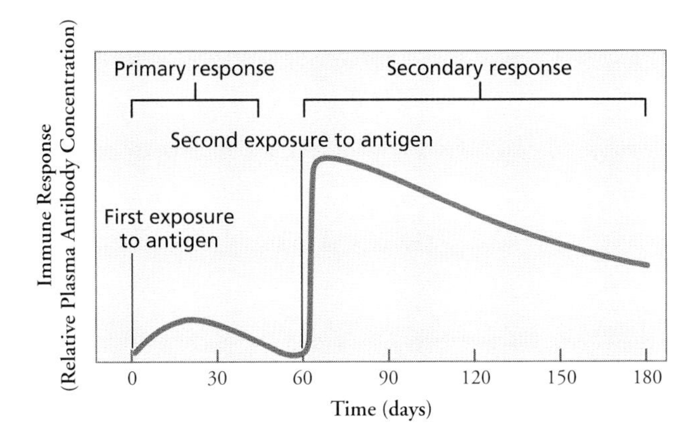

Fig. 13.16

RECOMMENDED IMMUNIZATION SCHEDULE IN 2003–2004	
VACCINE	AGE
Hepatitis B #1	Birth–2 months
Hepatitis B #2	1–4 months
Hepatitis B #3	6–18 months
DPT (Diphtheria, Tetanus, Pertussis)	2, 4, & 6 months
Booster	15–18 months
Booster	4–6 years
DT (Diphtheria Tetanus)	11–18 years
Haemophilus influenzae type b	2, 4, & 6 months
Booster	12–15 months
Polio	2 & 4 months
	6–18 months
	4–6 years
MMR (Measles, Mumps, Rubella)	12–15 months
Booster	4–6 years
Varicella (chickenpox)	12–18 months
PCV (pneumococcal conjugated vaccine)	2–15 months

Table 13.5

NOTES

Autoimmune Disorders		
AUTOIMMUNE DISORDER	**TARGET OF IMMUNE SYSTEM ATTACK**	**EFFECT**
Organ-specific		
Hashimoto's thyroiditis	Thyroid gland	Decreased production of thyroid hormone
Pernicious anemia	Cells in stomach lining that produce a chemical needed to absorb vitamin B_{12}, which is needed for red blood cell production	Decreased production of red blood cells
Addison's disease	Adrenal glands	Adrenal failure
Diabetes mellitus type 1	Insulin-producing cells in the pancreas	Elevated blood sugar
Graves' disease	Thyroid gland	Increased rate of chemical reactions in body
Multiple sclerosis	Myelin sheath of nerve cells	Short-circuiting of neural impulses resulting in sensory and/or motor defects
Ankylosing spondylitis	Joints between vertebrae	Spine bent and fused, inflammation
Myasthenia gravis	Connections between nerve and muscle	Muscle weakness
Glomerulonephritis	Kidney	Kidney failure
Encephalitis	Brain	Impaired brain functions, headaches, irritability, double vision, impaired speech
Non-organ-specific		
Lupus erythematosus	Connective tissue	Butterfly-shaped rash on face, skin lesions, joint pain
Rheumatoid arthritis	Collagen fibers of joints	Joint pain
Dermatomyositis	Skin inflammation	Body rash
Rheumatic fever	Heart valves, joints	Joint pain, kidney failure

Table 13.6

NOTES

TYPE OF ALLERGIC RESPONSE	COMMON CAUSES	LOCATION OF REACTIVE MAST CELLS	SYMPTOMS
Hay fever (allergic rhinitis)	Pollen, mold spores, animal dander (bits of skin and hair), feces of dust mites	Lining of nasal cavity	Sneezing, nasal congestion
Asthma	Pollen, mold spores, animal dander	Airways of lower respiratory tract	Difficulty breathing
Food allergy	Chicken, eggs, fish, milk, nuts (especially peanuts), shellfish, soybeans, and wheat	Lining of digestive system	Nausea, vomiting, abdominal cramps, and diarrhea
Hives	Foods (especially shellfish, strawberries, chocolate, nuts, and tomatoes); insect bites; certain drugs (especially penicillin and aspirin); chemicals such as food additives, dyes, and cosmetics	Skin	Patches of skin become red and swollen
Anaphylactic shock	Insect stings (especially from bees, wasps, hornets, yellow jackets, fire ants); medicines (especially penicillin and tetracycline); certain foods (especially eggs, seafood, nuts, and grains)	Throughout the body	Widening of blood vessels, causing blood to pool in capillaries and resulting in dizziness, nausea, diarrhea, and unconsciousness; death

COMMON ALLERGIES

Table 13.7

NOTES

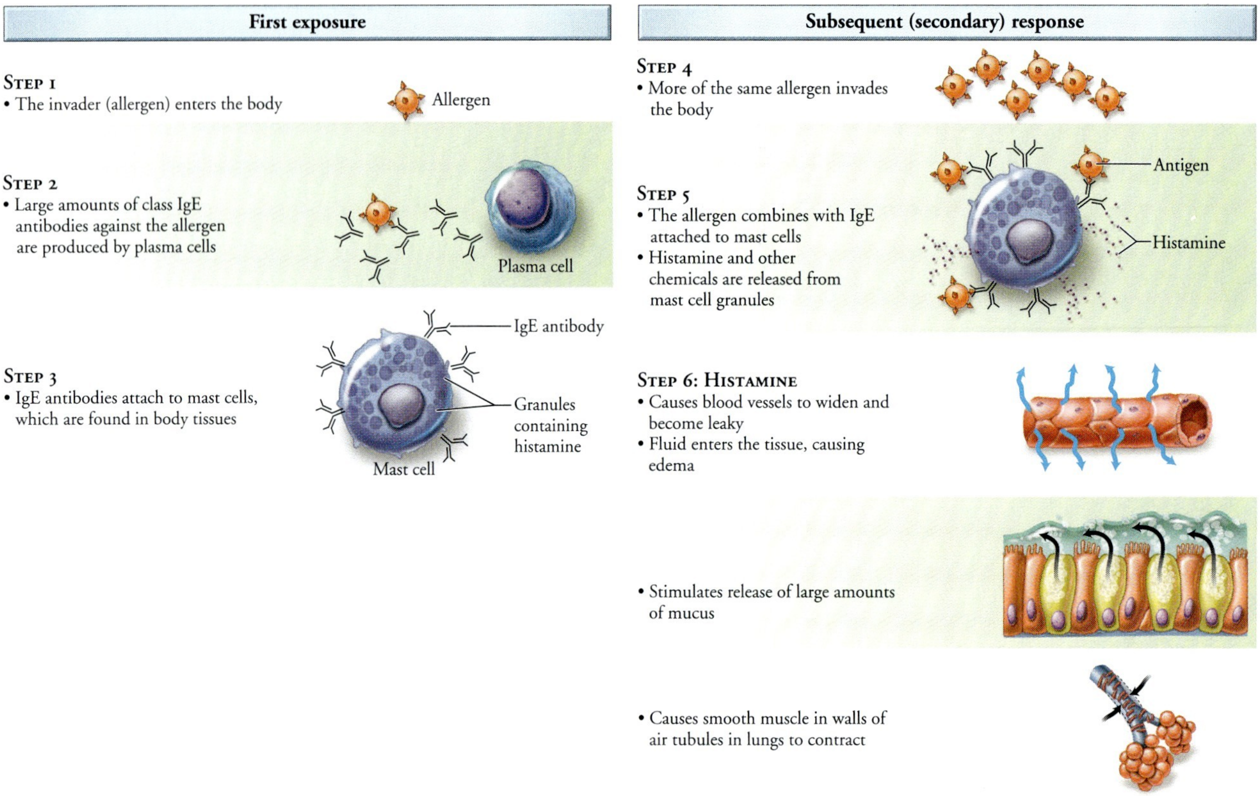
First exposure
Subsequent (secondary) response
STEP 1
• The invader (allergen) enters the body
Allergen
STEP 2
• Large amounts of class IgE antibodies against the allergen are produced by plasma cells
Plasma cell
STEP 3
• IgE antibodies attach to mast cells, which are found in body tissues
IgE antibody
Granules containing histamine
Mast cell
STEP 4
• More of the same allergen invades the body
STEP 5
• The allergen combines with IgE attached to mast cells
• Histamine and other chemicals are released from mast cell granules
Antigen
Histamine
STEP 6: HISTAMINE
• Causes blood vessels to widen and become leaky
• Fluid enters the tissue, causing edema
• Stimulates release of large amounts of mucus
• Causes smooth muscle in walls of air tubules in lungs to contract

Fig. 13.19

NOTES

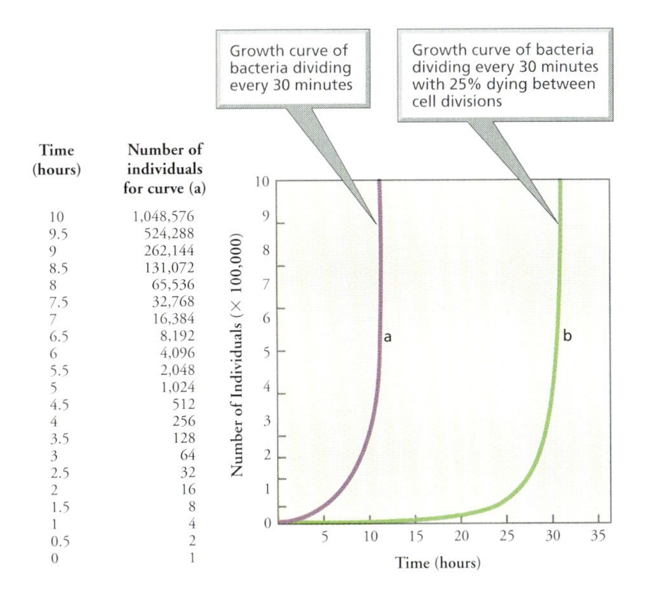

Time (hours)	Number of individuals for curve (a)
10	1,048,576
9.5	524,288
9	262,144
8.5	131,072
8	65,536
7.5	32,768
7	16,384
6.5	8,192
6	4,096
5.5	2,048
5	1,024
4.5	512
4	256
3.5	128
3	64
2.5	32
2	16
1.5	8
1	4
0.5	2
0	1

Fig. 13a.2

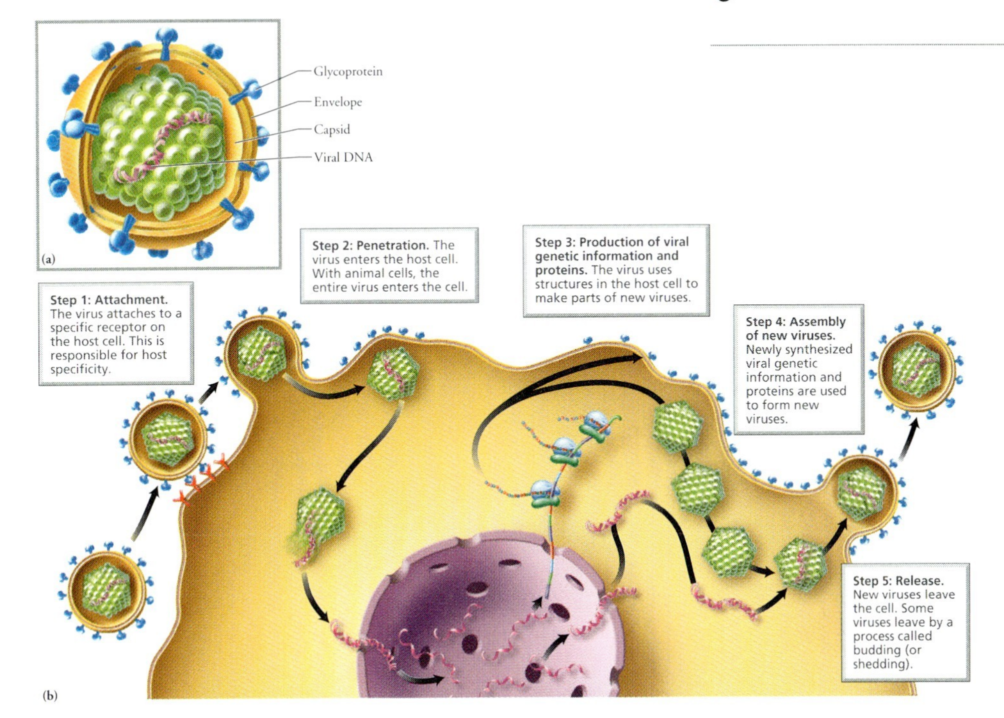

Fig. 13a.3

NOTES

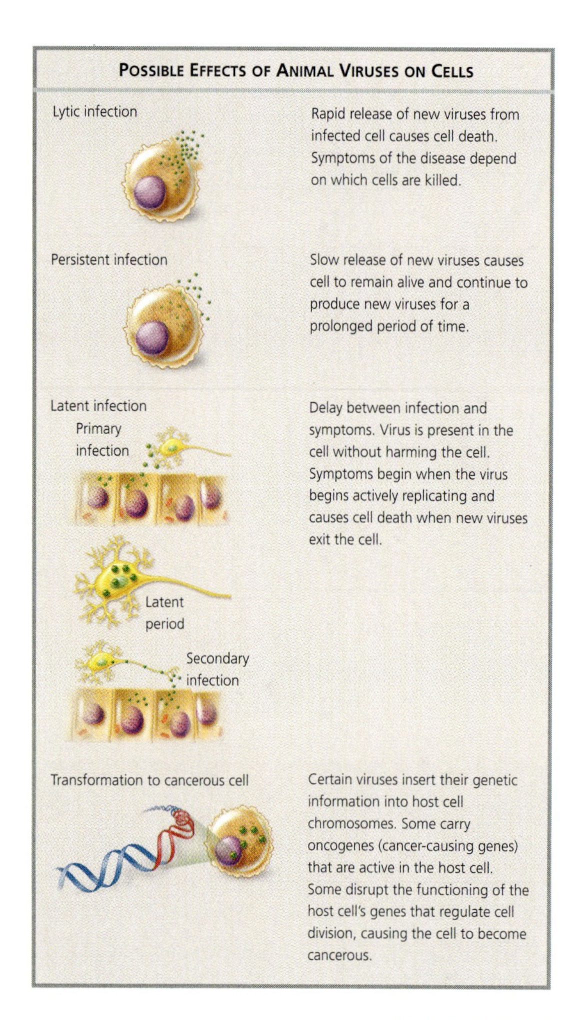

POSSIBLE EFFECTS OF ANIMAL VIRUSES ON CELLS	
Lytic infection	Rapid release of new viruses from infected cell causes cell death. Symptoms of the disease depend on which cells are killed.
Persistent infection	Slow release of new viruses causes cell to remain alive and continue to produce new viruses for a prolonged period of time.
Latent infection Primary infection Latent period Secondary infection	Delay between infection and symptoms. Virus is present in the cell without harming the cell. Symptoms begin when the virus begins actively replicating and causes cell death when new viruses exit the cell.
Transformation to cancerous cell	Certain viruses insert their genetic information into host cell chromosomes. Some carry oncogenes (cancer-causing genes) that are active in the host cell. Some disrupt the functioning of the host cell's genes that regulate cell division, causing the cell to become cancerous.

Table 13a.1

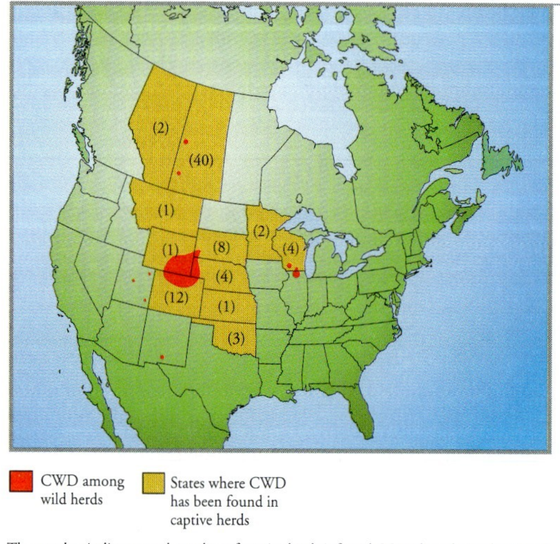

Fig. 13a.6

NOTES

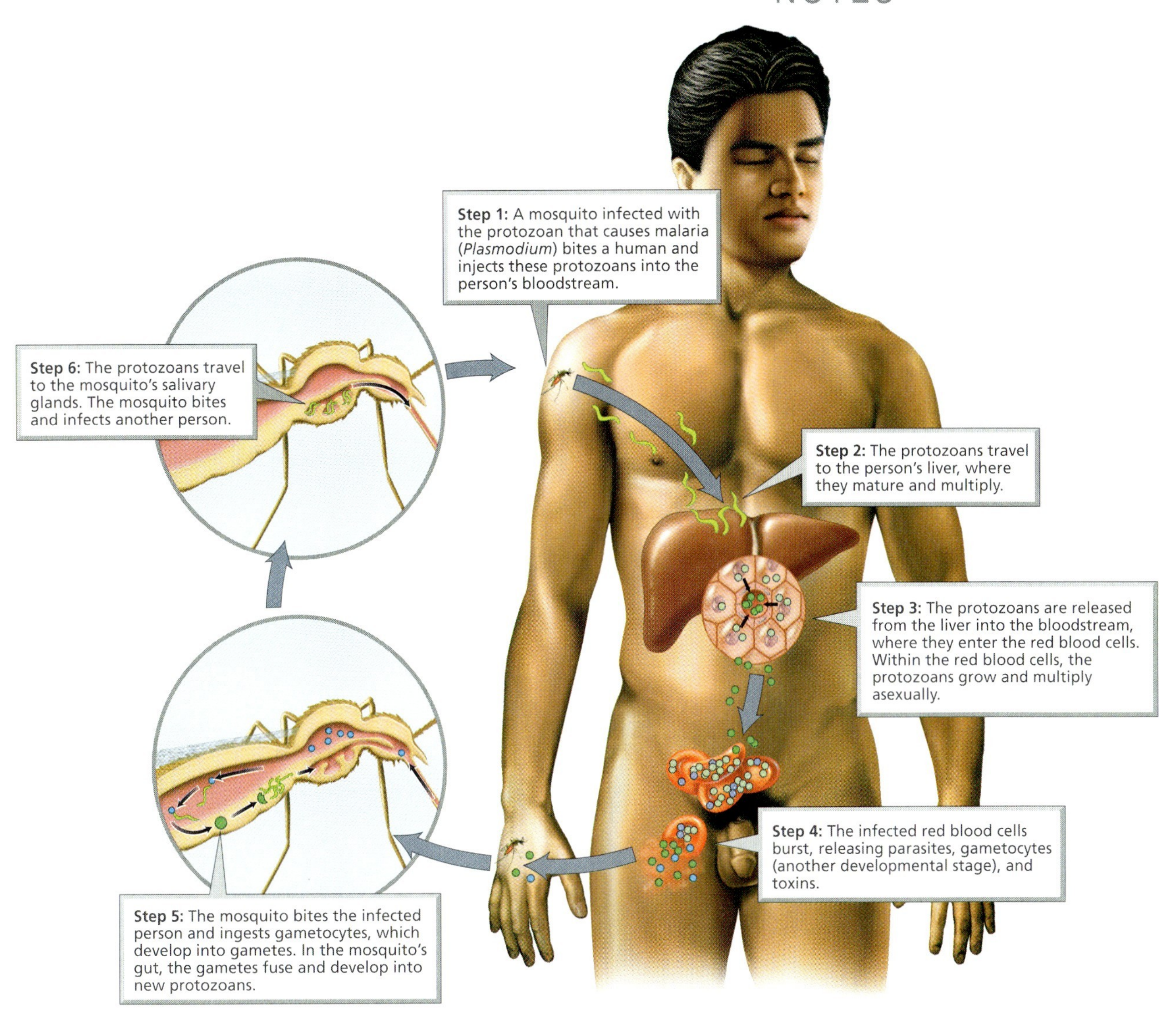

Fig. 13a.9

NOTES

EXAMPLES OF MODERN-DAY PLAGUES		
DISEASE	CAUSE	DISCUSSION
Meningitis	Bacterium or virus	Chapter 8
West Nile disease	Virus	Chapter 8
Malaria	Protozoan	Chapter 13a
Lyme disease	Bacterium	Chapter 13a
Transmissible spongiform encephalolopathies (TSEs; e.g., mad cow disease, chronic wasting syndrome, Creutzfeldt-Jakob disease)	Prion	Chapter 13a
Hantavirus pulmonary syndrome	Virus	Chapter 13a
Tuberculosis	Bacterium	Chapter 14
Severe acute respiratory syndrome (SARS)	Virus	Chapter 14
Influenza	Virus	Chapter 14
Hepatitis C	Virus	Chapter 15
Chlamydia	Bacterium	Chapter 17a
Gonorrhea	Bacterium	Chapter 17a
Genital herpes	Virus	Chapter 17a
Genital warts	Virus	Chapter 17a
HIV/AIDS	Virus	Chapter 17a

Table 13a.2

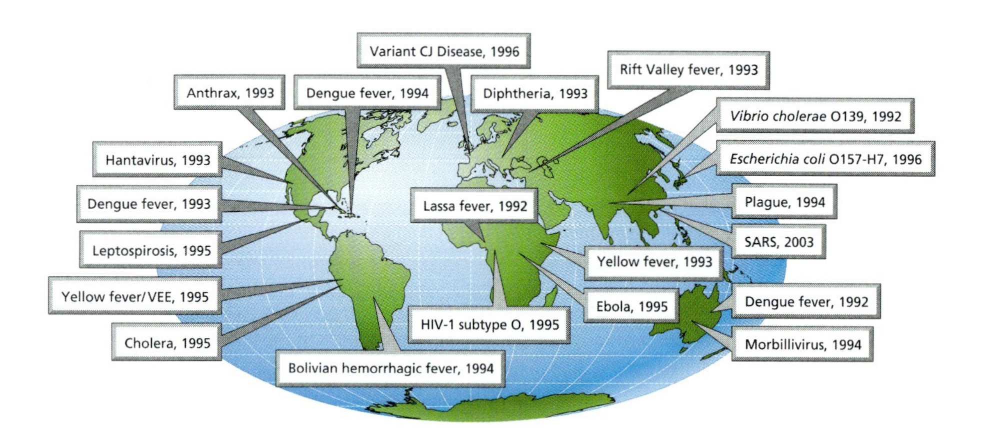

Fig. 13a.10

NOTES

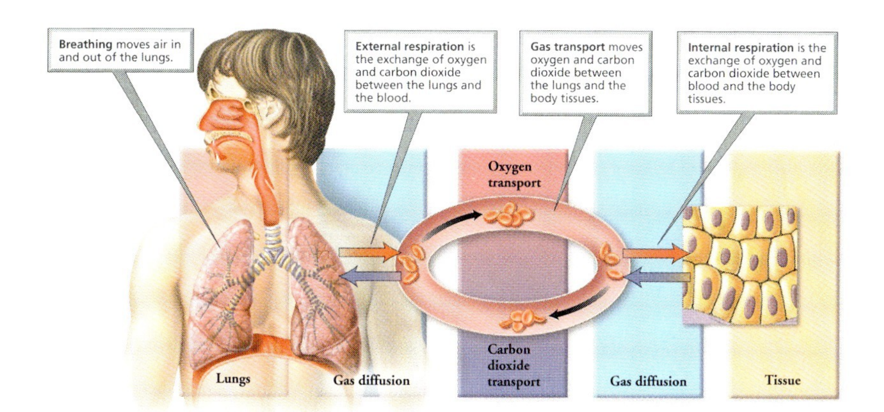

Fig. 14.1

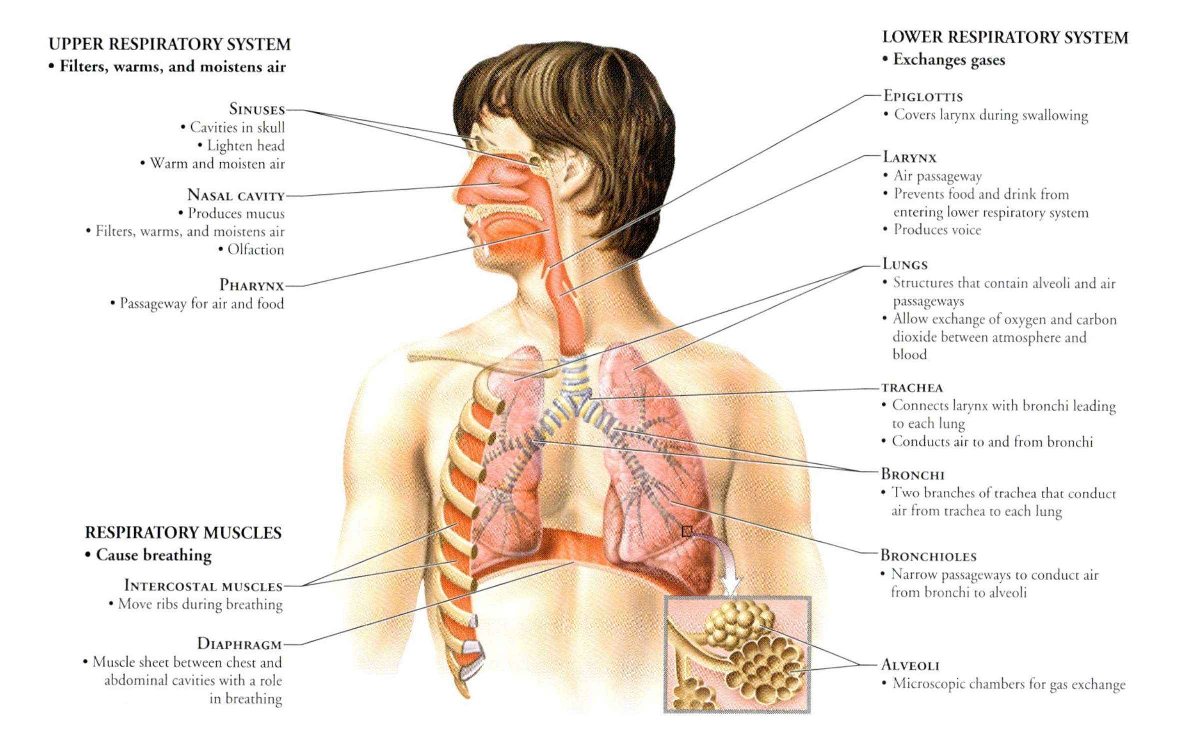

Fig. 14.2

NOTES

STRUCTURES OF THE RESPIRATORY SYSTEM		
STRUCTURE	DESCRIPTION	FUNCTION
Upper respiratory system		
Nasal cavity	Cavity within the nose, divided into right and left halves by nasal septum; has three shelflike bones	Filters and conditions (moistens and warms incoming air); olfaction (sense of smell)
Pharynx (throat)	Chamber connecting nasal cavities to esophagus and larynx	Common passageway for air, food, and drink
Lower respiratory system		
Larynx	Cartilaginous boxlike structure between the pharynx and trachea that contains the vocal cords and the glottis	Allows air but not other materials passage to the lower respiratory system; source of the voice
Epiglottis	Flap of tissue reinforced with cartilage	Covers the glottis during swallowing
Trachea	Tube reinforced with C-shaped rings of cartilage that leads from the larynx to the bronchi	The main airway; conducts air from larynx to bronchi
Bronchi (primary)	Two large branches of the trachea reinforced with cartilage	Conduct air from trachea to each lung
Bronchioles	Narrow passageways leading from bronchi to alveoli	Conduct air to alveoli; adjust airflow in lungs
Lungs	Two lobed, elastic structures within the thoracic (chest) cavity containing surfaces for gas exchange	Exchange oxygen and carbon dioxide between blood and air
Alveoli	Microscopic sacs within lungs, bordered by extensive capillary network	Provide immense, internal surface area for gas exchange

Table 14.1

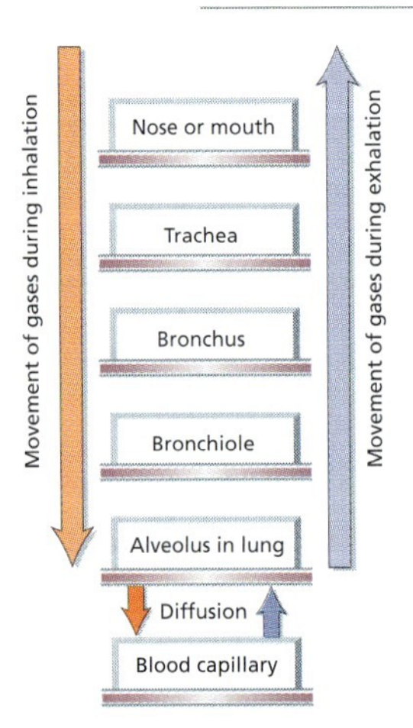

Fig. 14.3

NOTES

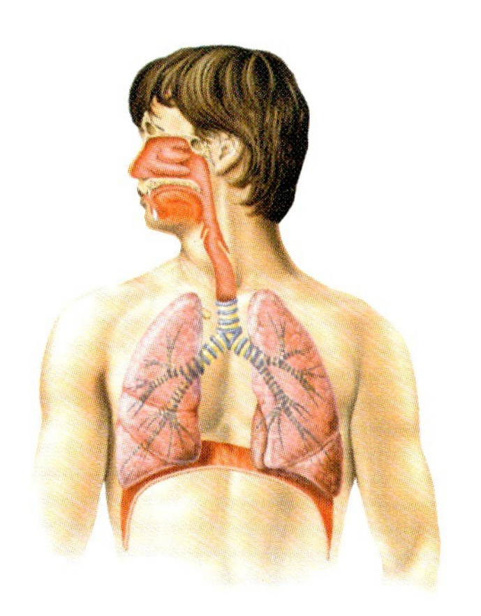

Fig. 14.4

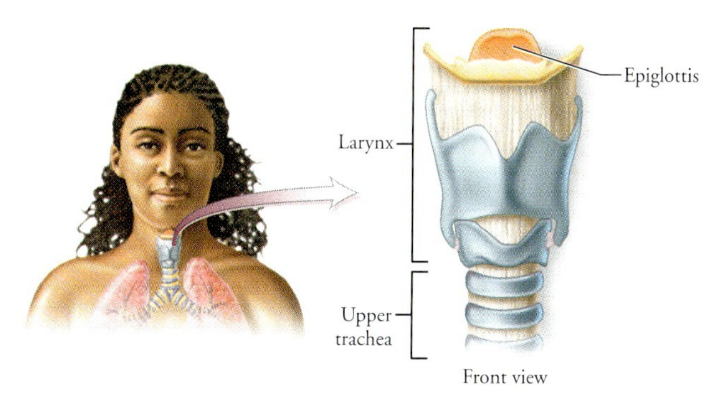

The epiglottis is open during breathing but covers the opening to the larynx during swallowing to prevent food or drink from entering the trachea.

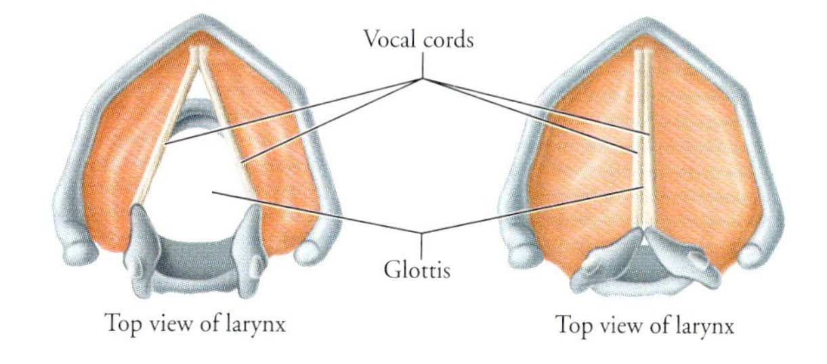

Fig. 14.5

NOTES

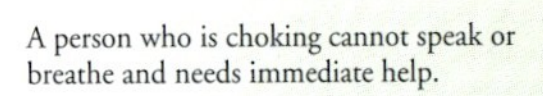

A person who is choking cannot speak or breathe and needs immediate help.

The **Heimlich maneuver** is a procedure intended to force a large burst of air out of the lungs and dislodge the object blocking air flow.

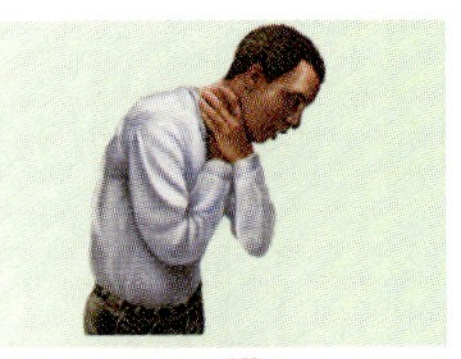

Step 1: Stand behind the choking person with arms around the waist.

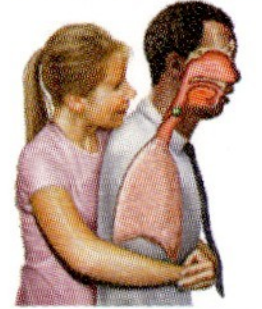

Step 2: Make a fist and place the thumb of the fist beneath the victim's rib cage about midway between the navel (belly button) and the breastbone.

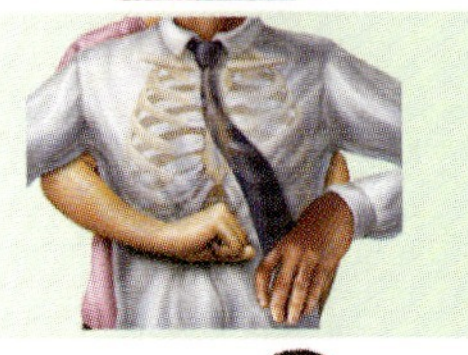

Step 3: Grasp the fist with your other hand and deliver a rapid "bear hug" up and under the rib cage with the clenched fist. Be careful not to press on the ribs or the breastbone because doing so could cause serious injury.

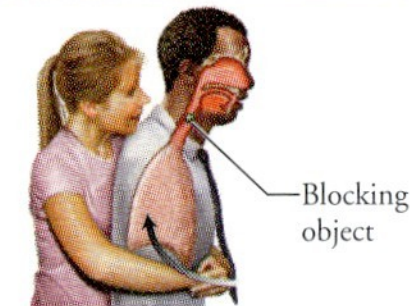

Step 4: Repeat until the object is dislodged.

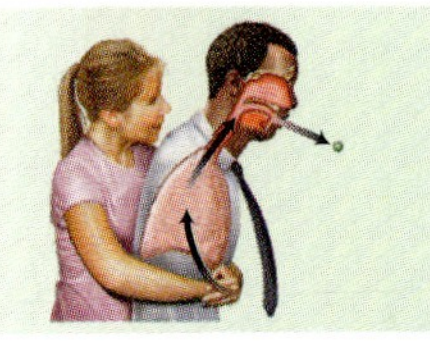

Fig. 14.6

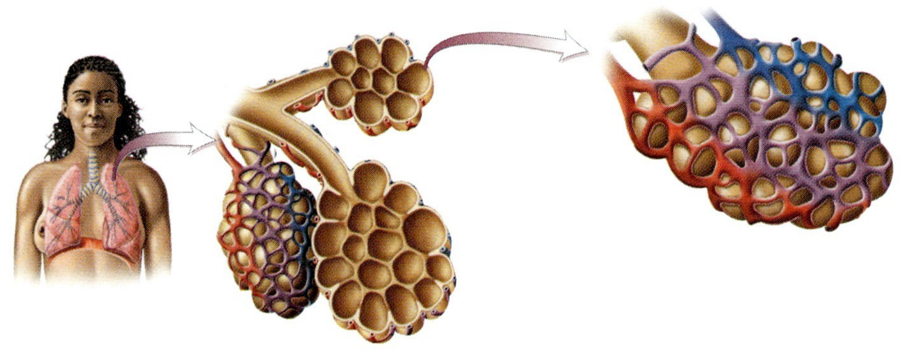

(a) Each alveolus is a cup-shaped chamber. In this section, some of the alveoli have been cut open and you can see into them.

(b) Much of the surface of each alveolus is covered with capillaries. The interface provides a vast surface area for the exchange of gases between the alveoli and the blood.

Fig. 14.8

NOTES

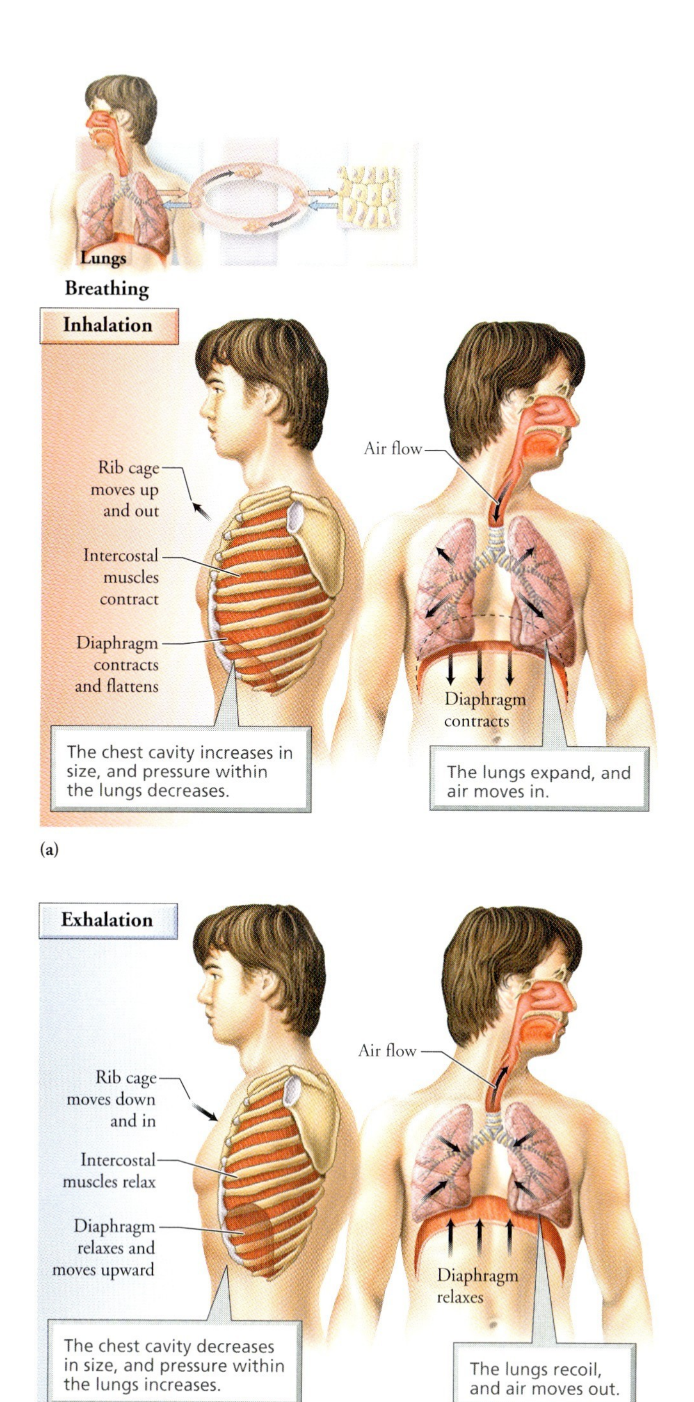

Fig. 14.9

NOTES

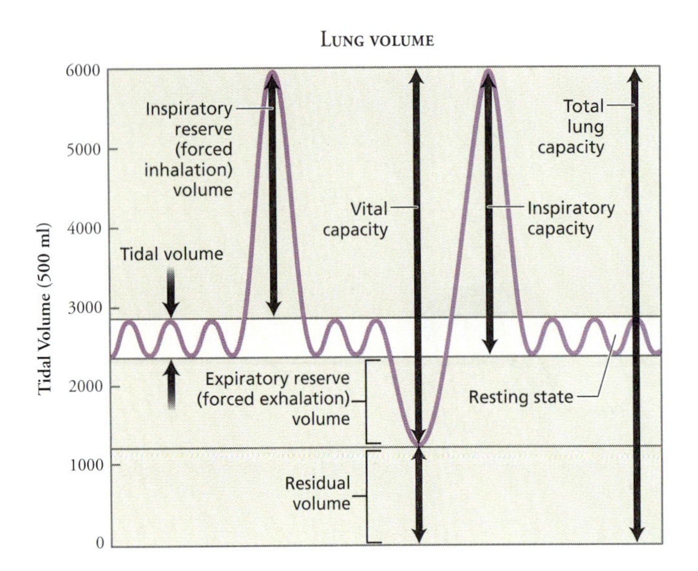

Tidal volume (~500 ml)	Amount of air inhaled or exhaled during an ordinary breath
Inspiratory reserve volume (~1900–3300 ml)	Amount of air that can be inhaled in addition to a normal breath
Expiratory reserve volume (~1000 ml)	Amount of air that can be exhaled in addition to a normal breath
Residual volume (~1100–1200 ml)	Amount of air remaining in the lungs after maximum exhalation
Vital capacity (~3400–4800 ml)	Amount of air that can be inhaled or exhaled in a single breath
Total lung capacity (4500–6000 ml)	Total amount of air in the lungs after maximal inhalation (vital capacity + residual volume)

Fig. 14.10

NOTES

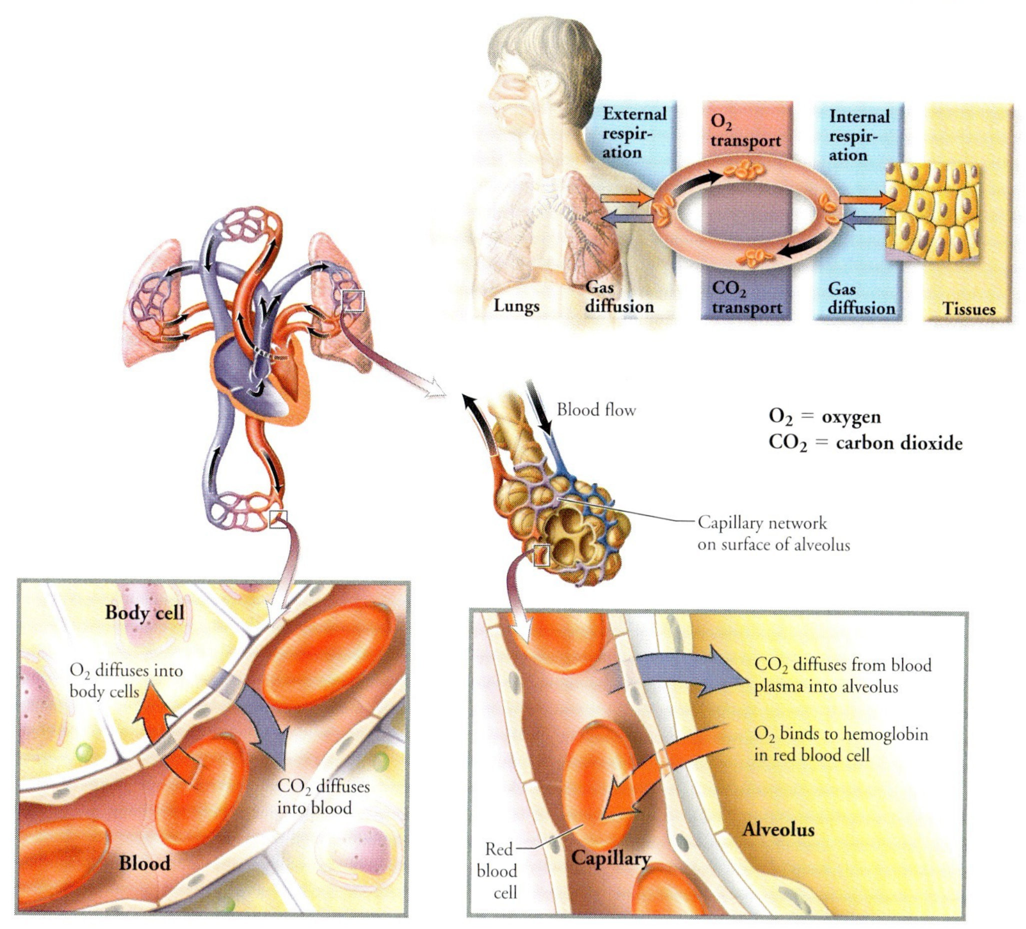

Internal Respiration: Gas exchange in capillary beds throughout body tissues

External Respiration: Gas exchange in lungs

Fig. 14.11

NOTES

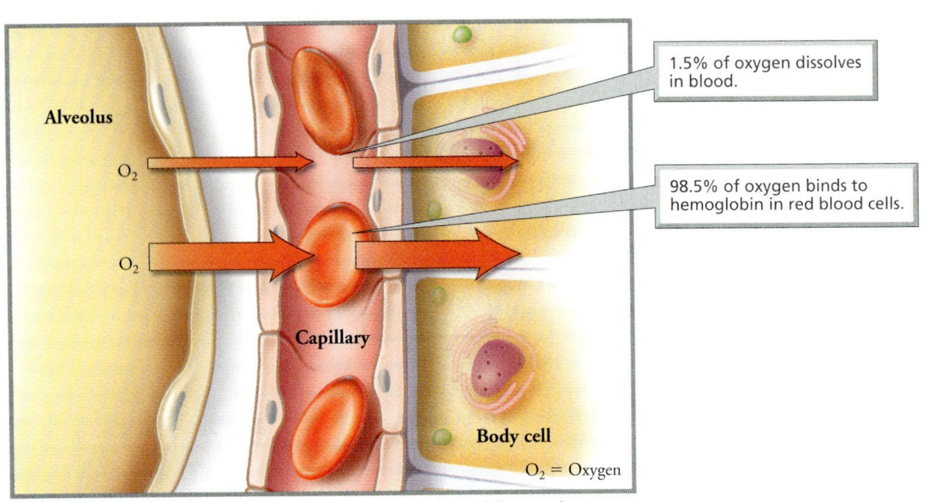

(a) Most oxygen is carried from the lungs to the cells bound to hemoglobin in red blood cells.

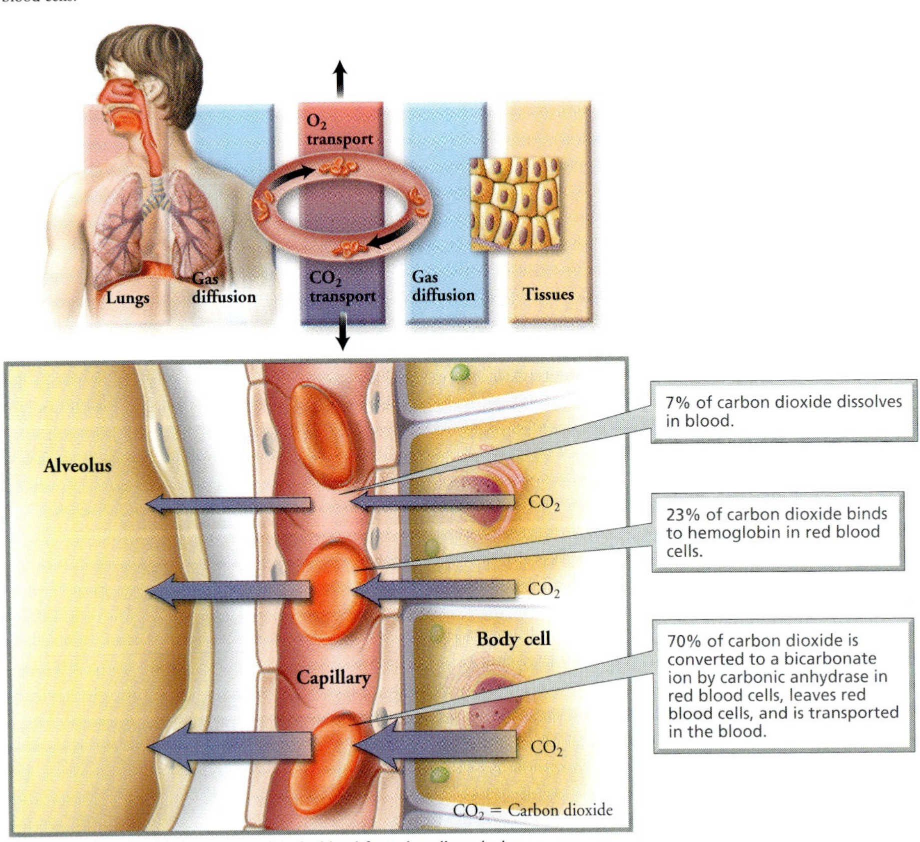

(b) Most carbon dioxide is transported in the blood from the cells to the lungs as bicarbonate ions that were formed in red blood cells by carbonic anhydrase.

Fig. 14.12

NOTES

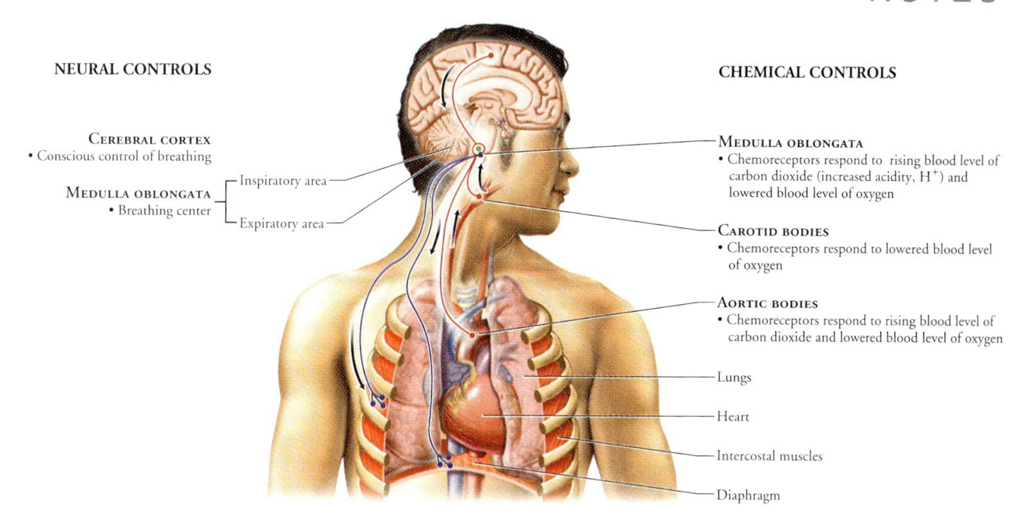

Fig. 14.13

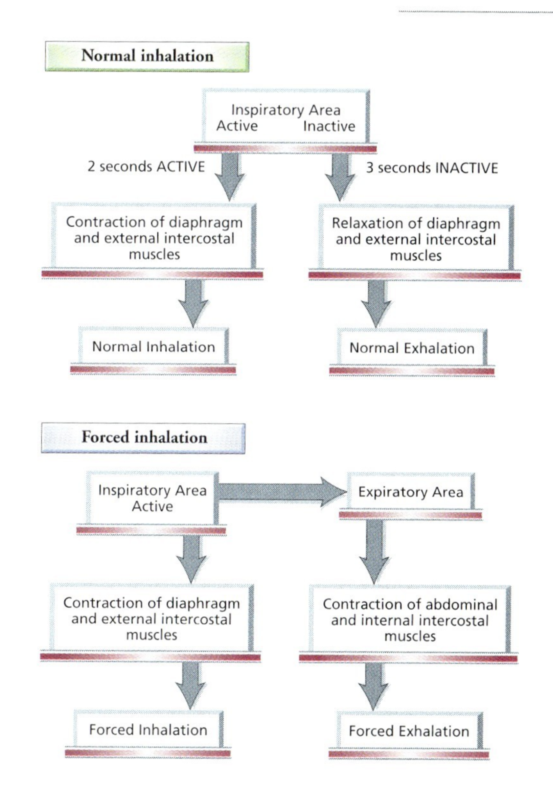

Fig. 14.14

NOTES

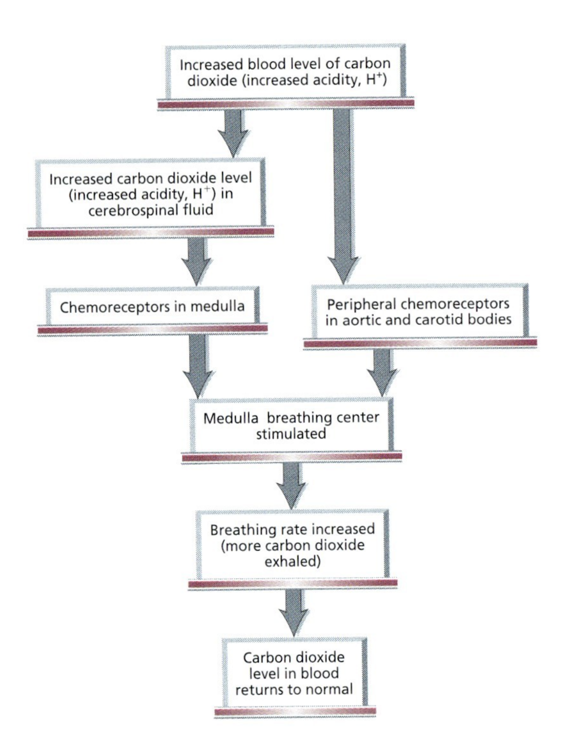

Fig. 14.15

(a) Normal alveoli

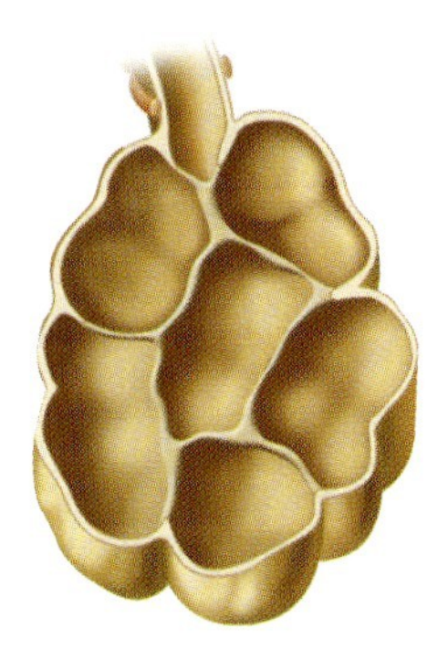

(b) Emphysema causes breakdown of alveolar walls.

Fig. 14.16

NOTES

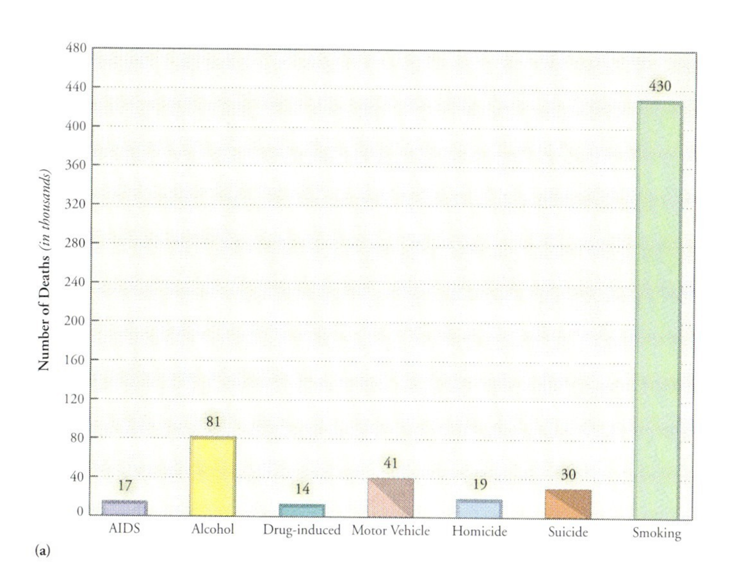

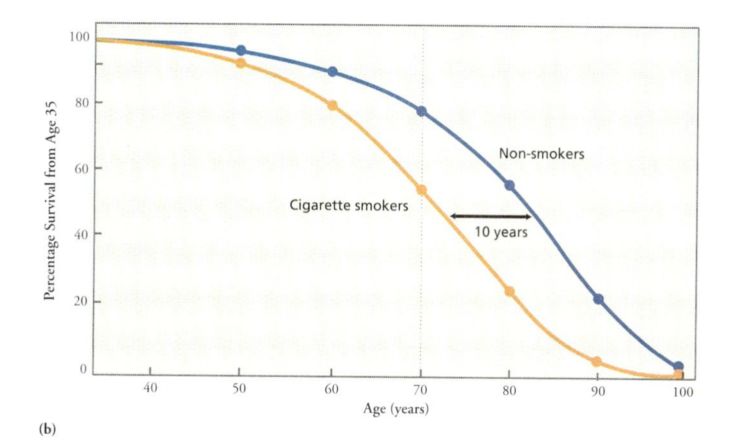

Fig. 14a.1

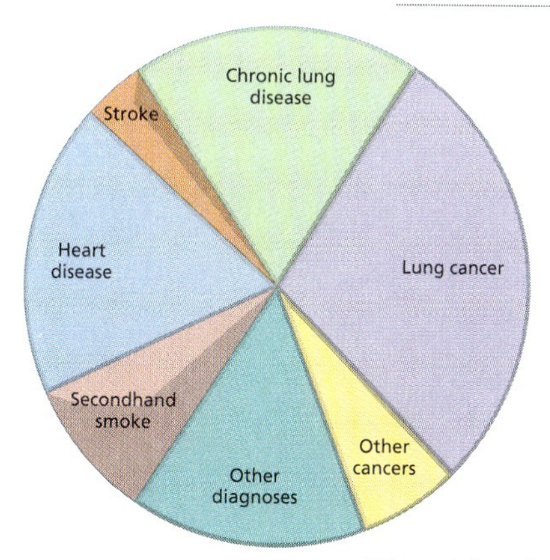

Fig. 14a.2

NOTES

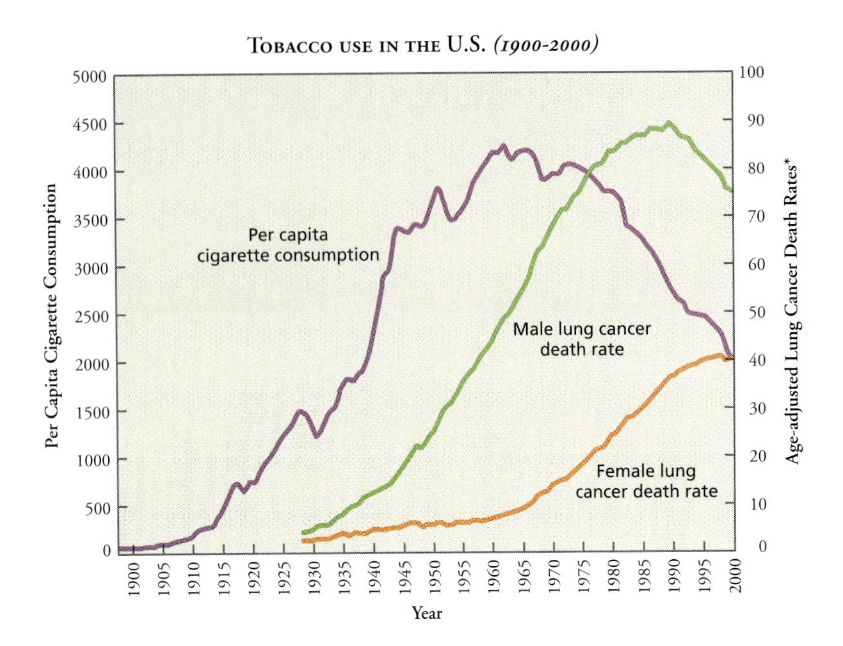

*Age-adjusted to 2000 U.S. standard population.

Fig. 14a.4

Types of Increased Cancer Risk Due to Smoking

TYPE	INCREASED RISK
Mouth and lips	4 times
Larynx	5 times in light smokers (less than 1 pack per day) 20–30 times in heavy smokers (more than 1 pack per day)
Esophagus	2–9 times
Kidney and bladder	2–10 times
Pancreas	2–5 times (especially if alcohol is also consumed)

Table 14a.1

NOTES

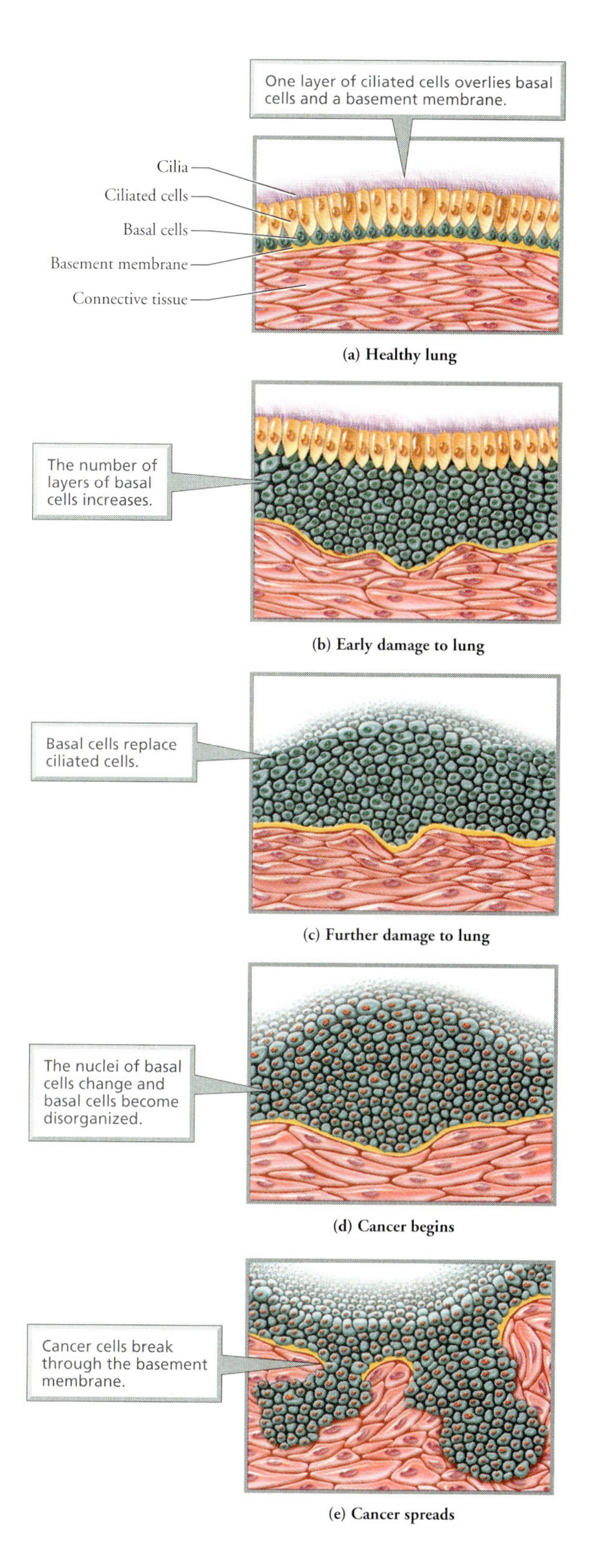

Fig. 14a.6

NOTES

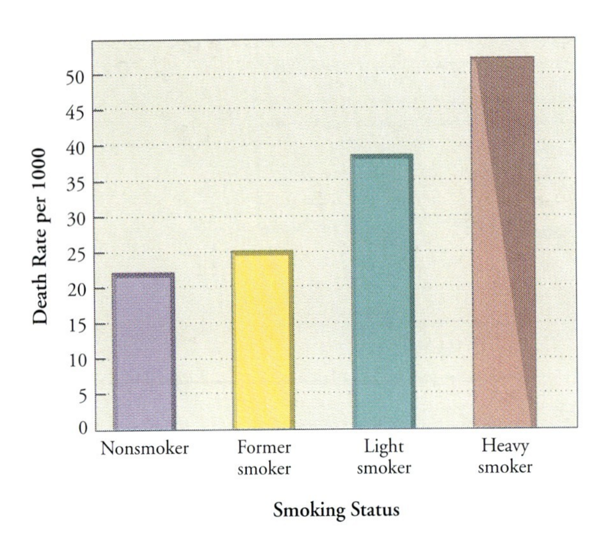

Fig. 14a.7

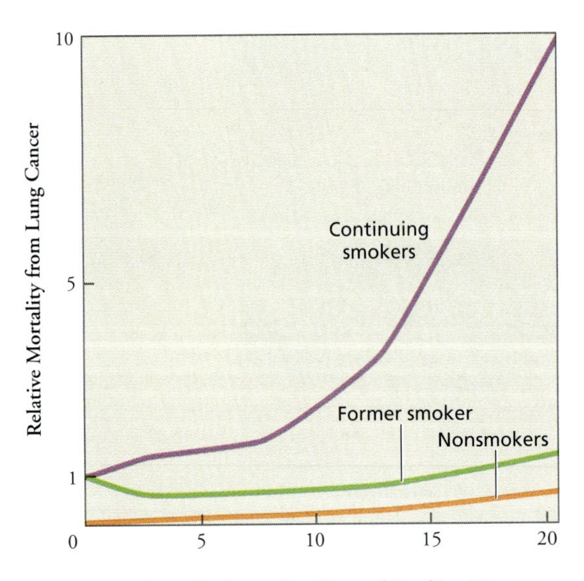

Fig. 14a.8

NOTES

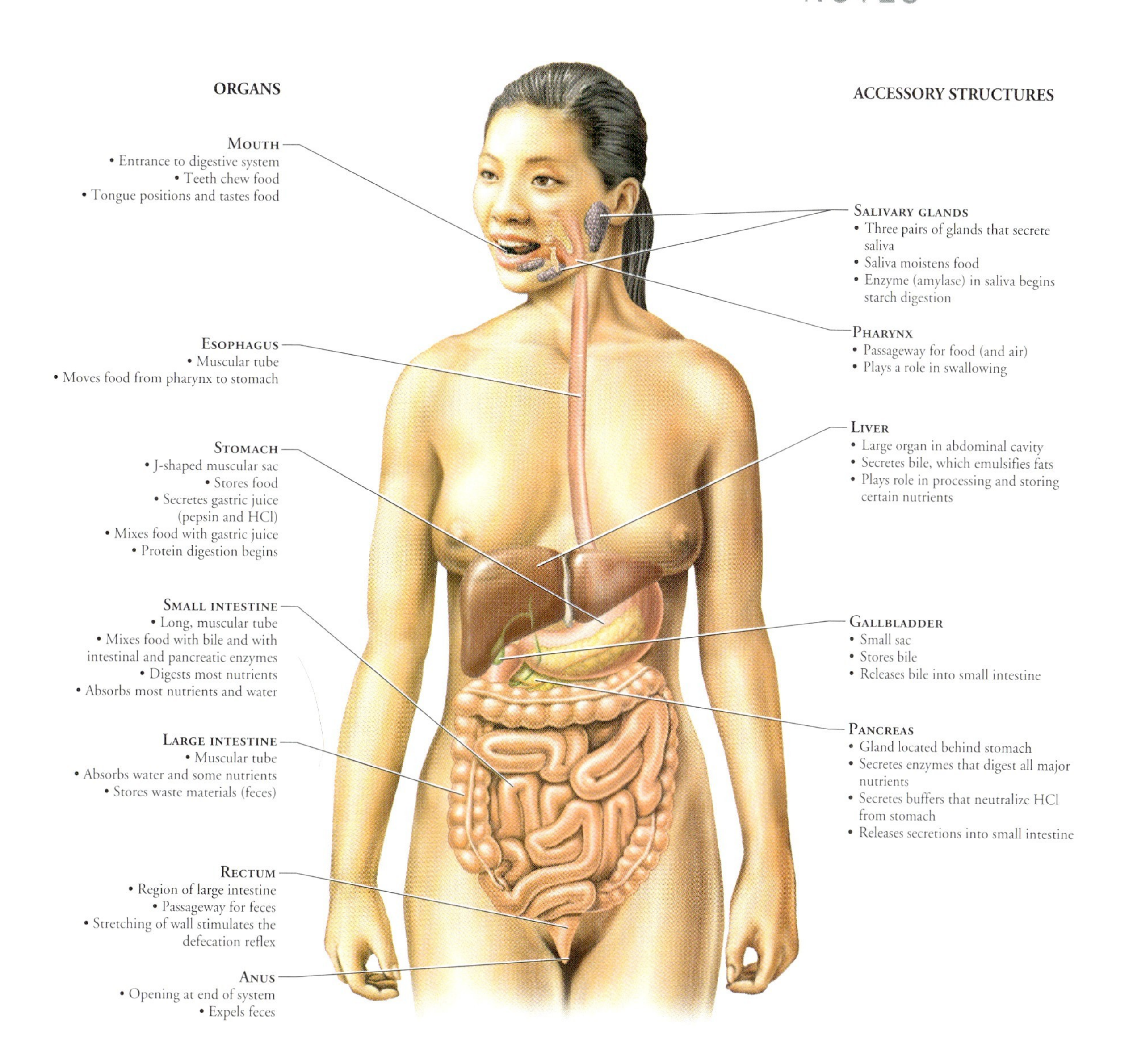
ORGANS
MOUTH
• Entrance to digestive system
• Teeth chew food
• Tongue positions and tastes food
ESOPHAGUS
• Muscular tube
• Moves food from pharynx to stomach
STOMACH
• J-shaped muscular sac
• Stores food
• Secretes gastric juice (pepsin and HCl)
• Mixes food with gastric juice
• Protein digestion begins
SMALL INTESTINE
• Long, muscular tube
• Mixes food with bile and with intestinal and pancreatic enzymes
• Digests most nutrients
• Absorbs most nutrients and water
LARGE INTESTINE
• Muscular tube
• Absorbs water and some nutrients
• Stores waste materials (feces)
RECTUM
• Region of large intestine
• Passageway for feces
• Stretching of wall stimulates the defecation reflex
ANUS
• Opening at end of system
• Expels feces
ACCESSORY STRUCTURES
SALIVARY GLANDS
• Three pairs of glands that secrete saliva
• Saliva moistens food
• Enzyme (amylase) in saliva begins starch digestion
PHARYNX
• Passageway for food (and air)
• Plays a role in swallowing
LIVER
• Large organ in abdominal cavity
• Secretes bile, which emulsifies fats
• Plays role in processing and storing certain nutrients
GALLBLADDER
• Small sac
• Stores bile
• Releases bile into small intestine
PANCREAS
• Gland located behind stomach
• Secretes enzymes that digest all major nutrients
• Secretes buffers that neutralize HCl from stomach
• Releases secretions into small intestine

Fig. 15.1

NOTES

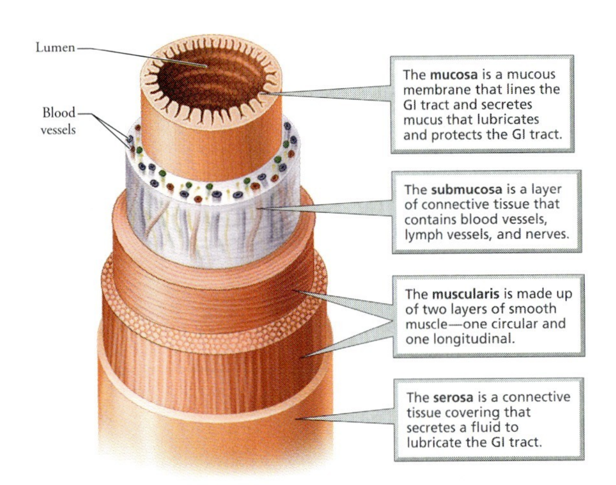

Fig. 15.2

Structures of the Digestive System

STRUCTURE	DESCRIPTION/FUNCTIONS	MECHANICAL DIGESTION	CHEMICAL DIGESTION
Gastrointestinal tract			
Mouth	Receives food; contains teeth and tongue; tongue manipulates food and monitors quality	Teeth tear and crush food into smaller pieces	Digestion of carbohydrates begins
Pharynx	Area that both food and air pass through	None	None
Esophagus	Tube that transports food from mouth to stomach	None	None
Stomach	J-shaped muscular sac for food storage	Churning of stomach mixes food with gastric juice, creating liquid chyme	Protein digestion begins
Small intestine	Long tube where digestion is completed and nutrients are absorbed	Segmental contractions mix food with intestinal enzymes, pancreatic enzymes, and bile	Carbohydrate, protein, and fat digestion completed
Large intestine	Final tubular region of GI tract; absorbs water and ions; houses bacteria; forms and expels feces	None	Some digestion is carried out by bacteria
Anus	Terminal outlet of digestive tract	None	None
Accessory structures			
Salivary glands (sublingual, submandibular, parotid)	Secrete saliva, a liquid that moistens food and contains an enzyme (amylase) for digesting carbohydrates	None	Saliva contains enzymes that begin carbohydrate digestion
Pancreas	Digestive secretions include bicarbonate ions that neutralize acidic chyme and enzymes that digest carbohydrates, proteins, fats, and nucleic acids	None	Pancreatic enzymes assist in digestion of carbohydrates, proteins, fats, and nucleic acids
Liver	Digestive function is to produce bile, a liquid that emulsifies fats, making chemical digestion easier and facilitating absorption	Bile emulsifies fats	Bile facilitates digestion and absorption of fats
Gallbladder	Stores bile and releases it into small intestine	None	None

Table 15.1

NOTES

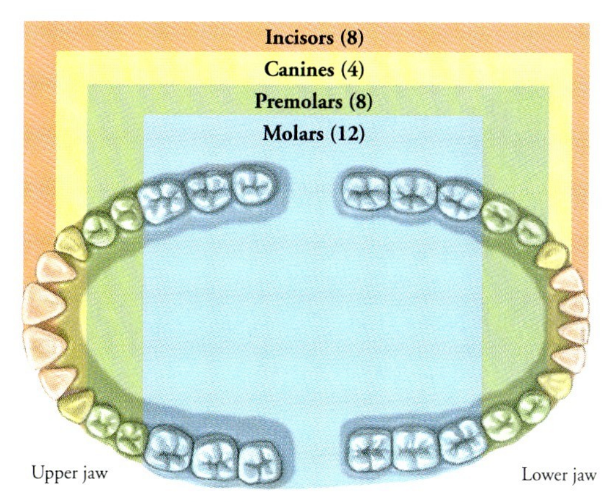

The teeth slice, tear, and grind food until it can be swallowed.

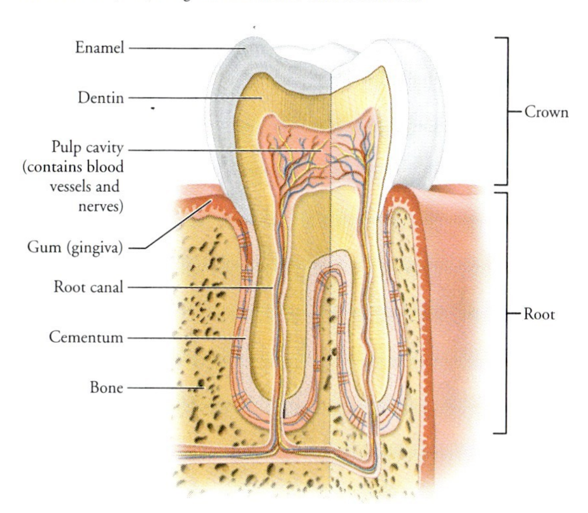

The structure of the human tooth is suited for its function of breaking food into smaller pieces.

Fig. 15.3

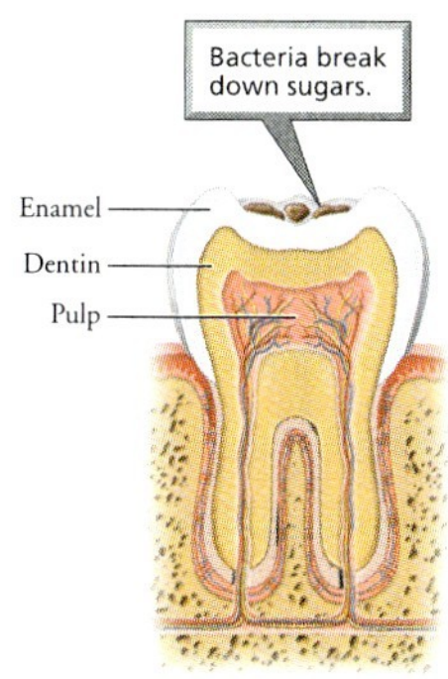

STEP 1
Tooth decay is caused by acid produced when bacteria living on the tooth surface break down sugars in food particles adhered to teeth.

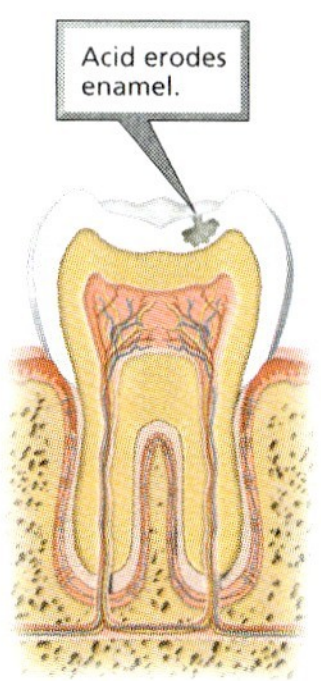

STEP 2
The acid erodes the tooth's enamel, causing a cavity to form.

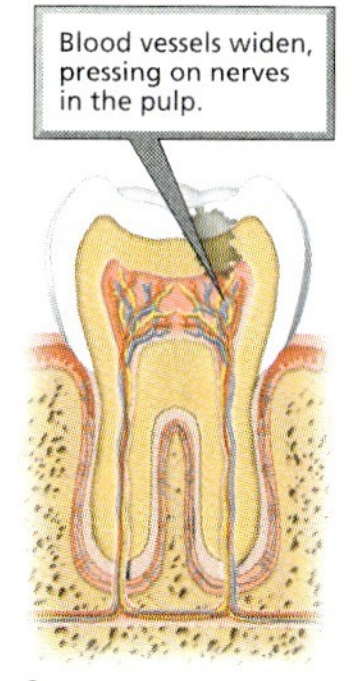

STEP 3
The body responds by widening blood vessels in the pulp, to increase delivery of white blood cells to fight the infection. When widened blood vessels press on nerves within the pulp, a toothache results.

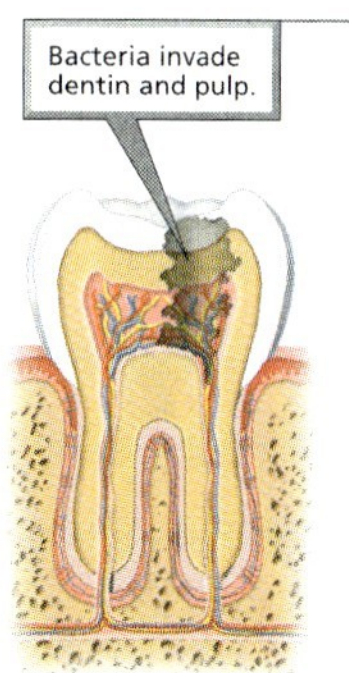

STEP 4
Bacteria can then infect the softer dentin beneath the enamel and later the pulp at the heart of the tooth.

Fig. 15.4

NOTES

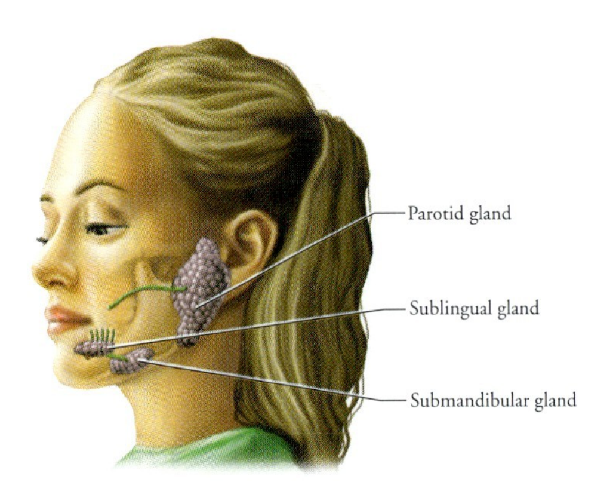

Fig. 15.5

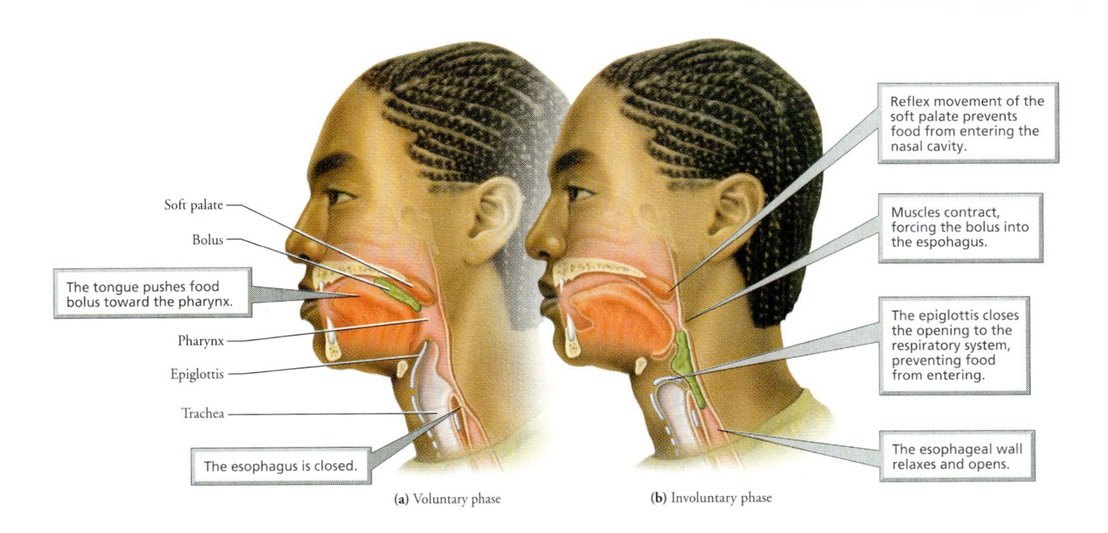

Fig. 15.6

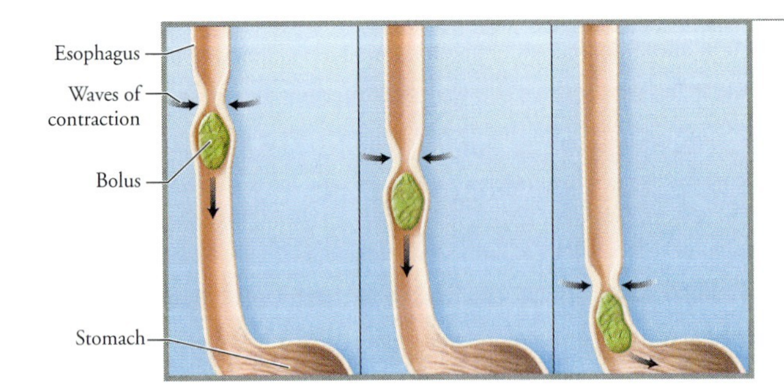

Fig. 15.7

NOTES

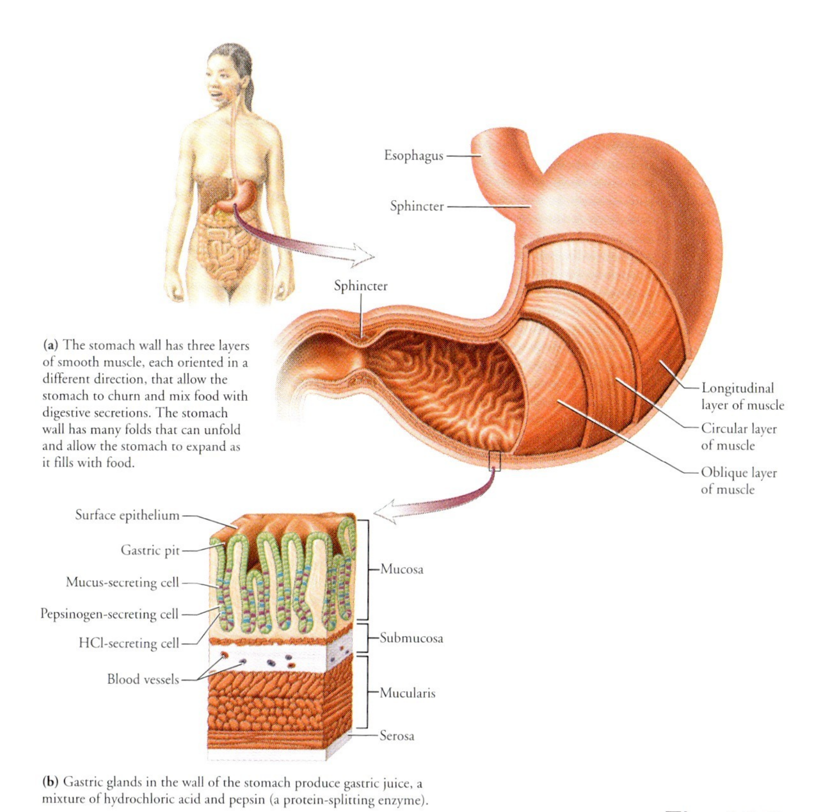

(a) The stomach wall has three layers of smooth muscle, each oriented in a different direction, that allow the stomach to churn and mix food with digestive secretions. The stomach wall has many folds that can unfold and allow the stomach to expand as it fills with food.

(b) Gastric glands in the wall of the stomach produce gastric juice, a mixture of hydrochloric acid and pepsin (a protein-splitting enzyme).

Fig. 15.8

MAJOR DIGESTIVE ENZYMES

SITE OF PRODUCTION	ENZYME	SITE OF ACTION	SUBSTRATE
Salivary glands	Salivary amylase	Mouth	Polysaccharides to shorter molecules
Stomach	Pepsin	Stomach	Proteins into protein fragments (polypeptides)
Pancreas	Trypsin	Small intestine	Proteins and polypeptides to smaller fragments
	Chymotrypsin	Small intestine	Proteins and polypeptides to smaller fragments
	Amylase	Small intestine	Polysaccharides to disaccharides
	Carboxypeptidase	Small intestine	Polypeptides to amino acids
	Lipase	Small intestine	Triglycerides (fats) to fatty acids and glycerol
	Nucleases (deoxyribonuclease and ribonuclease)	Small intestine	DNA or RNA to nucleotides
Small intestine	Maltase	Small intestine	Maltose to glucose units
	Sucrase	Small intestine	Sucrose to glucose and fructose
	Lactase	Small intestine	Lactose to glucose and galactose
	Aminopeptidase	Small intestine	Peptides to amino acids

Table 15.2

NOTES

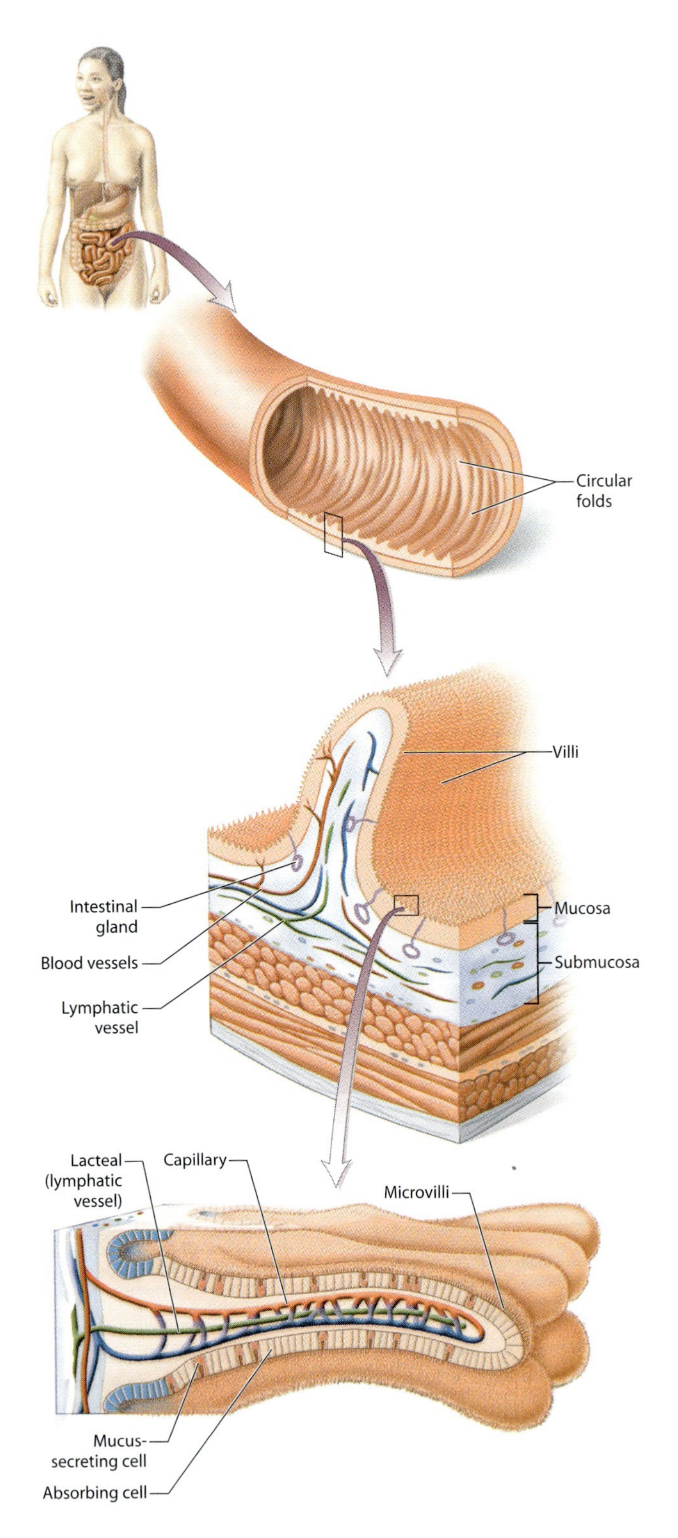

Fig. 15.9

NOTES

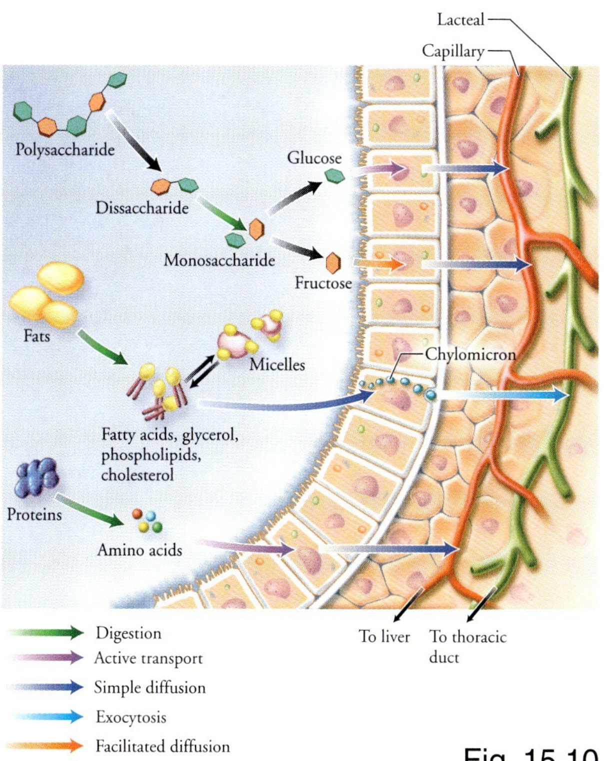

Fig. 15.10

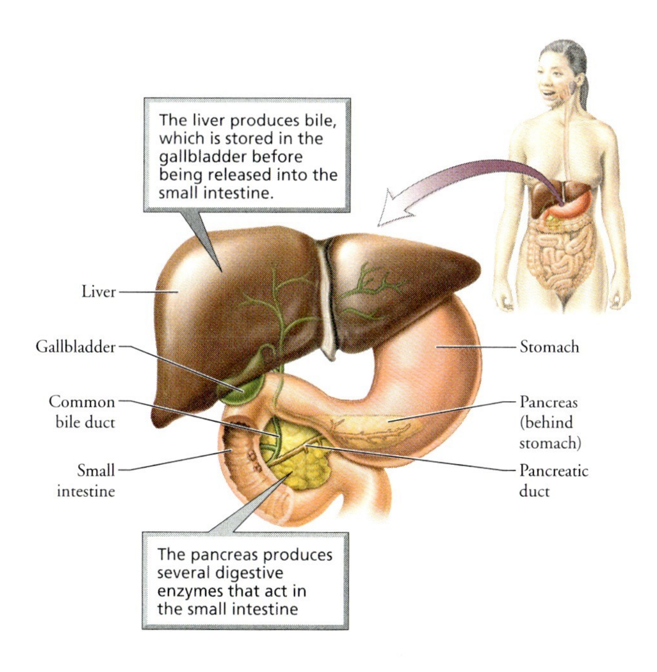

Fig. 15.11

NOTES

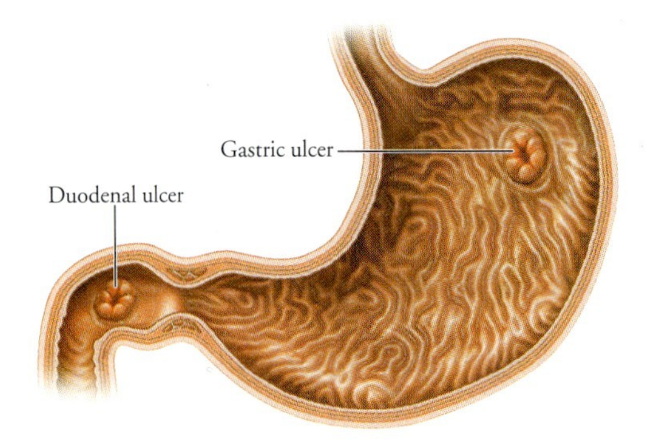

Fig. 15.A

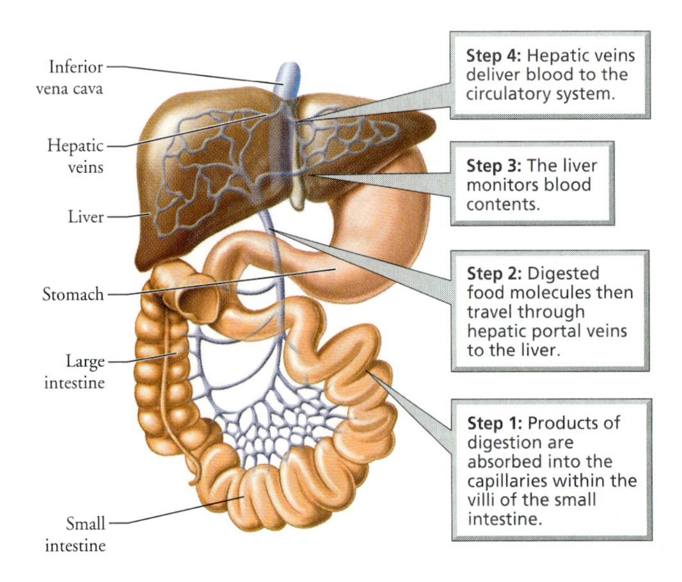

Fig. 15.12

NOTES

Forms of Viral Hepatitis

Hepatitis Virus	Means of Transmission	Symptoms
A	Sewage-contaminated water or food	Acute infection: fatigue, jaundice, fever, abdominal pain, loss of appetite Recovery within several months
B	Sexual contact, contact with contaminated blood, mother to newborn at birth	Acute infection: fatigue, jaundice, fever, abdominal pain, loss of appetite, joint pain Chronic, with possible liver failure and death Can lead to liver cancer
C	Contaminated blood or body fluids	Acute infection: fatigue, jaundice, dark urine, abdominal pain, nausea Chronic, with possible liver failure and death Can lead to liver cancer
D	Sexual contact, contact with contaminated blood, mother to newborn at birth	Causes symptoms only in persons already infected with hepatitis B Infection with B and D causes progressive and severe liver disease; can be acute or chronic
E	Sewage-contaminated food or water	Acute infection: fatigue and jaundice Potentially lethal in pregnant women
G	Contaminated blood? (appears in same populations as hepatitis C)	Recently described; little known

Table 15.3

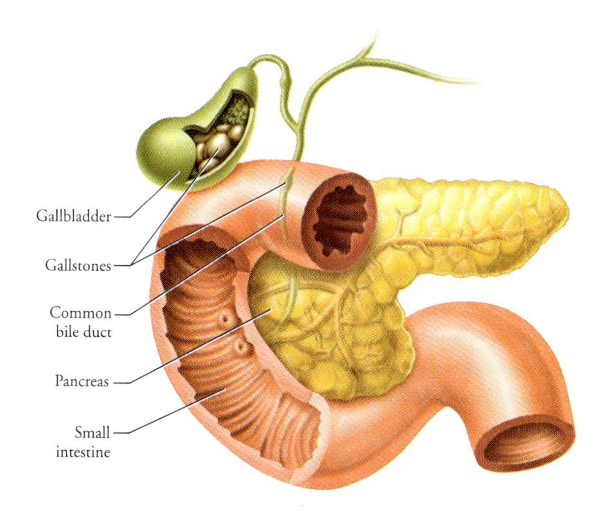

Fig. 15.13

NOTES

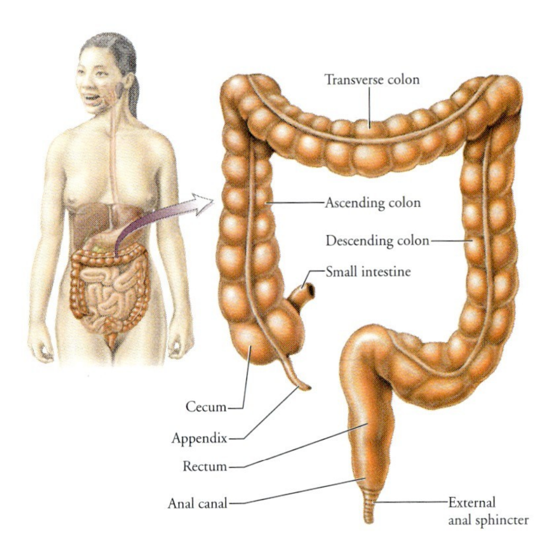

Fig. 15.14

NEURAL CONTROLS ON DIGESTIVE ACTIVITY	
STIMULUS	EFFECT
Sight of food, thought of food, presence of food in mouth	Release of saliva from salivary glands
Chewing food	Release of gastric juice (enzymes from stomach and HCl) and mucus from cells of stomach lining
Presence of acidic chyme in small intestine	Release of enzymes from small intestine into the small intestine; increased motility in small intestine

Table 15.4

HORMONAL CONTROLS ON DIGESTIVE ACTIVITY				
HORMONE	STIMULUS	ORIGIN	TARGET	EFFECTS
Gastrin	Distention of stomach by food; presence of partially digested proteins in stomach	Stomach	Stomach	Release of gastric juice (enzymes from stomach and HCl)
Vasoactive intestinal peptide (VIP)	Presence of acidic chyme in small intestine	Small intestine	Small intestine	Release of enzymes from small intestine
Secretin	Presence of acidic chyme in small intestine	Small intestine	Pancreas	Release of sodium bicarbonate into small intestine to neutralize acidic chyme
Cholecystokinin	Arrival of chyme containing lipids	Small intestine	Pancreas	Release of enzymes from pancreas
			Gall bladder	Contraction of gallbladder and release of bile

Table 15.5

NOTES

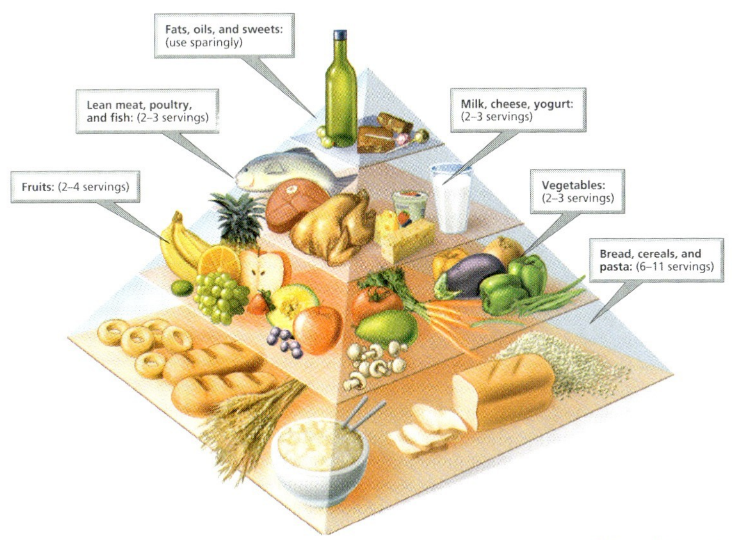

Fig. 15a.1

COMPONENTS OF FOOD

NUTRIENT	GOOD SOURCES	FUNCTIONS	RECOMMENDED QUANTITY*
Energy-containing nutrients			
Fats (lipids)	Milk, cheese, meat, vegetable oils, nuts	Provides 9 calories per gram; component of cell membranes; component of nerve sheaths; insulates body; forms protective cushions around vital body organs	20%–35% of daily calories
Carbohydrates	Cereal, bread, pasta, vegetables, fruits, sweets	Provides 4 calories per gram; primary fuel for all cells	45%–65% of daily calories (not more than 25% from added sugar)
Protein and amino acids	Meat, poultry, fish, legumes, nuts	Provides 4 calories per gram; important component of all cells; structural proteins including muscle fibers; regulatory proteins including enzymes and certain hormones	10%–35% of daily calories
Other nutrients			
Vitamins	Many vegetables, fruits, whole grain, meats, dairy products	Most function as co-enzymes that allow the cellular reactions of the body to take place fast enough to support life	Depends on the specific vitamin
Minerals	Many vegetables, fruits, whole grains, meats, seeds, nuts	Structural roles, including hardness of bones and teeth; functional roles including oxygen transport in blood; electrolyte balance; proper nerve and muscle function	Depends on the specific mineral
Water	Nearly every food and all beverages	Solvent; transport of materials; medium for all and participant in some chemical reactions; lubricant; protective cushion; regulation of body temperature	The equivalent of 8 8-ounce glasses
Fiber (not absorbed from intestines)	Whole grains, vegetables, fruits	Soluble fiber good for health of heart and blood vessels; insoluble fiber promotes intestinal health	Men 50 and younger: 38 grams/day Women 50 and younger: 25 grams/day Men over 50: 30 grams/day Women over 50: 21 grams/day

*The Food and Nutrition Board, the National Academy of Sciences, and the Institute of Medicine made these recommendations in 2002.

Table 15a.1

NOTES

Blood Cholesterol Levels and the Risk of Cardiovascular Disease

	Desirable	Borderline High	High
Total cholesterol/ 100 ml of blood	<200 mg/dl	200–239 mg/dl	≥240 mg/dl
LDL cholesterol/ 100 ml blood	100–129 mg/dl	130–159 mg/dl	≥160 mg/dl
Ratio of total cholesterol to HDL cholesterol	<4:1 mg/dl		

Table 15a.2

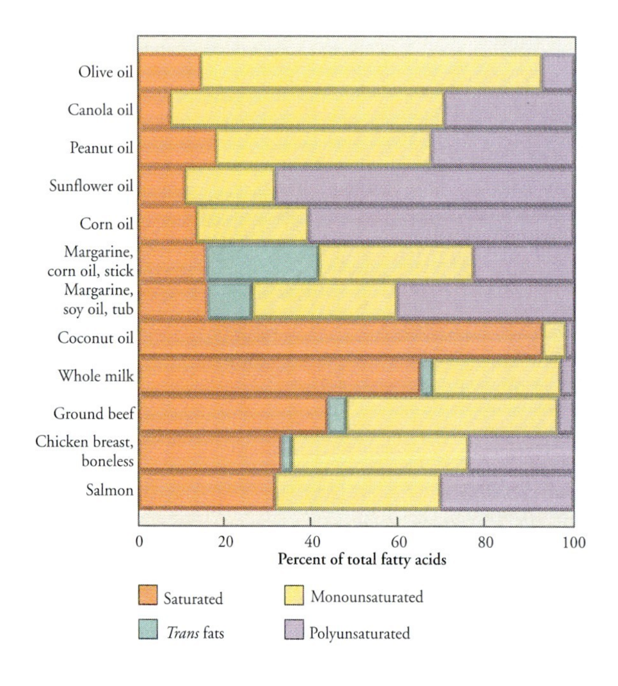

Fig. 15a.3

NOTES

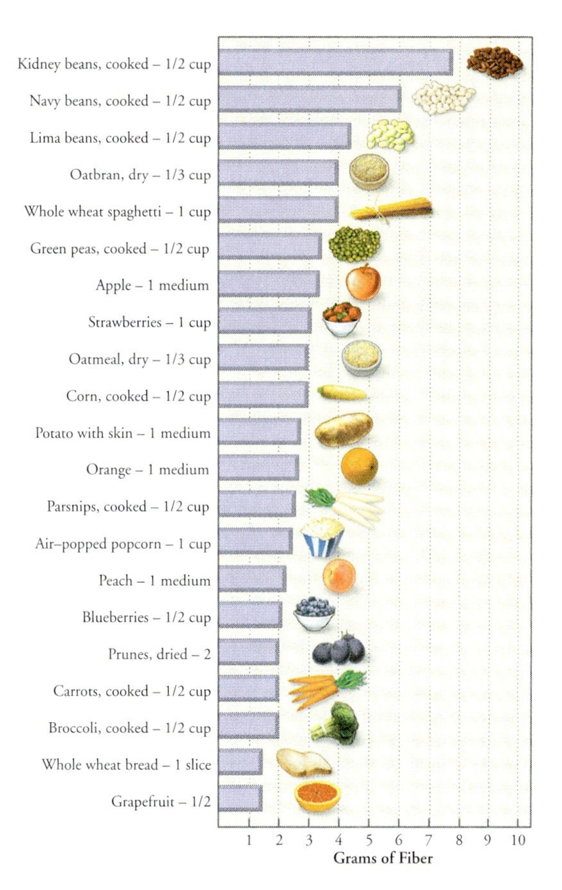

Fig. 15a.5

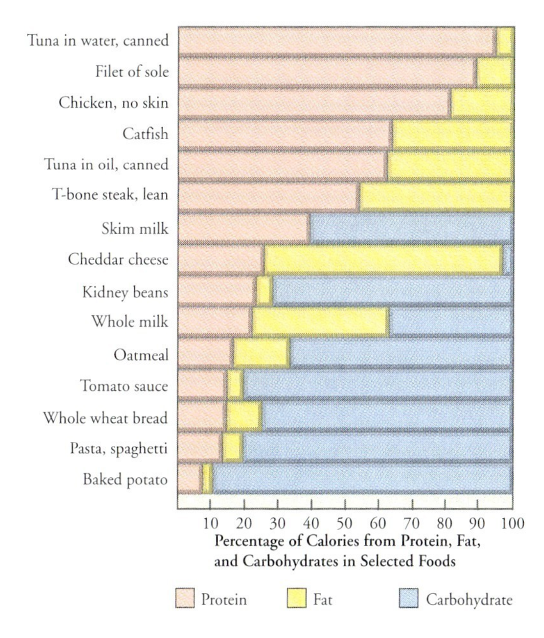

Fig. 15a.8

NOTES

Vitamins				
VITAMIN	**GOOD SOURCES**	**FUNCTION**	**EFFECTS OF DEFICIENCY**	**EFFECTS OF EXCESS**
Fat-soluble vitamins				
A	Liver, egg yolk, fat-containing and fortified dairy products; formed from carotene (found in deep yellow and deep green leafy vegetables)	Components of rhodopsin, the eye pigment responsible for black-and-white vision; maintains epithelia; cell differentiation	Night-blindness; dry, scaly skin; dry hair; skin sores; increased respiratory, urogenital, and digestive infections; xerophthalmia (the leading cause of preventable blindness worldwide); most common vitamin deficiency in world	Drowsiness; headache; dry, coarse, scaly skin; hair loss; itching; brittle nails; abdominal and bone pain
D	Fortified milk, fish liver oil, egg yolk; formed in skin when exposed to ultraviolet light	Increases absorption of calcium; enhances bone growth and calcification	Bone deformities in children, rickets, bone softening in adults	Calcium deposits in soft tissues, kidney damage, vomiting, diarrhea, weight loss
E	Whole grains, dark green vegetables, vegetable oils, nuts, seeds	May inhibit effects of free radicals; helps maintain cell membranes; prevents oxidation of vitamins A and C in gut	Rare; possible anemia and nerve damage	Muscle weakness, fatigue, nausea
K	Primary source from bacteria in large intestine; leafy green vegetables, cabbage, cauliflower	Important in forming proteins involved in blood clotting	Easy bruising, abnormal blood clotting, severe bleeding	Liver damage and anemia
Water-soluble vitamins				
C (ascorbic acid)	Citrus fruits, cantaloupe, strawberries, tomatoes, broccoli, cabbage, green pepper	Collagen synthesis; may inhibit free radicals; improves iron absorption	Scurvy, poor wound healing, impaired immunity	Diarrhea, kidney stones; may alter results of certain diagnostic lab tests
Thiamin (B_1)	Pork, legumes, whole grains, leafy green vegetables	Coenzyme in energy metabolism; nerve function	Water retention in tissues, nerve changes leading to poor coordination, heart failure, beriberi	None known
Riboflavin (B_2)	Dairy products such as milk; whole grains, meat, liver, egg whites, leafy green vegetables	Coenzyme used in energy metabolism	Skin lesions	None known
Niacin (B_3)	Nuts, green leafy vegetables, potatoes; can be formed from tryptophan found in meats	Coenzyme used in energy metabolism	Contributes to pellagra (damage to skin, gut, nervous system)	Flushing of skin on face, neck, and hands; possible liver damage
B_6	Meat, poultry, fish, spinach, potatoes, tomatoes	Coenzyme used in amino acid metabolism	Nervous, skin, and muscular disorders; anemia	Numbness in feet, poor coordination
Pantothenic acid	Widely distributed in foods, animal products, and whole grains	Coenzyme in energy metabolism	Fatigue, numbness and tingling of hands and feet, headaches, nausea	Diarrhea, water retention
Folic acid (folate)	Dark green vegetables, orange juice, nuts, legumes, grain products	Coenzyme in nucleic acid and amino acid metabolism	Anemia (megaloblastic and pernicious), gastrointestinal disturbances, nervous system damage, inflamed tongue, neural tube defects	High doses mask vitamin B_{12} deficiency
B_{12}	Poultry, fish, red meat, dairy products except butter	Coenzyme in nucleic acid metabolism	Anemia (megaloblastic and pernicious), impaired nerve function	None known
Biotin	Legumes, egg yolk; widely distributed in foods; bacteria of large intestine	Coenzyme used in energy metabolism	Scaly skin (dermatitis), sore tongue, anemia	None known

Table 15a.3

NOTES

MINERALS

MINERAL	GOOD SOURCES	FUNCTION	EFFECTS OF DEFICIENCY	EFFECTS OF EXCESS
Major minerals				
Calcium	Milk, cheese, dark green vegetables, legumes	Hardness of bones, tooth formation, blood clotting, nerve and muscle action	Stunted growth, loss of bone mass, osteoporosis, convulsions	Impaired absorption of other minerals, kidney stones
Phosphorus	Milk, cheese, red meat, poultry, whole grains	Bone and tooth formation; component of nucleic acids, ATP, and phospholipids; acid-base balance	Weakness, demineralized bone	Impaired absorption of some minerals
Magnesium	Whole grains, green leafy vegetables, milk, dairy products, nuts, legumes	Component of enzymes	Muscle cramps, neurologic disturbances	Neurologic disturbances
Potassium	Available in many foods including meats, fruits, vegetables, and whole grains	Body water balance, nerve function, muscle function, role in protein synthesis	Muscle weakness	Muscle weakness, paralysis, heart failure
Sulfur	Protein-containing foods including meat, legumes, milk, and eggs	Component of body proteins	None known	None known
Sodium	Table salt	Body water balance, nerve function	Muscle cramps, reduced appetite	High blood pressure in susceptible people
Chloride	Table salt, processed foods	Formation of hydrochloric acid in stomach, role in acid-base balance	Muscle cramps, reduced appetite, poor growth	High blood pressure in susceptible people
Trace minerals				
Iron	Meat, liver, shellfish, egg yolk, whole grains, green leafy vegetables, nuts, dried fruit	Component of hemoglobin, myoglobin, and cytochrome (transport chain enzyme)	Iron-deficiency anemia, weakness, impaired immune function	Liver damage, heart failure, shock
Iodine	Marine fish and shellfish, iodized salt, dairy products	Thyroid hormone function	Enlarged thyroid	Enlarged thyroid
Fluoride	Drinking water, tea, seafood	Bone and tooth maintenance	Tooth decay	Digestive upsets, mottling of teeth, deformed skeleton
Copper	Nuts, legumes, seafood, drinking water	Synthesis of melanin, hemoglobin, and transport chain components; collagen synthesis; immune function	Rare; anemia, changes in blood vessels	Nausea, liver damage
Zinc	Seafood, whole grains, legumes, nuts, meats	Component of digestive enzymes; required for normal growth, wound healing, and sperm production	Difficulty in walking, slurred speech, scaly skin, impaired immune function	Nausea, vomiting, diarrhea, impaired immune function
Manganese	Nuts, legumes, whole grains, leafy green vegetables	Role in synthesis of fatty acids, cholesterol, urea, and hemoglobin; normal neural function	None known	Nerve damage

Table 15a.4

NOTES

APPROXIMATE NUMBER OF CALORIES BURNED PER HOUR BY VARIOUS ACTIVITIES			
ACTIVITY	100-LB PERSON	150-LB PERSON	200-LB PERSON
Bicycling, 6 mph	160	240	312
Bicycling, 12 mph	270	410	534
Jogging, 5.5 mph	440	660	962
Jogging, 10 mph	850	1280	1664
Jumping rope	500	750	1000
Swimming, 25 yd/min	185	275	358
Swimming, 50 yd/min	325	500	650
Walking, 2 mph	160	240	312
Walking, 4.5 mph	295	440	572
Tennis (singles)	265	400	535

Source: American Heart Association

Table 15a.5

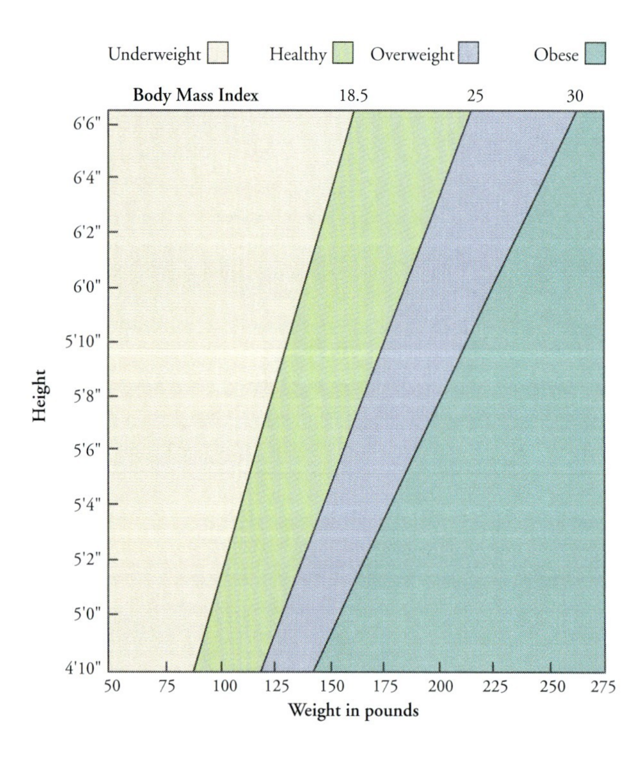

Fig. 15a.9

NOTES

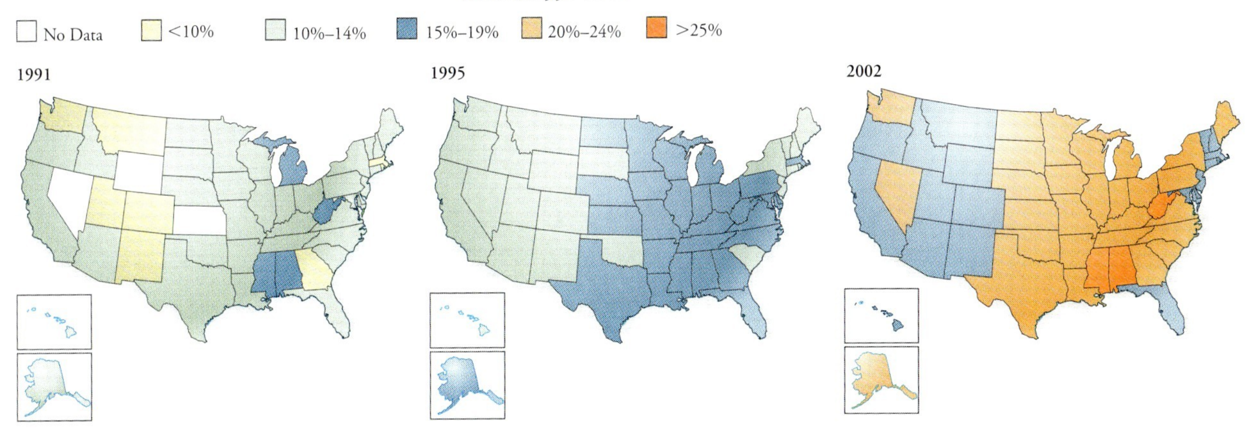

Fig. 15a.10

Nutrition and Your Health
Dietary guidelines for Americans

Aim For Fitness...

- Aim for a healthy weight
- Be physically active each day

Build A Healthy Base...

- Let the pyramid guide your food choices
- Choose a variety of grains daily, especially whole grains
- Choose a variety of fruits and vegetables daily
- Keep food safe to eat

Choose Sensibly...

- Choose a diet that is low in saturated fat and cholesterol and moderate in total fat
- Choose beverages and foods to moderate your intake of sugars
- Choose and prepare foods with less salt
- If you drink alcoholic beverages, do so in moderation

...for good health

Fig. 15a.11

NOTES

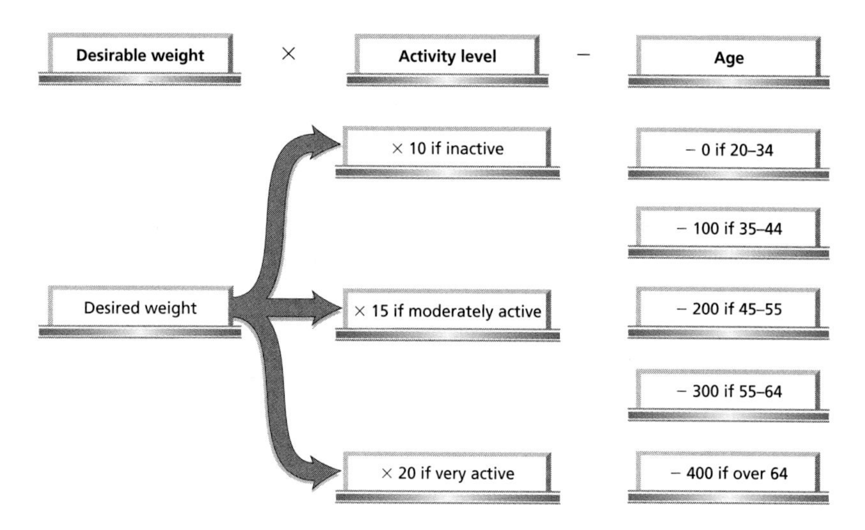

Fig. 15a.12

NOTES

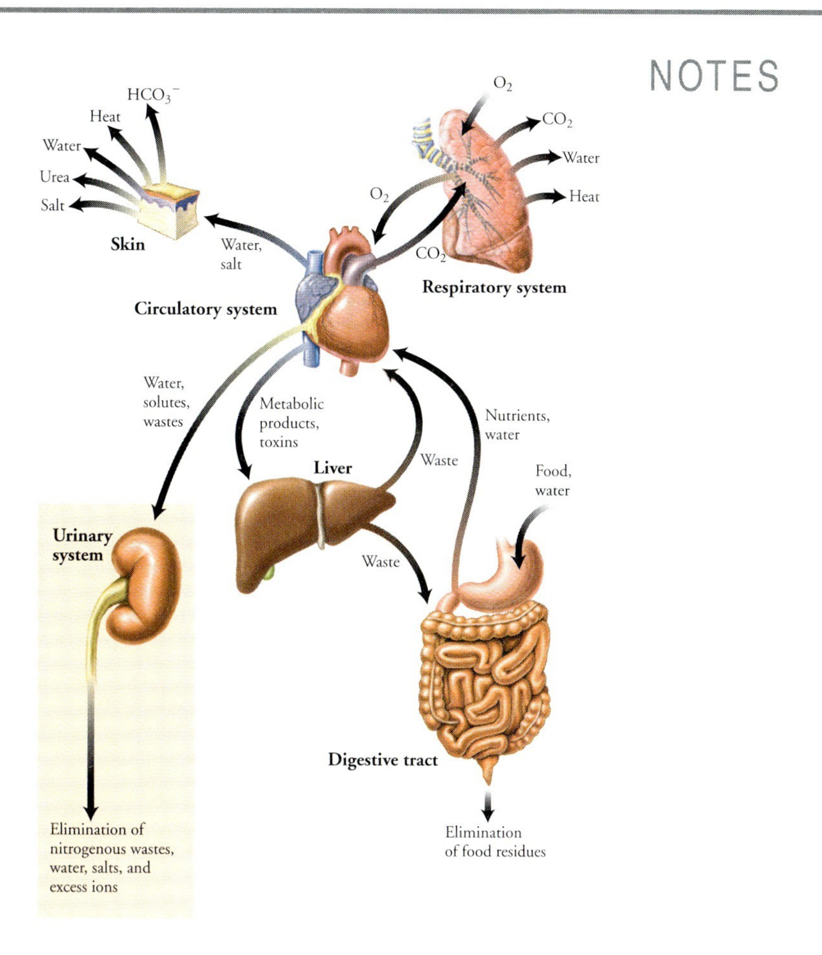

Fig. 16.2

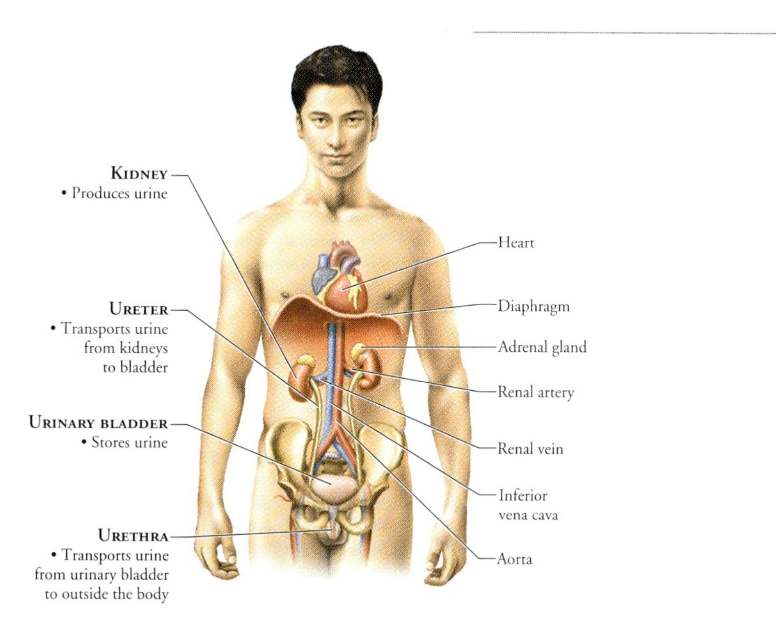

Fig. 16.3

NOTES

Components of the Urinary System and Their Functions	
Component	**Function**
Kidneys	Filter wastes and excess material from the blood
	Help regulate blood pressure and pH
	Maintain fluid balance by regulating the volume and composition of blood and urine
	Release erythropoietin, which stimulates production of red blood cells
	Transform vitamin D into its active form
Ureters	Transport urine from kidneys to urinary bladder
Urinary bladder	Stores urine
	Contracts and expels urine into urethra
Urethra	Transports urine from urinary bladder to outside the body
	In males, transports semen to outside the body

Table 16.1

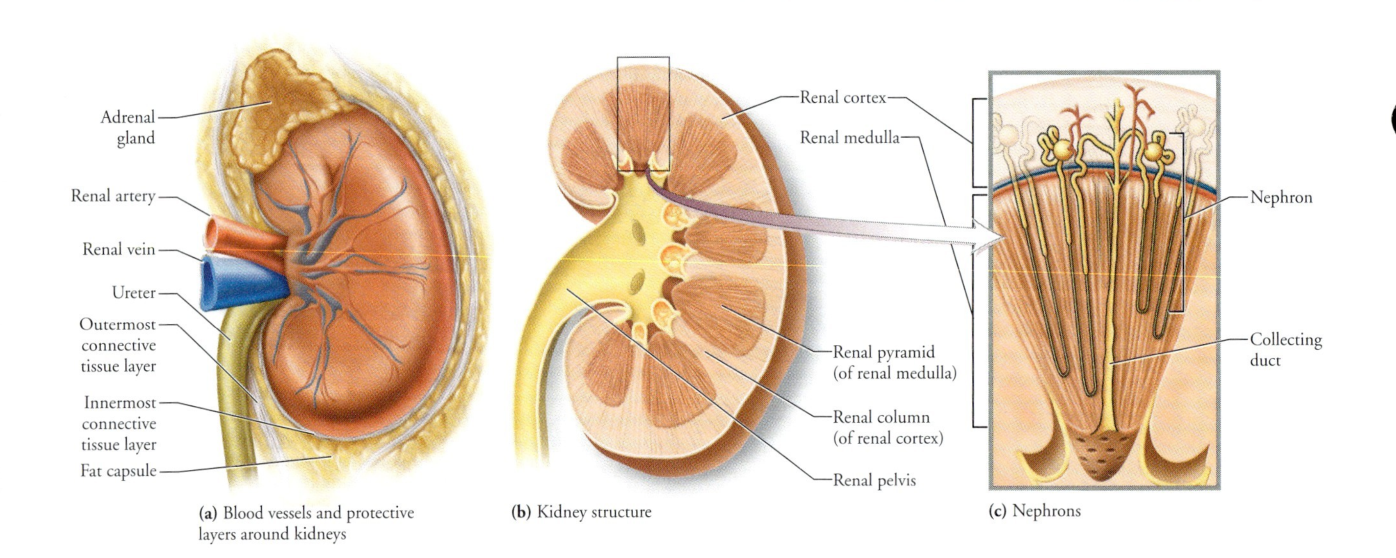

(a) Blood vessels and protective layers around kidneys

(b) Kidney structure

(c) Nephrons

Fig. 16.4

NOTES

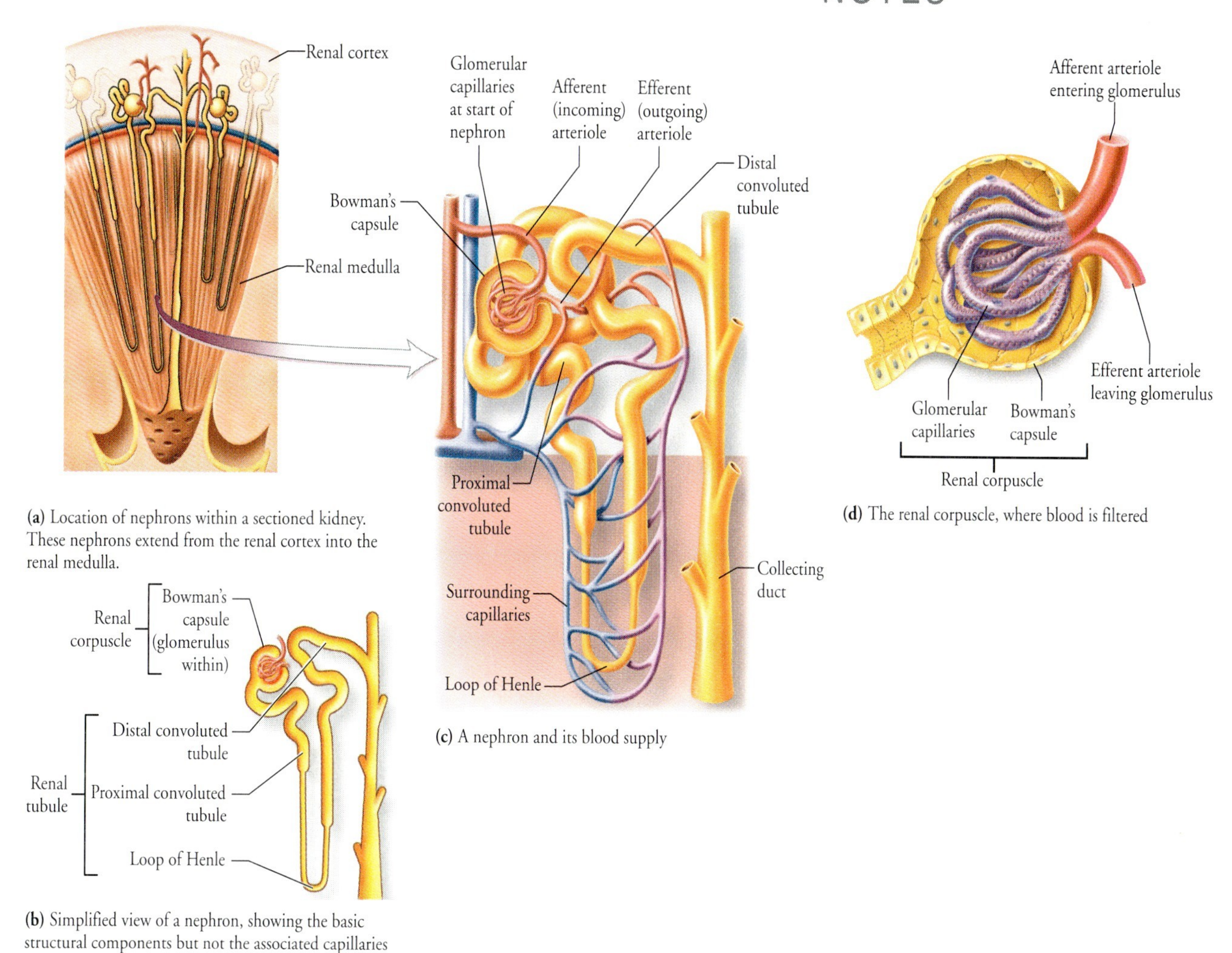

(a) Location of nephrons within a sectioned kidney. These nephrons extend from the renal cortex into the renal medulla.

(b) Simplified view of a nephron, showing the basic structural components but not the associated capillaries

(c) A nephron and its blood supply

(d) The renal corpuscle, where blood is filtered

Fig. 16.5

NOTES

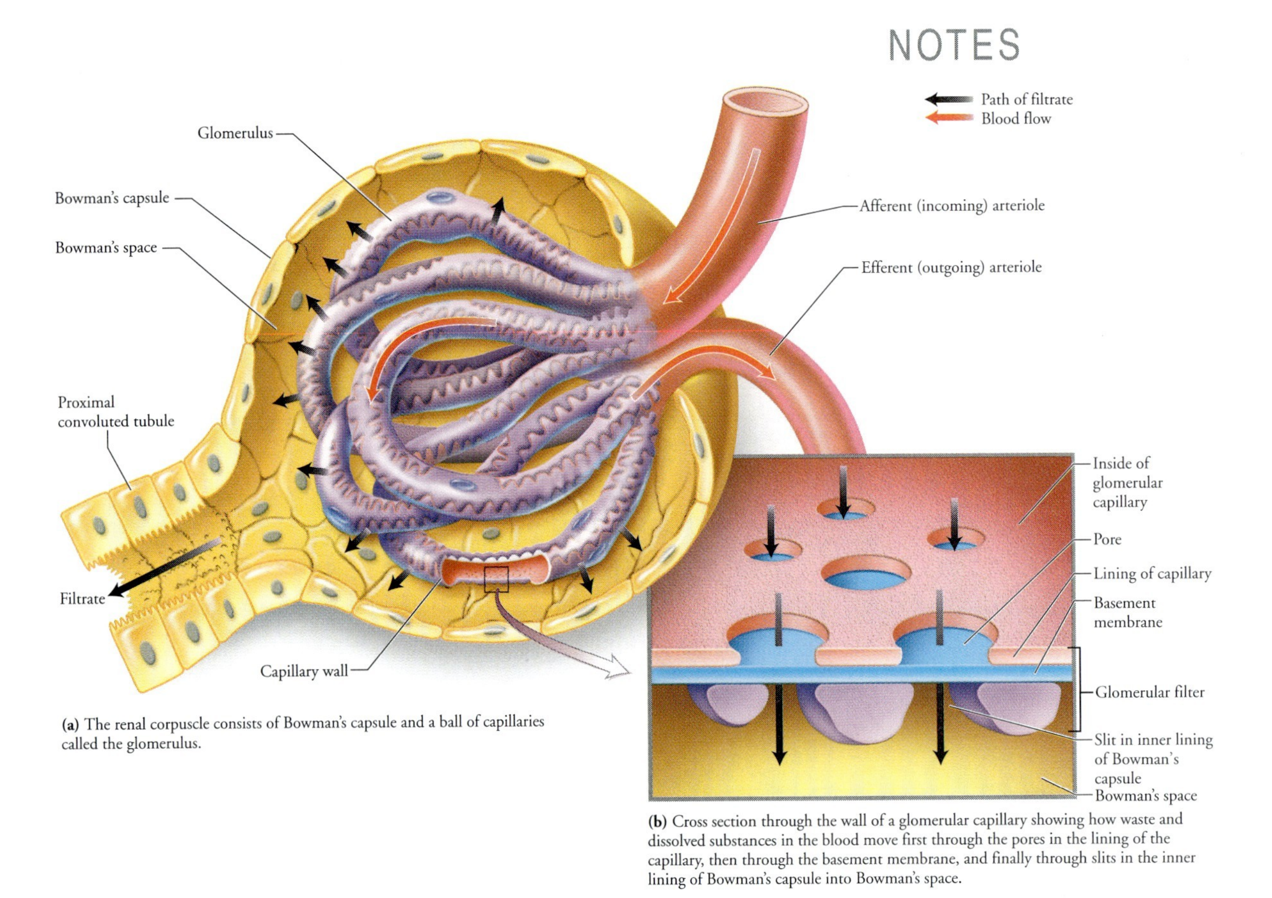

(a) The renal corpuscle consists of Bowman's capsule and a ball of capillaries called the glomerulus.

(b) Cross section through the wall of a glomerular capillary showing how waste and dissolved substances in the blood move first through the pores in the lining of the capillary, then through the basement membrane, and finally through slits in the inner lining of Bowman's capsule into Bowman's space.

Fig. 16.6

NOTES

REGIONS OF THE NEPHRON AND THEIR ROLES IN FILTRATION, REABSORPTION, AND SECRETION	
REGION OF NEPHRON	ROLE
Renal corpuscle (Bowman's capsule and glomerulus)	Filters the blood, removing water, glucose, amino acids, ions, nitrogenous wastes, and other small molecules
Proximal convoluted tubule	Reabsorbs water, glucose, amino acids, some urea, Na^+, Cl^-, and HCO_3^- Secretes drugs, H^+, NH_4^+
Loop of Henle	Reabsorbs water, Na^+, Cl^-, and K^+
Distal convoluted tubule	Reabsorbs water, Na^+, Cl^-, and HCO_3^- Secretes drugs, H^+, K^+, NH_4^+

Table 16.2

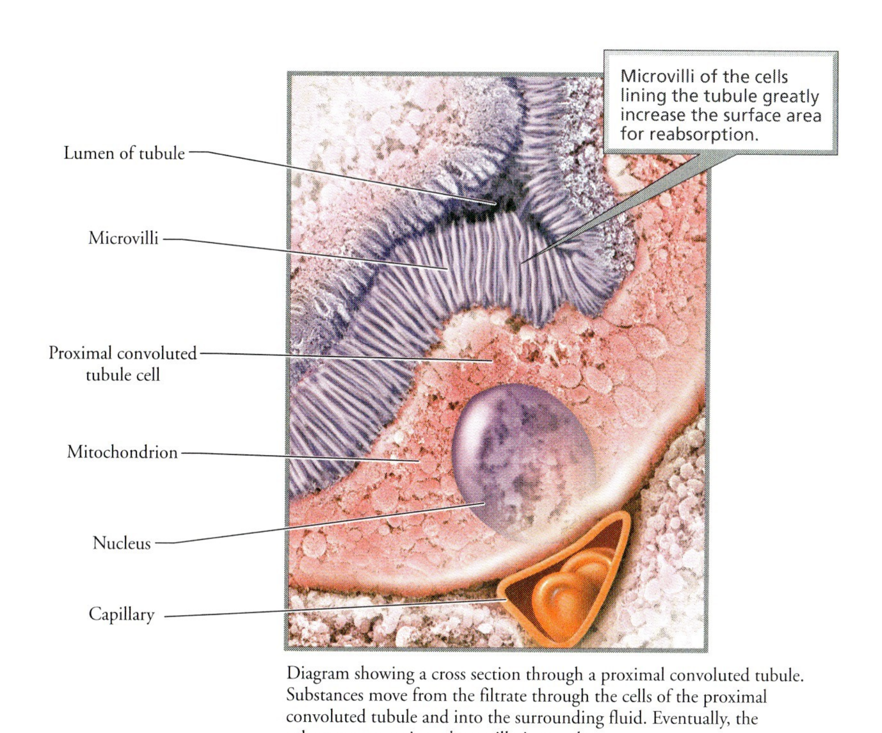

Diagram showing a cross section through a proximal convoluted tubule. Substances move from the filtrate through the cells of the proximal convoluted tubule and into the surrounding fluid. Eventually, the substances move into the capillaries nearby.

Fig. 16.7

NOTES

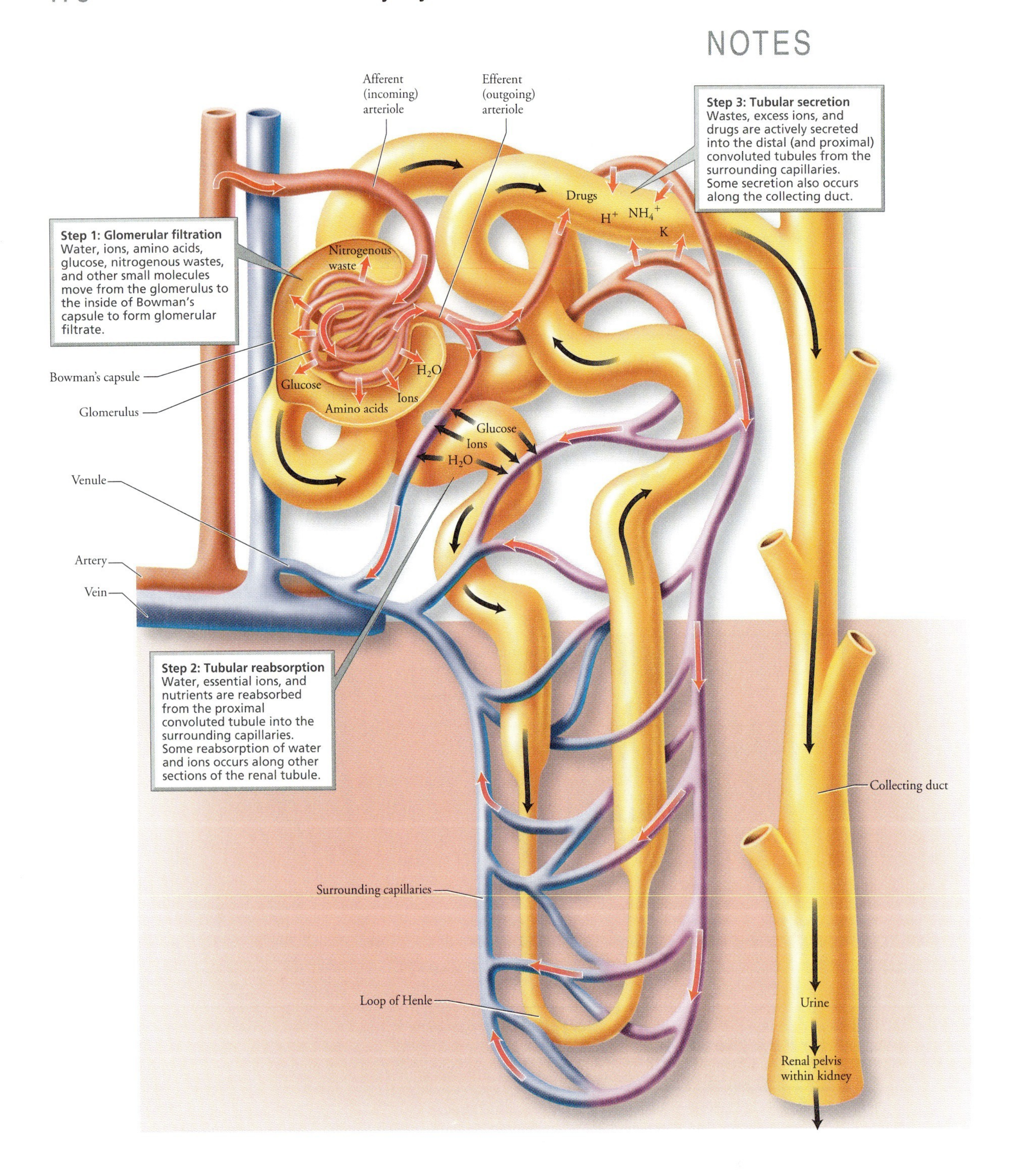

Fig. 16.8

NOTES

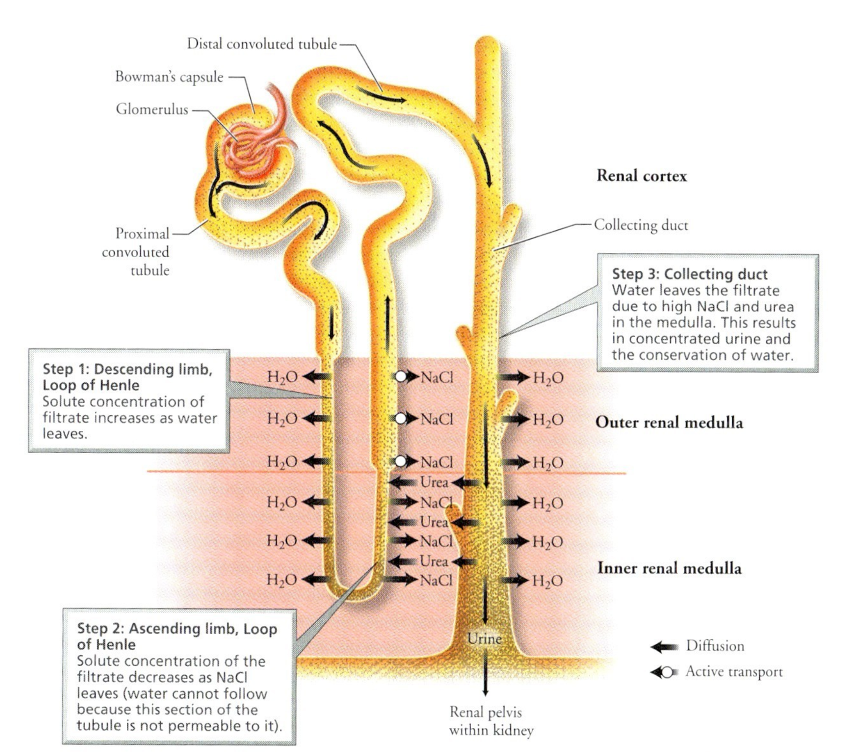

Fig. 16.9

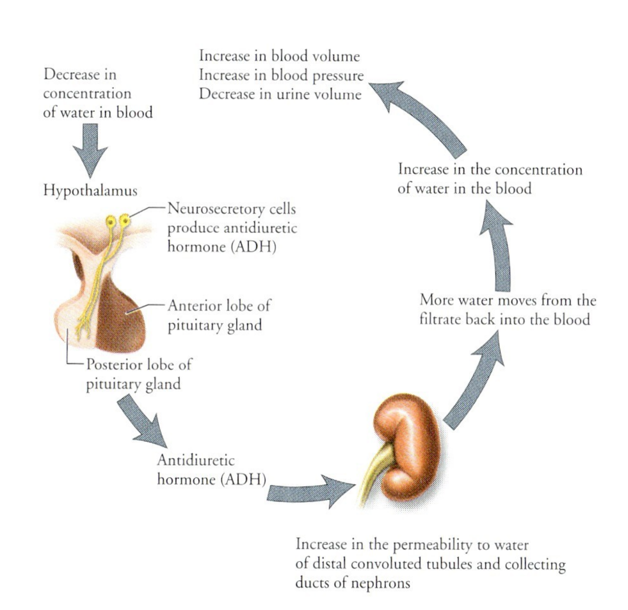

Fig. 16.10

NOTES

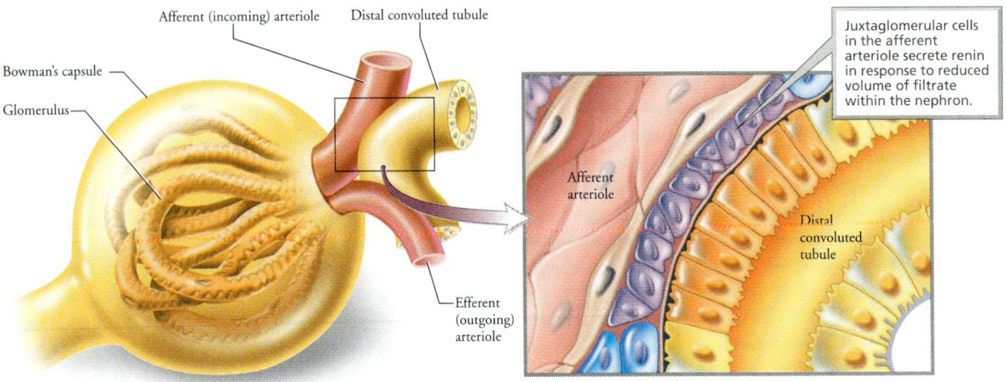

Afferent (incoming) arteriole

Distal convoluted tubule

Bowman's capsule

Glomerulus

Efferent (outgoing) arteriole

Juxtaglomerular cells in the afferent arteriole secrete renin in response to reduced volume of filtrate within the nephron.

Afferent arteriole

Distal convoluted tubule

(a) The juxtaglomerular apparatus (within the square) is a section of the nephron where the distal convoluted tubule contacts the afferent arteriole. The nearby renal corpuscle is shown in ghosted view to reveal its components, Bowman's capsule and the glomerulus.

(b) Close-up view of the juxtaglomerular apparatus

Fig. 16.11

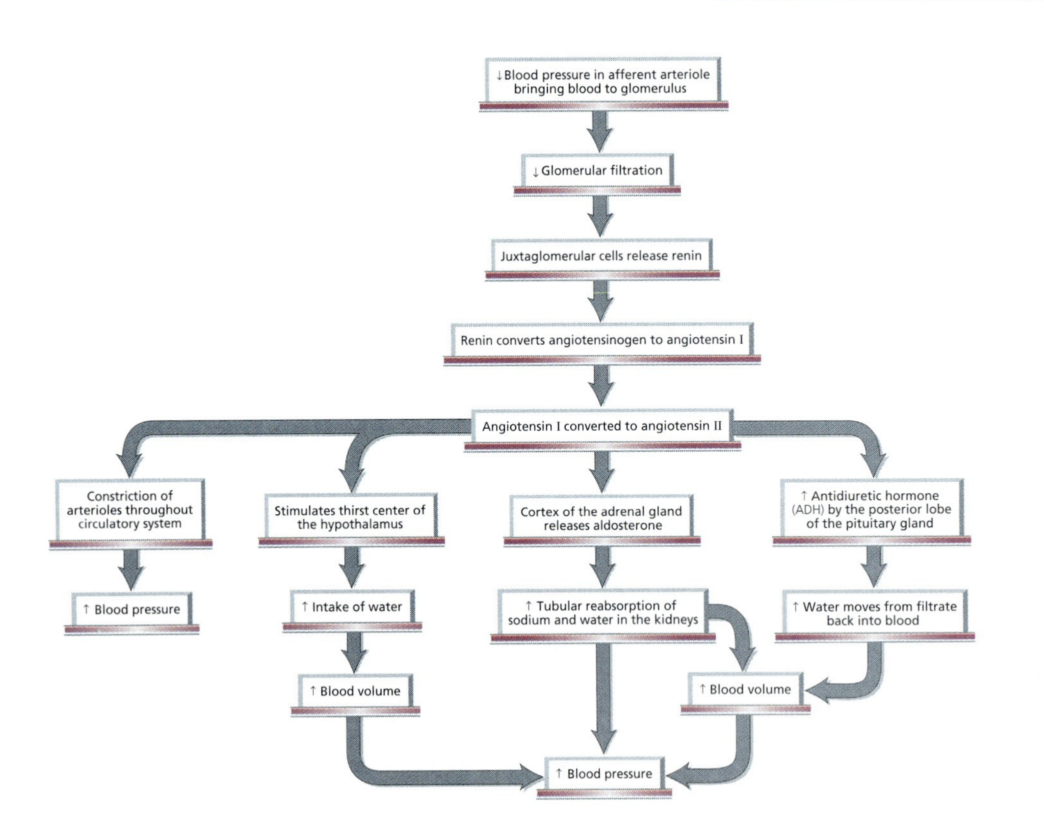

Fig. 16.12

NOTES

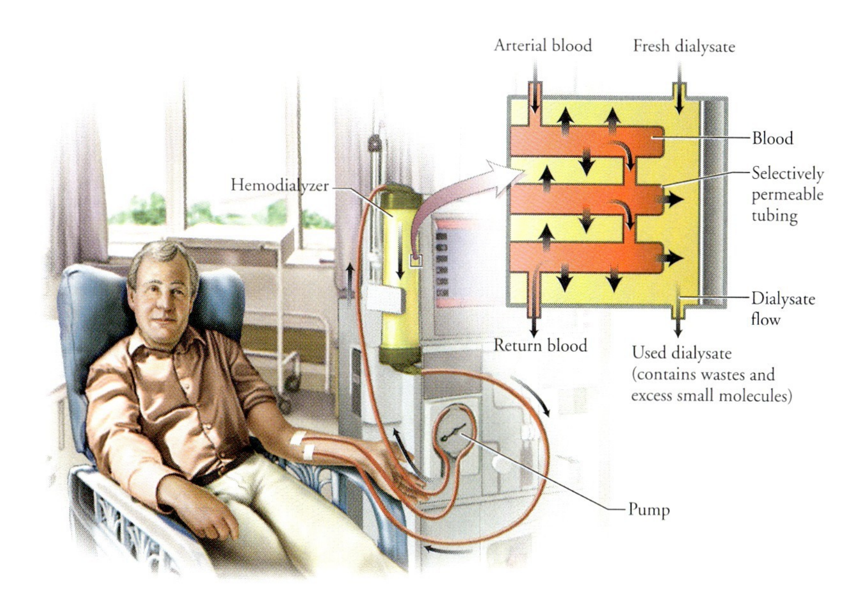

Fig. 16.13

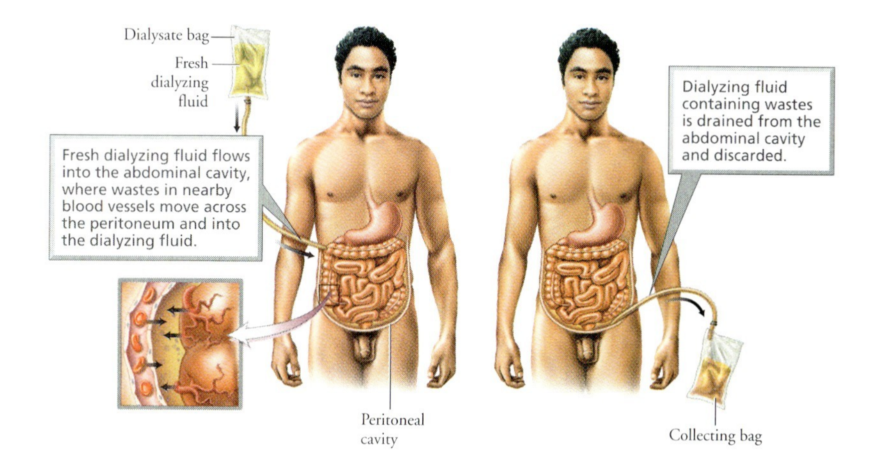

Fig. 16.14

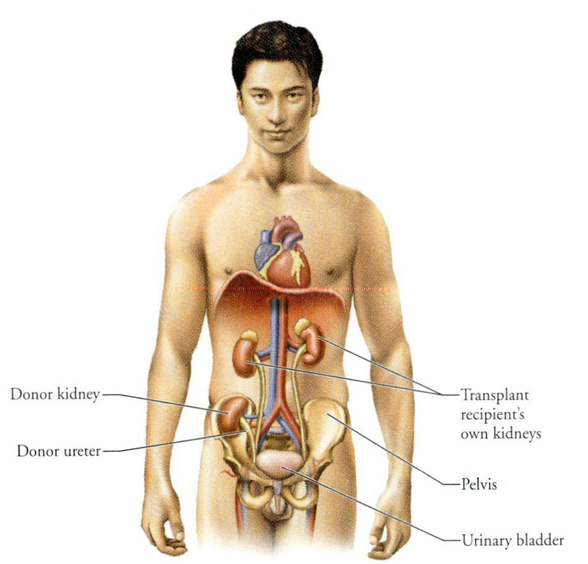

Fig. 16.15

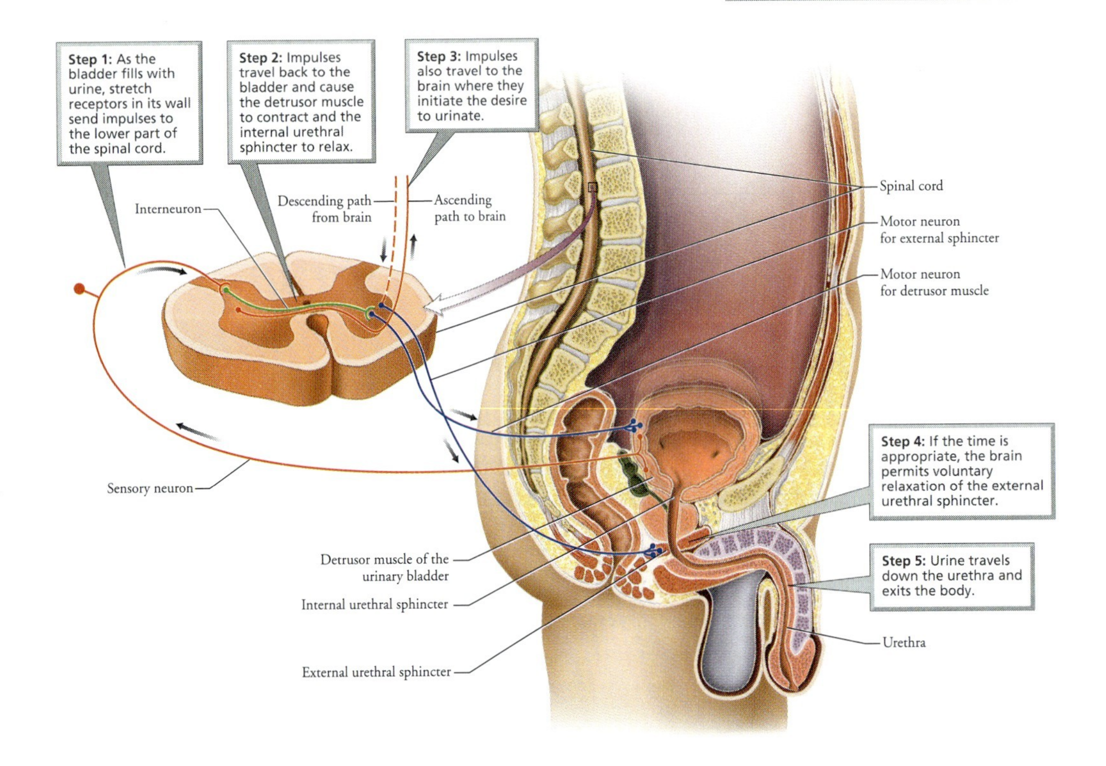

Fig. 16.17

NOTES

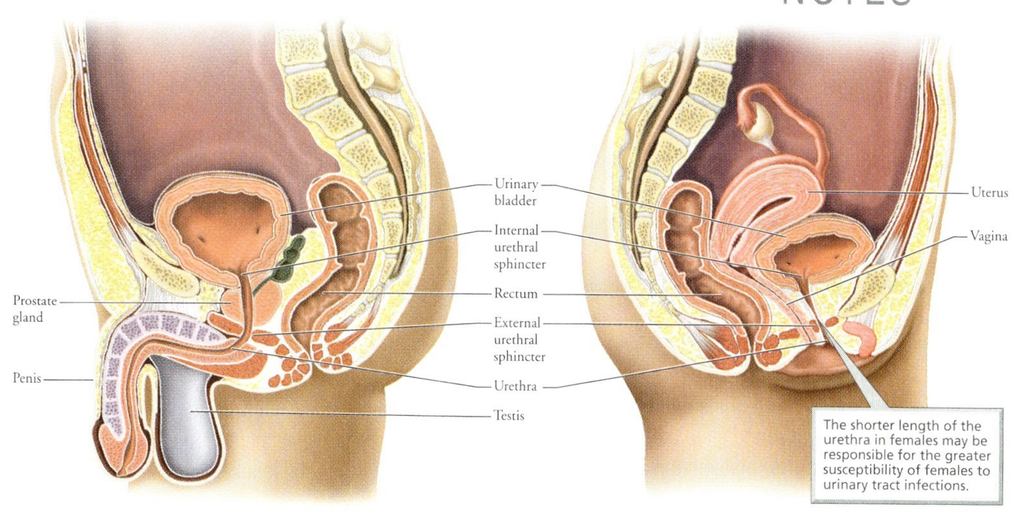

Fig. 16.19

Some Normal Constituents of Urine and Their Clinical Implications When Present in Abnormal Amounts

CONSTITUENT	CLINICAL IMPLICATIONS
Calcium (Ca^{2+})	Values depend on dietary intake
	Increases signal overactive parathyroid glands or cancer of breast or lungs
	Decreases signal vitamin D deficiency or underactive parathyroid glands
Chloride (Cl^-)	Values depend on dietary salt intake
	Increases signal Addison's disease (undersecretion of glucocorticoids by adrenal cortex), dehydration, or starvation
	Decreases signal diarrhea or emphysema
Creatinine	Increases signal infection
	Decreases signal muscular atrophy, anemia, or kidney disease
Potassium (K^+)	Increases signal chronic renal failure, dehydration, starvation, or Cushing's syndrome (oversecretion of glucocorticoids by adrenal cortex)
	Decreases signal diarrhea or underactive adrenal cortex
Sodium (Na^+)	Values depend on dietary salt intake
	Increases signal dehydration, starvation, or low blood pH from diabetes
	Decreases signal diarrhea, acute renal failure, emphysema, or Cushing's syndrome
Urea	Increases signal high protein intake
	Decreases signal impaired kidney function
Uric acid	Increases signal gout (excessive uric acid in blood causes crystals to form in joints, soft tissues, and kidneys), leukemia (cancer of white blood cells), or liver disease
	Decreases signal kidney disease

Table 16.A

NOTES

SOME NORMAL CONSTITUENTS OF URINE AND THEIR CLINICAL IMPLICATIONS WHEN PRESENT IN ABNORMAL AMOUNTS

CONSTITUENT	CLINICAL IMPLICATIONS
Calcium (Ca^{2+})	Values depend on dietary intake
	Increases signal overactive parathyroid glands or cancer of breast or lungs
	Decreases signal vitamin D deficiency or underactive parathyroid glands
Chloride (Cl^{-})	Values depend on dietary salt intake
	Increases signal Addison's disease (undersecretion of glucocorticoids by adrenal cortex), dehydration, or starvation
	Decreases signal diarrhea or emphysema
Creatinine	Increases signal infection
	Decreases signal muscular atrophy, anemia, or kidney disease
Potassium (K^{+})	Increases signal chronic renal failure, dehydration, starvation, or Cushing's syndrome (oversecretion of glucocorticoids by adrenal cortex)
	Decreases signal diarrhea or underactive adrenal cortex
Sodium (Na^{+})	Values depend on dietary salt intake
	Increases signal dehydration, starvation, or low blood pH from diabetes
	Decreases signal diarrhea, acute renal failure, emphysema, or Cushing's syndrome
Urea	Increases signal high protein intake
	Decreases signal impaired kidney function
Uric acid	Increases signal gout (excessive uric acid in blood causes crystals to form in joints, soft tissues, and kidneys), leukemia (cancer of white blood cells), or liver disease
	Decreases signal kidney disease

Table 16.B

NOTES

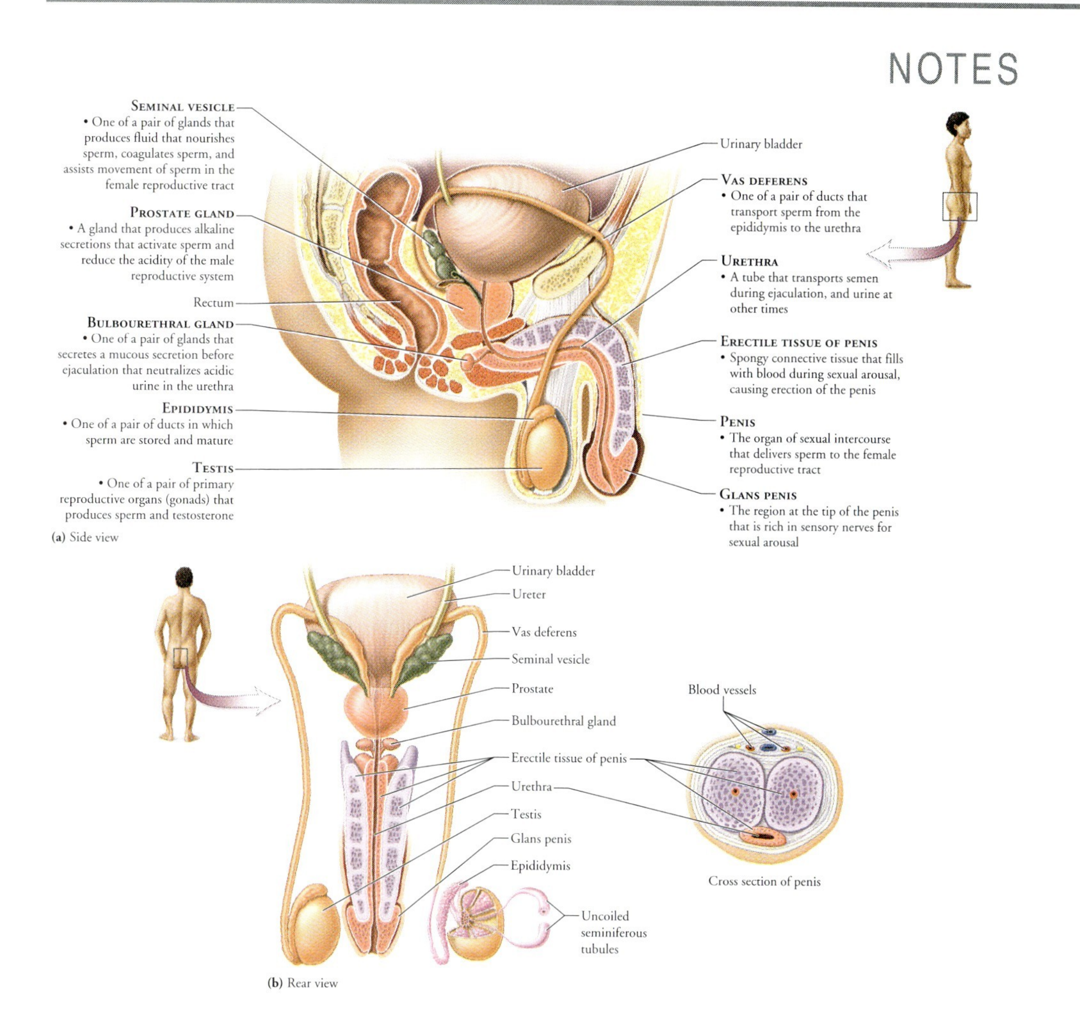

Fig. 17.1

THE MALE REPRODUCTIVE SYSTEM

STRUCTURE	FUNCTION
Testes	Produce sperm and testosterone
Epididymis	Location of sperm storage and maturation
Vas deferens	Conducts sperm from epididymis to urethra
Urethra	Tube through which sperm or urine leaves the body
Prostate gland	Produces secretions that make sperm mobile and that counteract the acidity of the female reproductive tract
Seminal vesicles	Produce secretions that make up most of the volume of semen
Bulbourethral glands	Produce secretions just before ejaculation; may lubricate; may rinse urine from urethra
Penis	Delivers sperm to female reproductive tract

Table 17.1

NOTES

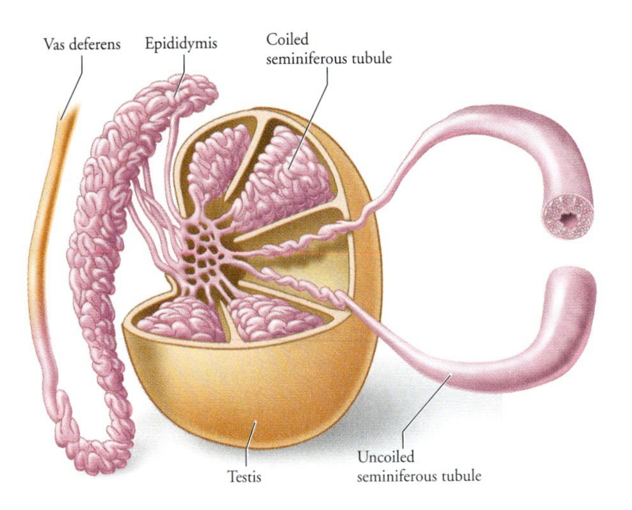

Fig. 17.2

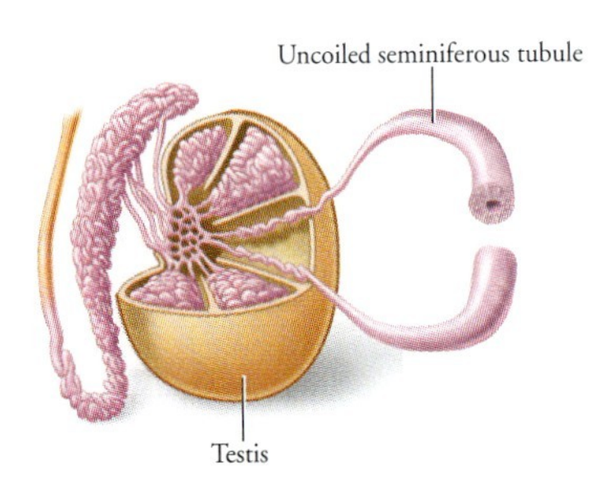

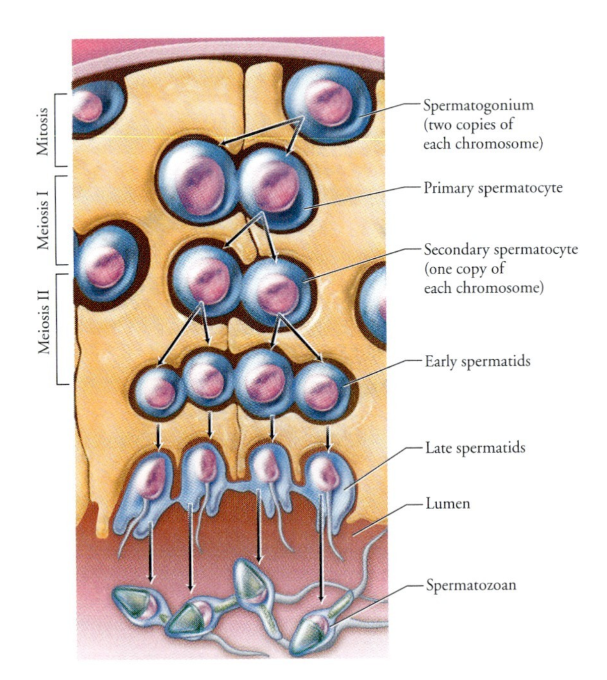

Fig. 17.3

NOTES

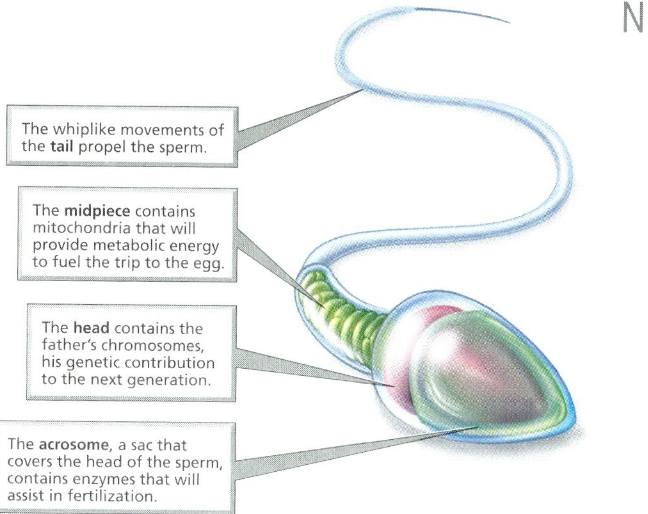

Fig. 17.4

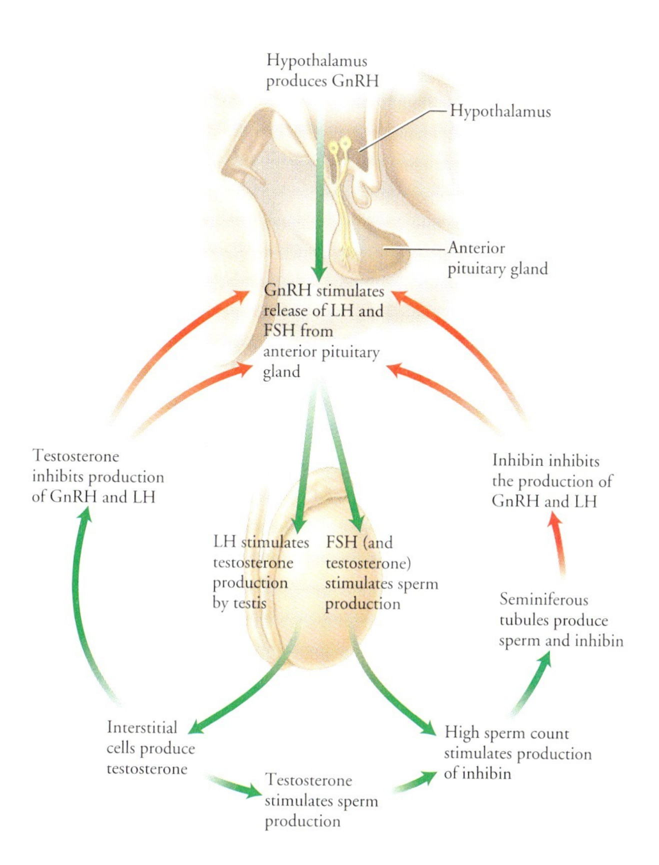

Fig. 17.5

NOTES

HORMONES IMPORTANT IN THE REGULATION OF MALE REPRODUCTIVE PROCESSES		
HORMONE	**SOURCE**	**EFFECTS**
Testosterone	Interstitial cells in testes	Sperm production; development and maintenance of male reproductive structures, male secondary sex characteristics; sex drive
Gonadotropin-releasing hormone (GnRH)	Hypothalamus (in brain)	Stimulates the anterior pituitary gland to release LH
Luteinizing hormone (LH)	Anterior pituitary gland (in brain)	Stimulates interstitial cells of testis to produce testosterone
Follicle-stimulating hormone (FSH)	Anterior pituitary gland (in brain)	Enhances sperm formation
Inhibin	Seminiferous tubules in testes	Inhibits FSH secretion by anterior pituitary gland, causing a decrease in sperm production and testosterone production

Table 17.2

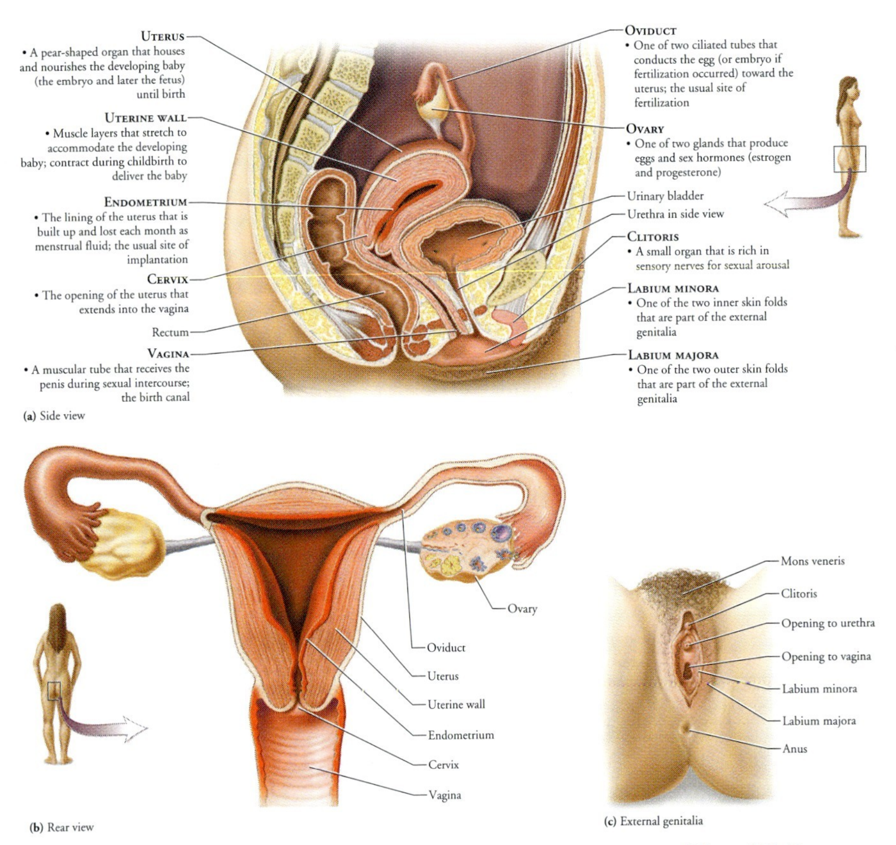

Fig. 17.6

NOTES

THE FEMALE REPRODUCTIVE SYSTEM	
STRUCTURE	FUNCTION
Ovary	Produces eggs and the hormones estrogen and progesterone
Oviducts	Transport ovulated egg (or embryo if fertilization occurred) from ovary to uterus; the usual site of fertilization
Uterus	Receives and nourishes embryo
Vagina	Receives penis during intercourse; serves as birth canal
Clitoris	Contributes to sexual arousal
Breasts	Produce milk

Table 17.3

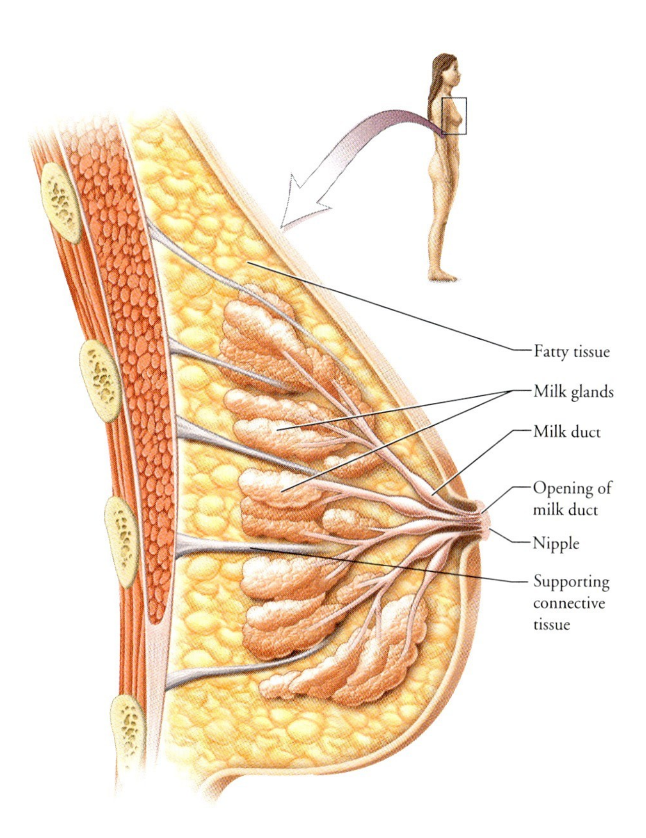

Fig. 17.7

NOTES

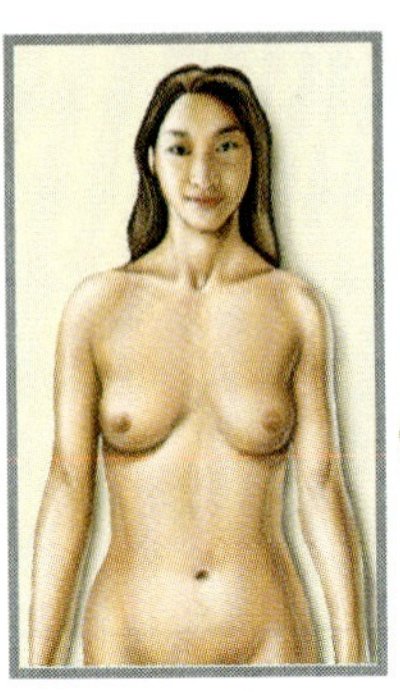

Step 1: Stand in front of the mirror and look at each breast to see if there is a lump, a depression, a difference in texture, or any other change in appearance.

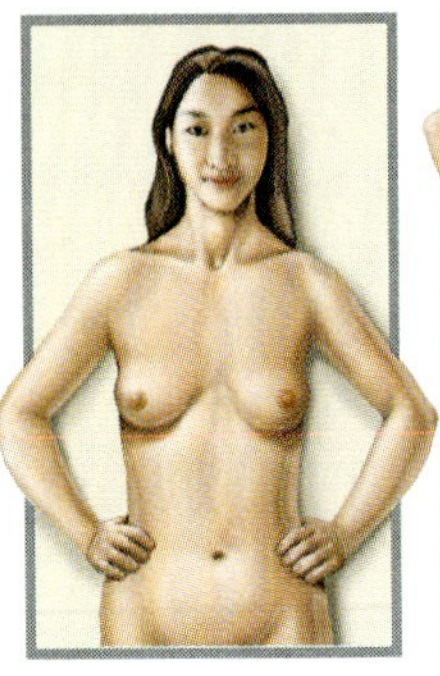

Step 2: Get to know how your breasts look. Be especially alert for any changes in the nipples' appearance.

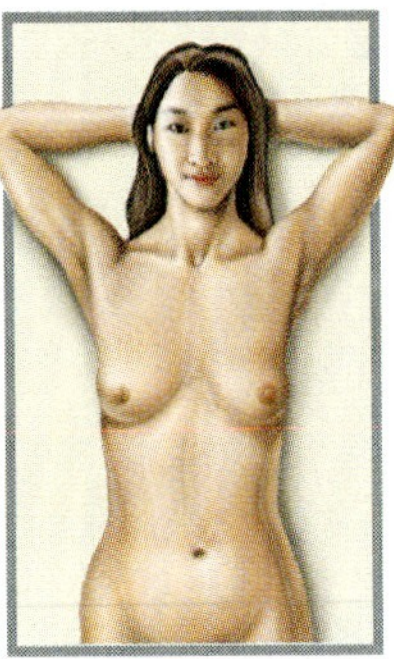

Step 3: Raise both arms and check for any swelling or dimpling in the skin of your breasts.

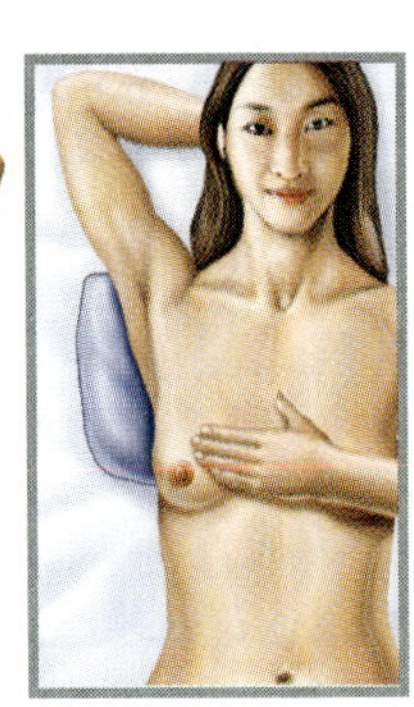

Step 4: Lie down with a pillow under your right shoulder and put your right arm behind your head. Perform a manual breast examination. With the nipple as the center, divide your breast into imaginary quadrants.

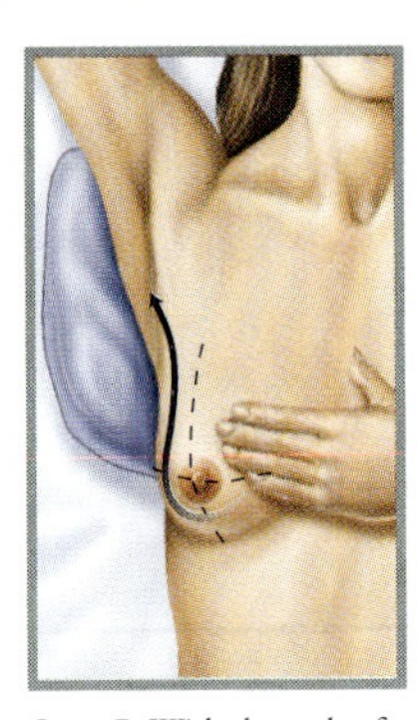

Step 5: With the pads of the fingers of the left hand, make firm circular movements over each quadrant, feeling for unusual lumps or areas of tenderness. When you reach the upper, outer quadrant of your breast, continue toward your armpit. Press down in all directions.

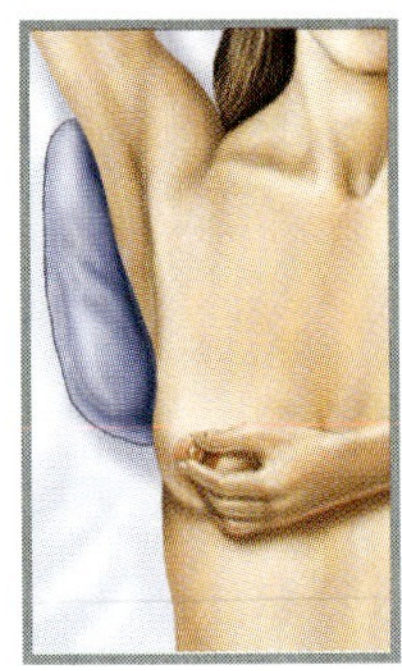

Step 6: Feel your nipple for any change in size and shape. Squeeze your nipple to see if there is any discharge. Repeat from step 4 on the other breast.

Fig. 17.A

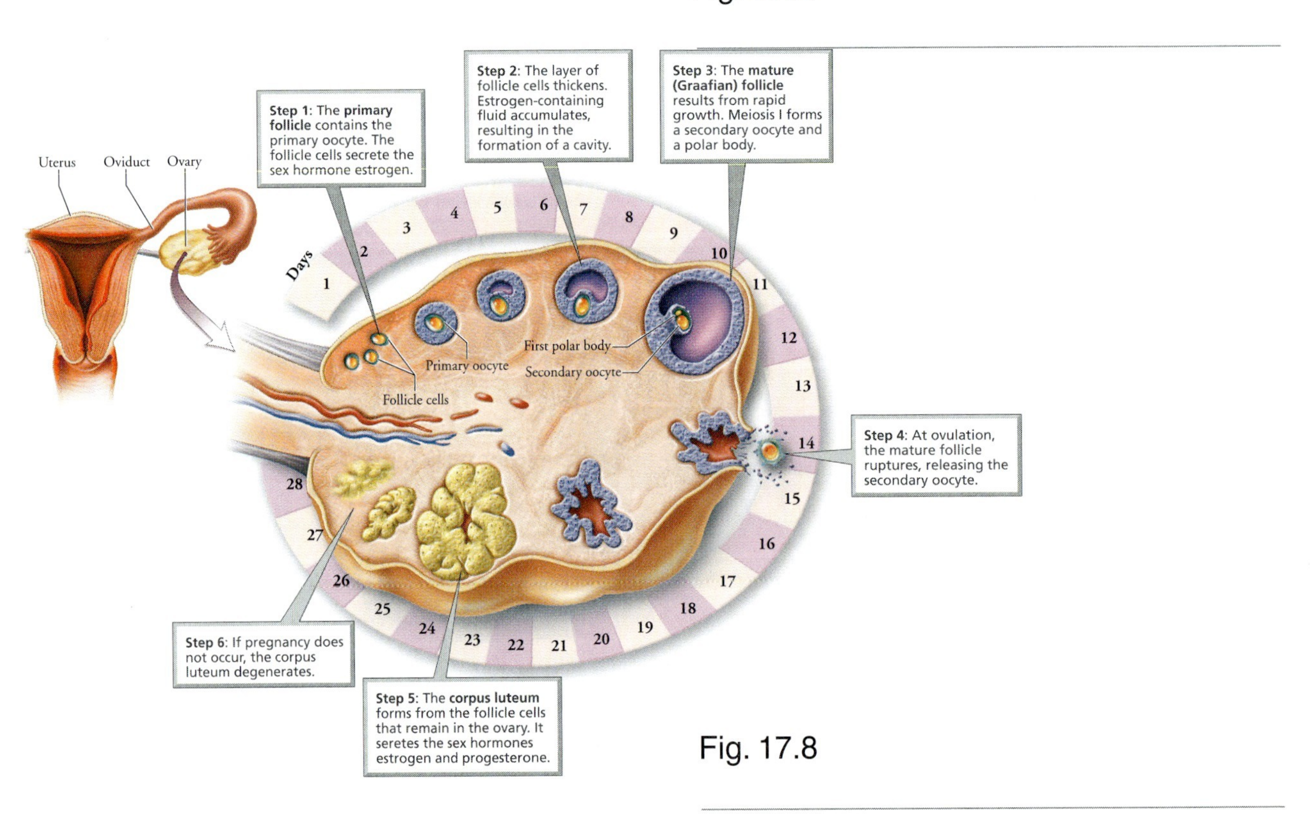

Fig. 17.8

NOTES

HORMONES INVOLVED IN THE REGULATION OF FEMALE REPRODUCTIVE PROCESSES		
HORMONE	**SOURCE**	**EFFECTS**
Estrogen	Ovaries (follicle cells and corpus luteum)	Maturation of the egg; development and maintenance of female reproductive structures, secondary sex characteristics; thickens endometrium of uterus in preparation for implantation of embryo; cell division in breast tissue
Progesterone	Ovaries (corpus luteum)	Further prepares uterus for implantation of embryo; maintains endometrium
Follicle-stimulating hormone (FSH)	Anterior pituitary gland (in brain)	Stimulates development of a follicle in the ovary
Luteinizing hormone (LH)	Anterior pituitary gland (in brain)	Triggers ovulation; causes formation of the corpus luteum

Table 17.4

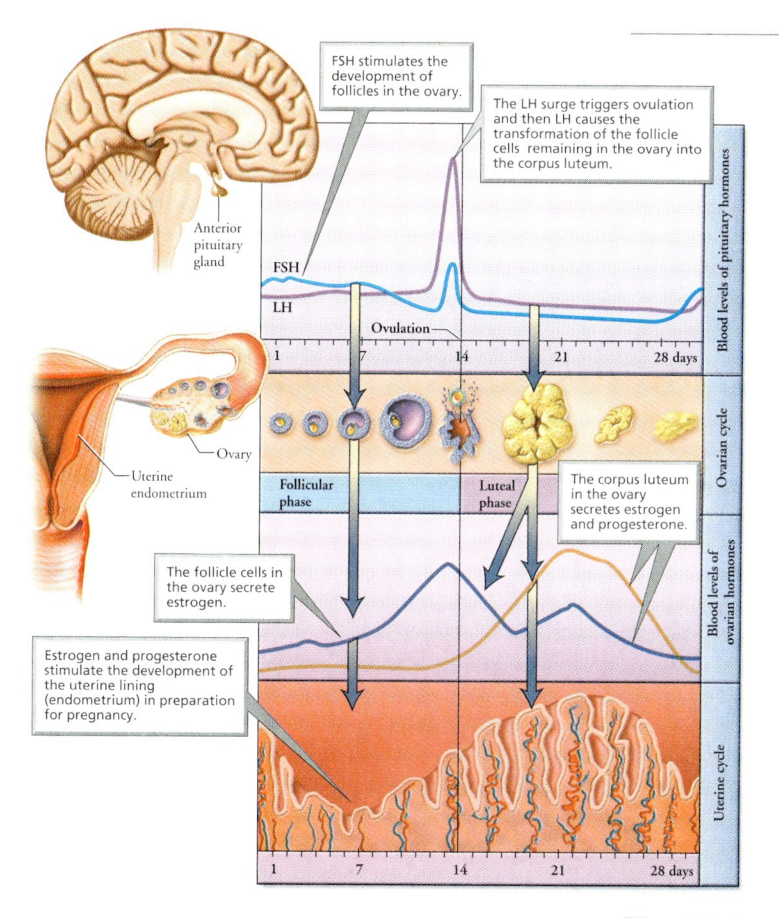

Fig. 17.9

NOTES

OVARIAN AND UTERINE CYCLES			
OVARIAN CYCLE		**UTERINE CYCLE**	
Approximate timing in 28-day cycle	Events	Approximate timing in 28-day cycle	Events
Days 1–13	Follicle develops, caused by FSH	Day 1	Onset of menstrual flow (breakdown of endometrium)
	Follicle cells produce estrogen	Day 6	Endometrium begins to get thicker
Day 14	Ovulation is triggered by LH surge		
Days 15–21	Corpus luteum forms and secretes estrogen and progesterone	Days 15–23	Endometrium is further prepared for implantation of the embryo by estrogen and progesterone
Days 22–28	Corpus luteum degenerates, causing estrogen and progesterone level to decline	Days 24–28	Endometrium begins to degenerate owing to declining maintenance by progesterone

Table 17.5

NOTES

METHODS OF BIRTH CONTROL AVAILABLE IN THE UNITED STATES

METHOD	PROCEDURE	HOW IT WORKS	PERCENTAGE FAILURE (TYPICAL USE)	PERCENTAGE FAILURE (PERFECT USE)	RISKS	PROTECTION FROM STDs
Abstinence	Abstain from sexual activity	Sperm never contacts egg	0	0	None	Yes
Sterilization						
Vasectomy	Cut and seal each vas deferens	No sperm in semen	<1	<1	No risks presently known	None
Tubal ligation	Oviducts blocked	Sperm cannot reach egg	<1	<1	Infection from surgery	None
Hormonal Methods						
Combination estrogen and progesterone						
Oral (the pill)	Hormone pill taken daily	Prevents egg development and release	3	0.1	Problems with heart or blood vessels; blood clot formation; stroke	None
Injection (Lunelle®)	Estrogen and progesterone injection every month	Same as pill	Better than pill	0.1	Assumed same as pill	None
Vaginal ring (NuvaRing®)	Ring inserted in vagina by woman for 3 weeks and removed for 1 week	Same as pill	Better than pill	1–2	Assumed same as pill	None
Skin patch (Ortho Evra®)	Patch adheres to skin of abdomen or buttocks	Same as pill	1.24	0.99	Assumed same as pill	None
Progesterone only						
Minipill	Hormone pill taken daily	Thickens cervical mucus; endometrium not properly prepared; sperm movement impaired; egg does not mature	13.2	1.1	Ovarian cysts	None
Injection (Depro-Provera®)	Progesterone injection every 3 months	Same as minipill	0.3	0.3	Possible link to osteoporosis	None
Intrauterine device (IUD)	Small plastic device inserted into uterus by physician	Interferes with both fertilization and implantation	<2	<2	Increased risk of pelvic inflammatory disease following insertion	None
Barrier Methods						
Diaphragm	Inserted into vagina before intercourse	Covers cervix and prevents sperm from entering	18	9	No risks presently known	Some for woman
Cervical cap	Inserted into vagina before intercourse	Same as diaphragm (partly by suction)	18	9	No risks presently known	Some for woman
Male condom	Fits over penis before intercourse	Prevents sperm/penis from contacting vagina	12	3	No risks presently known	Latex, excellent; "skin," poor
Female condom	Held in vagina by flexible rim rings	Prevents sperm/penis from contacting vagina	21	5	No risks presently known	Very good
Spermicides	Inserted into vagina before intercourse	Kill sperm for 1 hour after application	21	6	No risks presently known	Increased for women

Table 17.6

NOTES

Table 17.6 (continued)

METHODS OF BIRTH CONTROL AVAILABLE IN THE UNITED STATES						
METHOD	PROCEDURE	HOW IT WORKS	PERCENTAGE FAILURE (TYPICAL USE)	PERCENTAGE FAILURE (PERFECT USE)	RISKS	PROTECTION FROM STDs
Fertility Awareness						
	Abstain from sex on days that eggs and sperm may meet	Sperm never come into contact with egg	20	2–9	None	None
Emergency contraception (morning-after pill)						
Combined estrogen and progesterone (Preven®)	First dose taken within 72 hours of unprotected intercourse; second dose 12 hours later	Inhibit or delay ovulation; prevent fertilization; thicken the cervical mucus; alter the endometrium		Reduces likelihood of pregnancy by 75%	Nausea; menstrual cycle disturbance	None
Progesterone-only (Plan B®)	Same as combination	Same as combination		Same as combination	Same as combination	None

Table 17.6 (continued)

NOTES

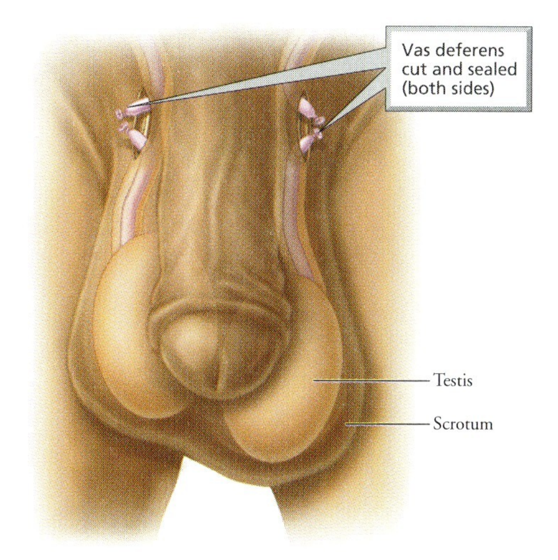

(a) Vasectomy

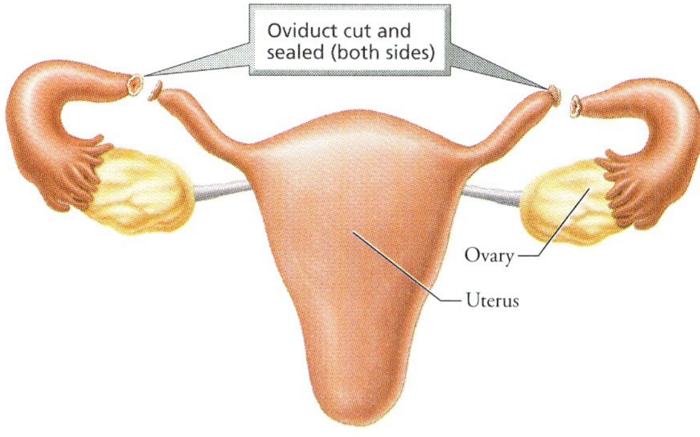

(b) Tubal ligation

Fig. 17.10

Fig. 17.12

NOTES

NOTES

SEXUALLY TRANSMITTED DISEASES

DISEASE	SYMPTOMS	DIAGNOSIS AND TREATMENT	EFFECTS
STDs caused by bacteria			
Chlamydia	First symptoms occur 7–21 days after contact Up to 75% of women and 50% of men show no symptoms *Women:* Vaginal discharge Vaginal bleeding between periods Pain during urination and intercourse Abdominal pain accompanied by fever and nausea *Men:* Urethral discharge Pain during urination	*Diagnosis:* Urine test for chlamydial DNA *Treatment:* Antibiotics	Long-term reproductive consequences such as sterility Infection can pass to infant during childbirth Can cause rupture of the protective membrane surrounding the fetus
Gonorrhea	First symptoms occur 2–21 days after contact About 30%–40% of men and women show no symptoms *Women:* Vaginal discharge Pain during urination and bowel movement Cramps and pain in lower abdomen More pain than usual during menstruation *Men:* Thick yellow or white discharge from penis Inflammation of the urethra Pain during urination and bowel movements	*Diagnosis:* Examination of penile discharge or cervical secretions Urine test for DNA of the bacterium that causes gonorrhea Cell culture *Treatment:* Antibiotics	Can cause long-term reproductive consequences such as sterility Infection can pass to infant during childbirth Can cause heart trouble, arthritis, and blindness
Syphilis	*Stage 1:* Occurs 2–8 weeks after contact. Chancre forms at site of contact. Lymph nodes in groin area swell *Stage 2:* Occurs 6 weeks to 6 months after contact Reddish brown rash appears anywhere on the body Flulike symptoms present Ulcers or warty growth may appear Patches of hair may be lost *Stage 3:* Lesions appear on skin and internal organs May affect nervous system Blindness Brain damage	*Diagnosis:* Identification of the bacterium from a chancre Blood test to detect antibodies to the bacterium that causes syphilis *Treatment:* Large doses of antibiotics over a prolonged period of time	Infection can pass to fetus during pregnancy Can cause heart disease, brain damage, blindness, and death

Table 17a.1

NOTES

Table 17a.1 (continued)

SEXUALLY TRANSMITTED DISEASES			
DISEASE	SYMPTOMS	DIAGNOSIS AND TREATMENT	EFFECTS
STDs caused by viruses			
Genital herpes	First symptoms appear 2–20 days after contact Some people have no symptoms Flulike symptoms present Small, painful blisters that can leave painful ulcers appear Blisters go away but the virus remains Symptoms recur periodically	*Diagnosis:* Examination of blisters Laboratory tests on the fluid from the sore to detect the presence of the virus Blood test for antibodies *Treatment:* Antiviral drugs can ease symptoms	Cannot be cured Reoccurrences of blisters Infection can pass to fetus, causing miscarriage or stillbirth Can cause brain damage in newborns
Genital warts	First symptoms appear 1–6 months after exposure Small warts appear on sex organs May cause itching, burning, irritation, discharge, bleeding	*Diagnosis:* Appearance of growth In women, Pap test may help *Treatment:* For removal: freezing, burning, laser surgery	Formation of additional warts Closely associated with cervical and penile cancer Infection can pass to infant during childbirth

Table 17a.1 (continued)

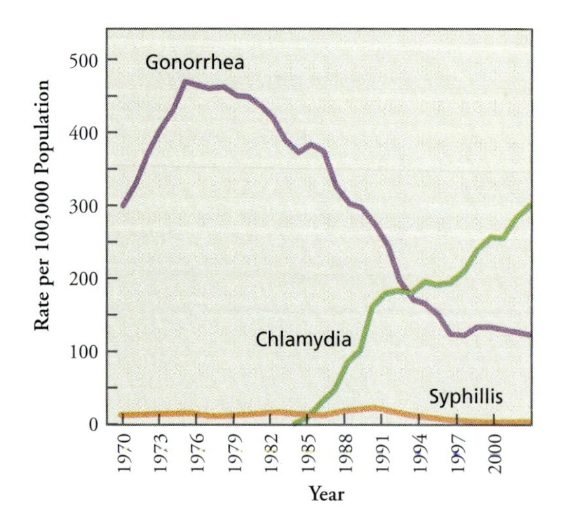

Fig. 17a.1

NOTES

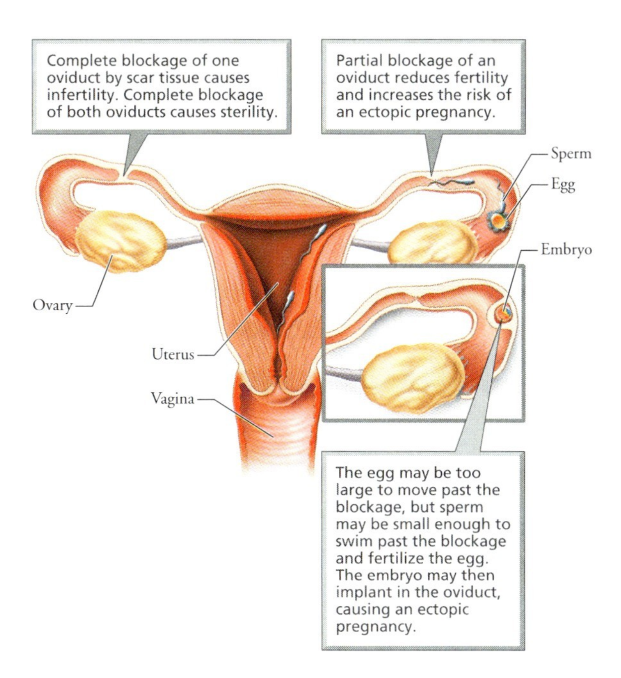

Fig. 17a.2

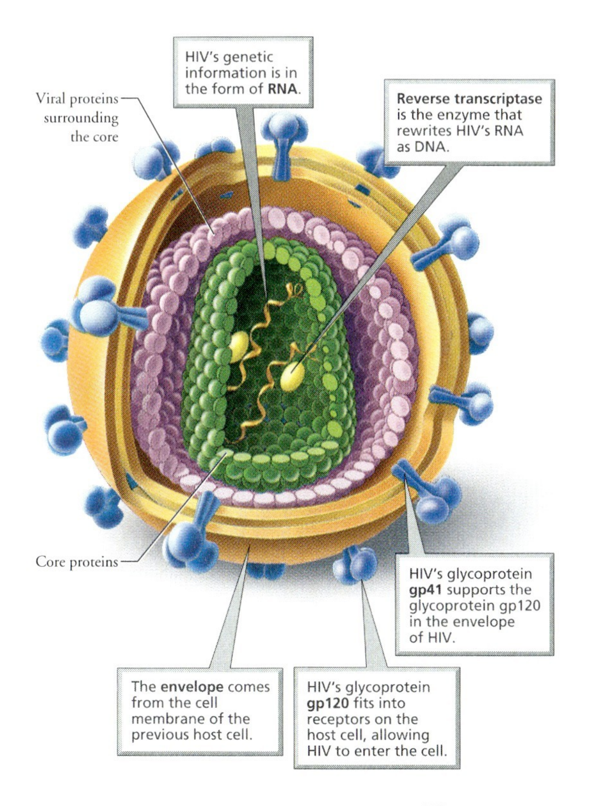

Fig. 17a.7

NOTES

Step 1: HIV attaches to a CD4 receptor on the cell membrane of the host cell.

Step 2: HIV fuses with the cell membrane and releases its contents into the host cell.

Step 3: HIV's protein coat degrades. HIV's RNA and reverse transcriptase are released into cell cytoplasm.

Step 4: Reverse transcriptase rewrites HIV's single-stranded RNA as HIV's double-stranded DNA.

Step 5: HIV's DNA enters the host cell's nucleus and is then inserted into the DNA of a host cell chromosome.

Step 6: Many copies of single-stranded HIV RNA are made. They direct the synthesis of HIV's proteins.

Step 7: HIV RNA and proteins gather at the cell membrane and leave the host cell as an immature virus.

Step 8: HIV RNA and proteins are reassembled to form a mature virus.

Reverse transcriptase

HIV RNA

HIV DNA

HIV provirus

HIV RNA

Fig. 17a.8

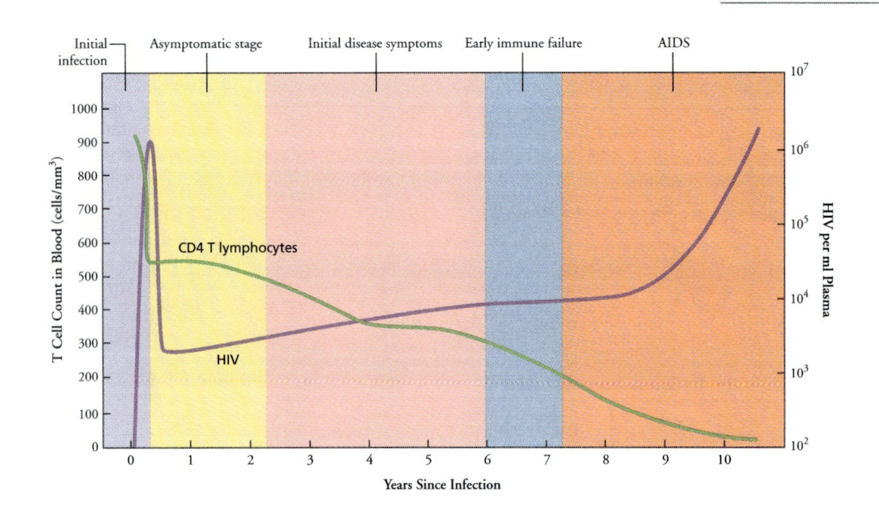

Fig. 17a.10

NOTES

THE STAGES OF AN HIV INFECTION				
STAGE	HIV	IMMUNE SYSTEM	T CELL COUNT/mm^3	SYMPTOMS
Initial infection	Enters body and replicates	Fights back; produces antibodies	~1000–800	• None • Flulike • Neurological
Asymptomatic	Replicating in lymph nodes	Vigorously produces helper T cells	~800–600	• None • Considered HIV positive when antibodies against HIV detected in blood
Initial disease symptoms	• Replicating • Viral load gradually increasing	Production of helper T cells cannot keep pace with destruction of helper T cells	~600–400	• Wasting syndrome • Lymphadenopathy • Neurological
Early immune failure	• Replicating • Viral load gradually increasing	Helper T cell number continues to decline	~400–200	• Thrush (and vaginal yeast infections) • Shingles • Hairy leukoplakia • Herpes simplex recurrences
AIDS	• Replicating • Viral load rapidly increasing	Helper T cell number continues to decline	200–below	

Table 17a.2

NOTES

NOTES

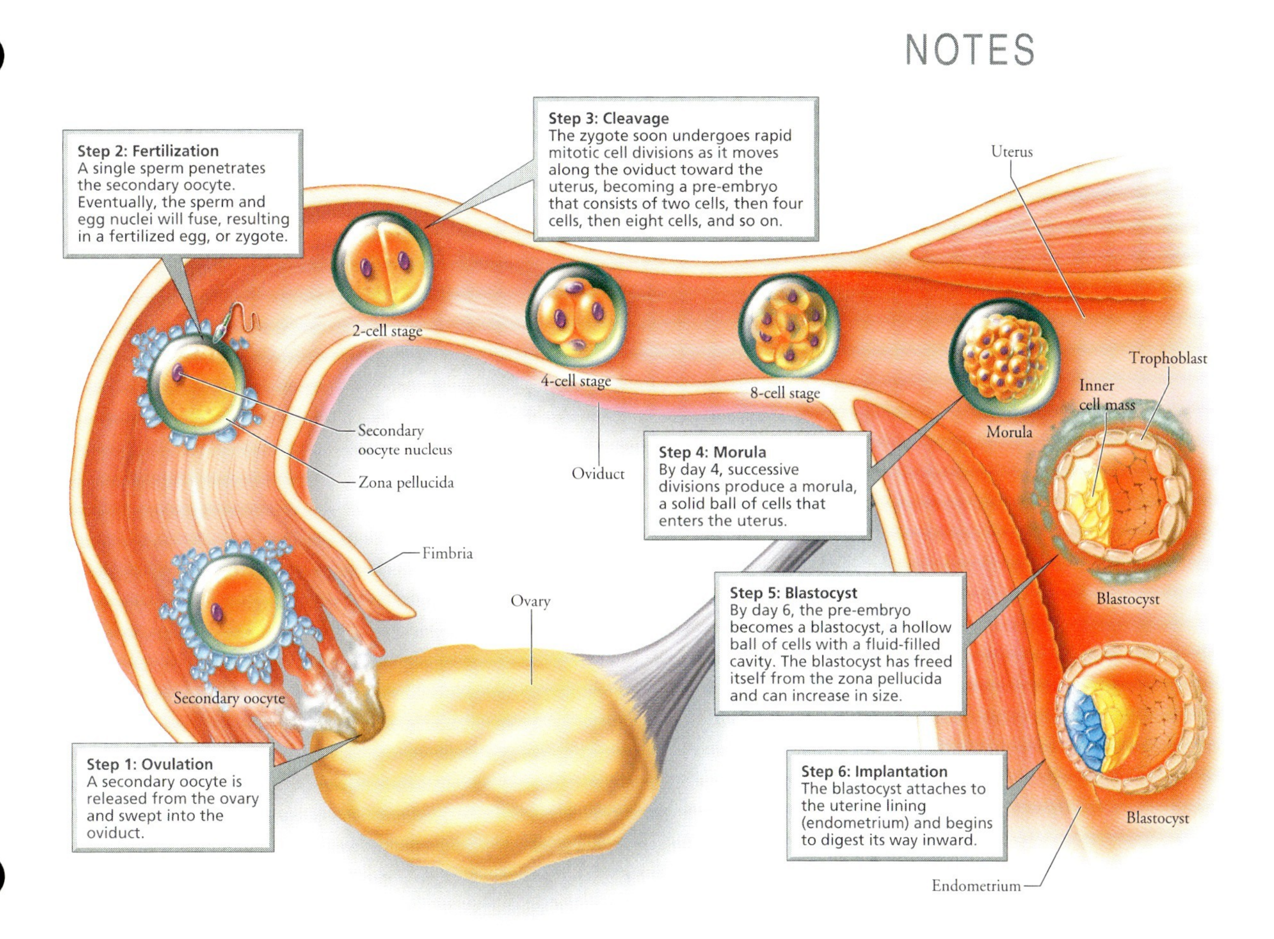

Fig 18.2

NOTES

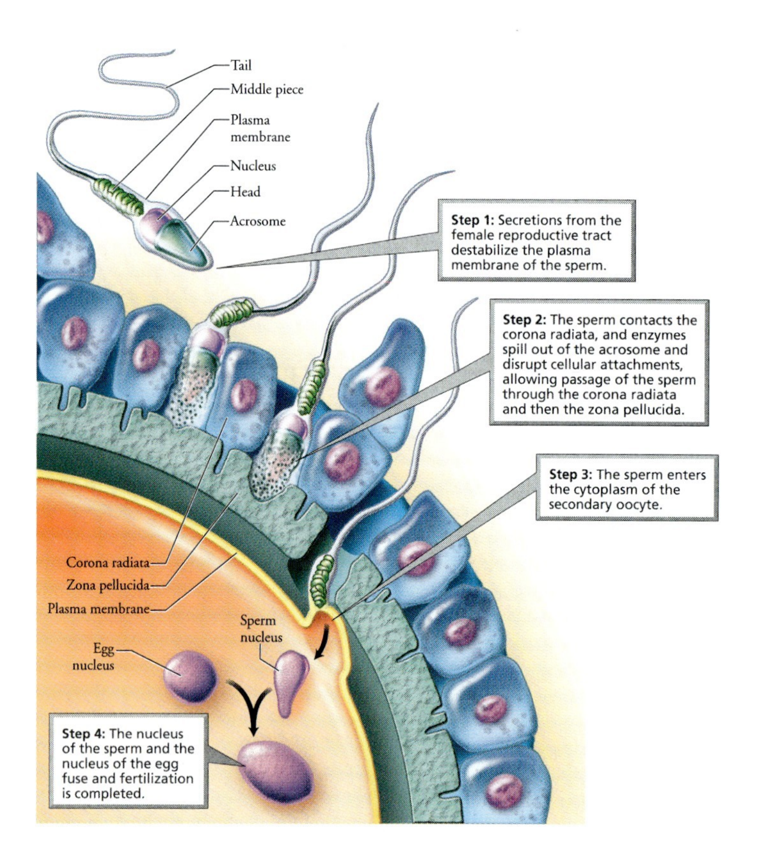

Fig. 18.3

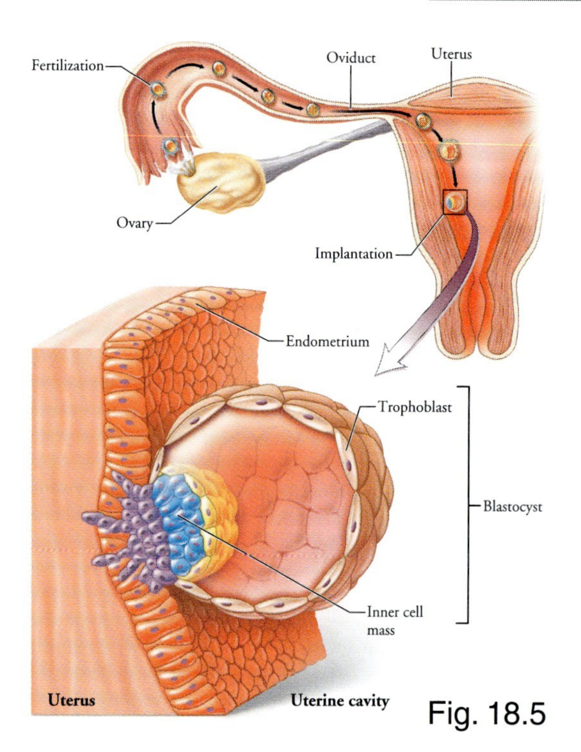

Fig. 18.5

NOTES

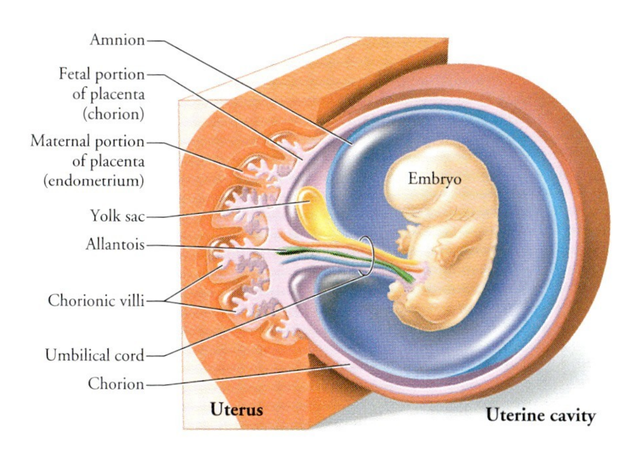

Fig. 18.6

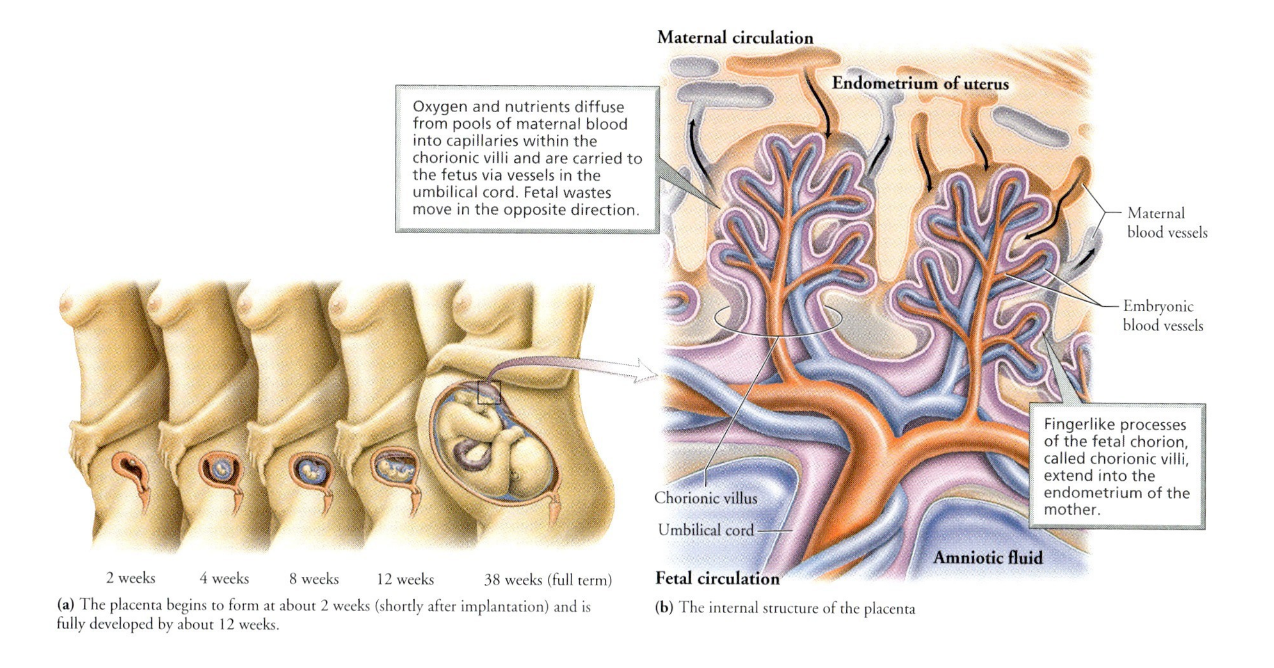

(a) The placenta begins to form at about 2 weeks (shortly after implantation) and is fully developed by about 12 weeks.

(b) The internal structure of the placenta

Fig. 18.7

NOTES

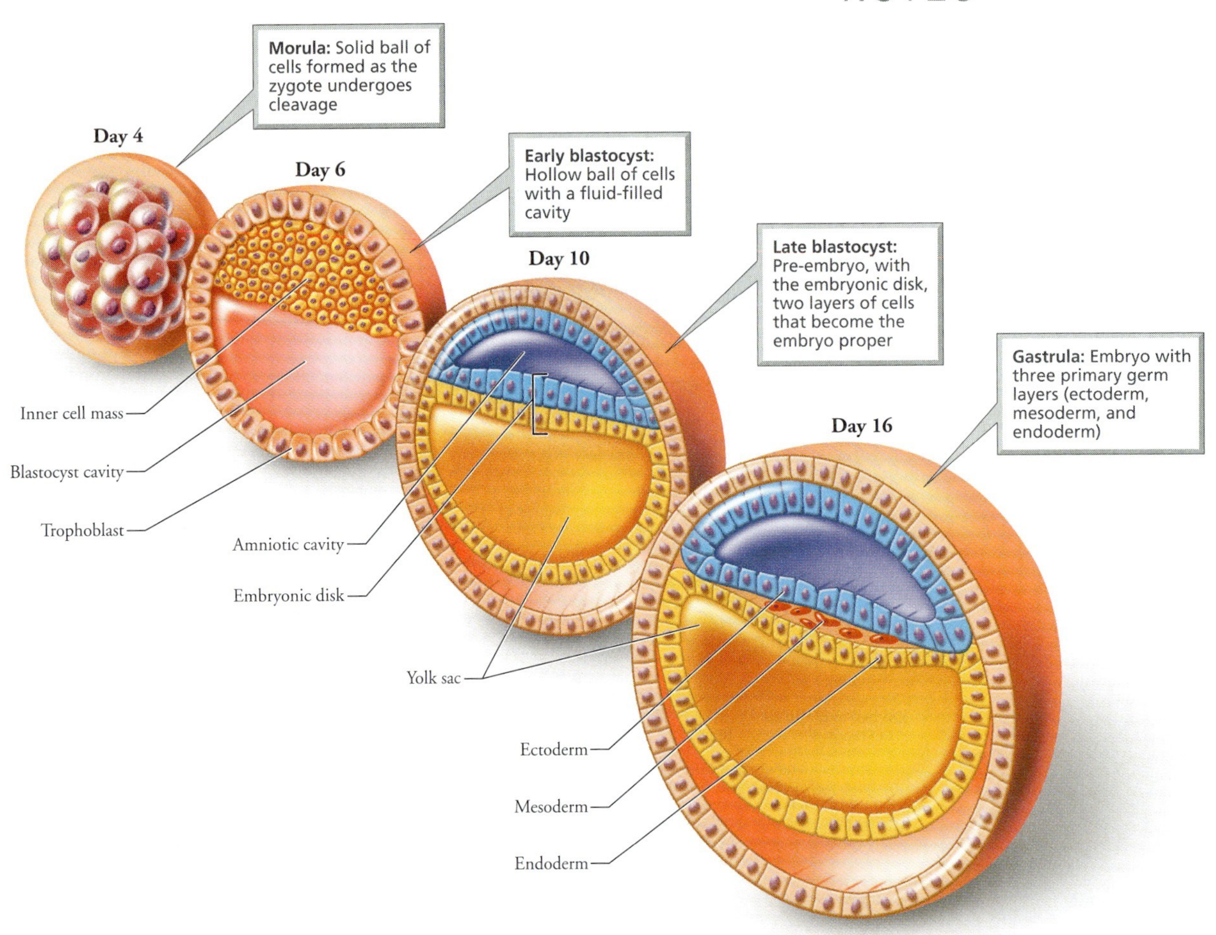

Fig. 18.8

NOTES

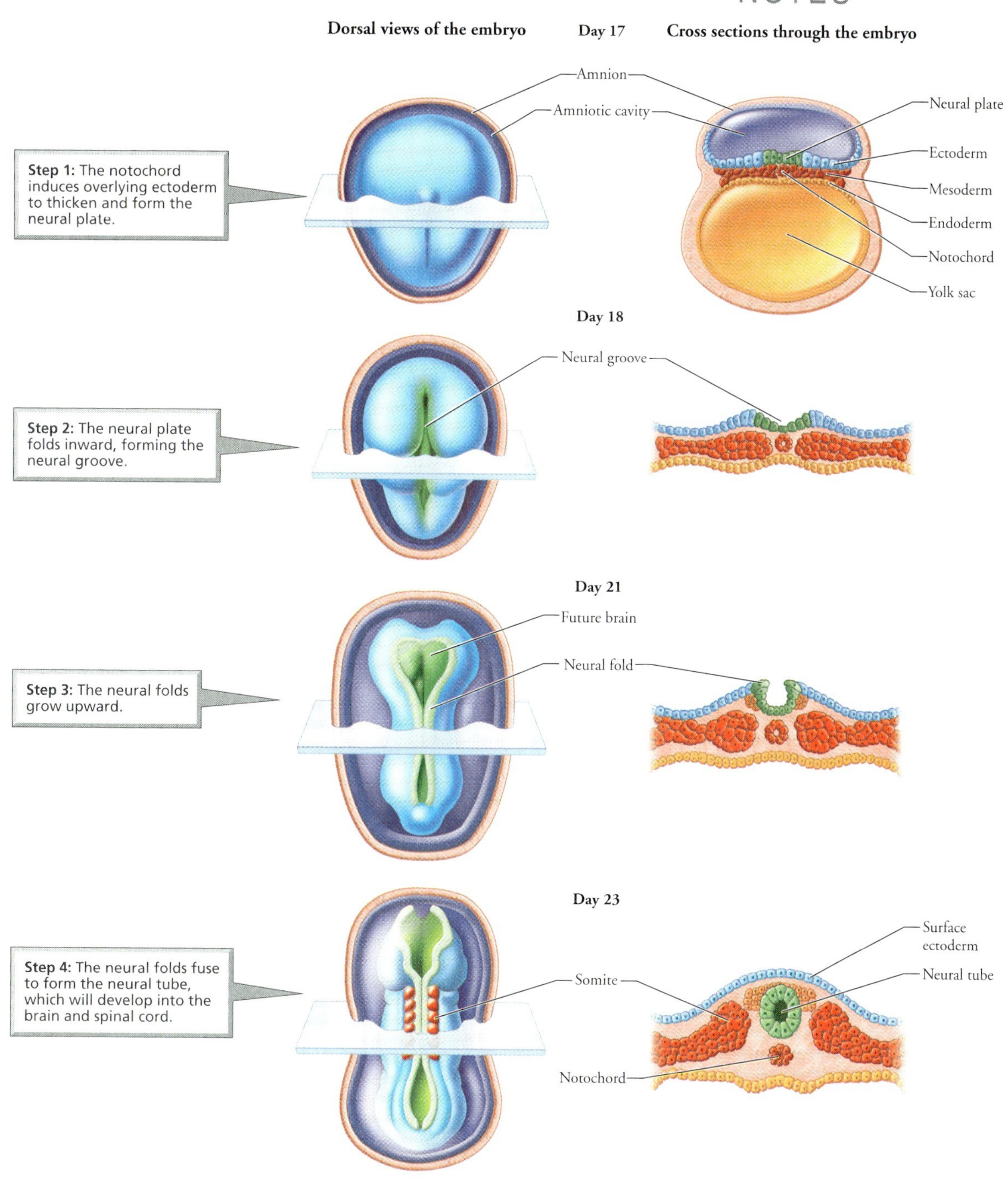

Fig. 18.9

NOTES

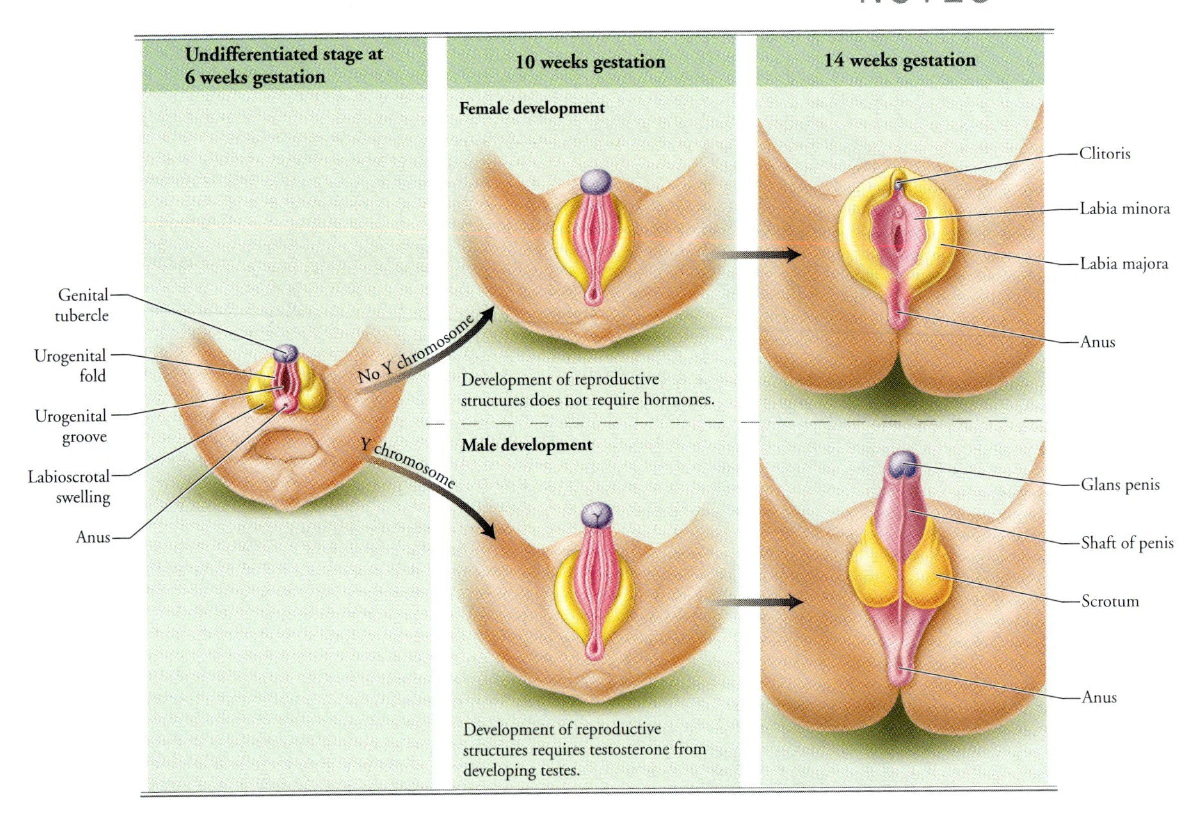

Fig. 18.10

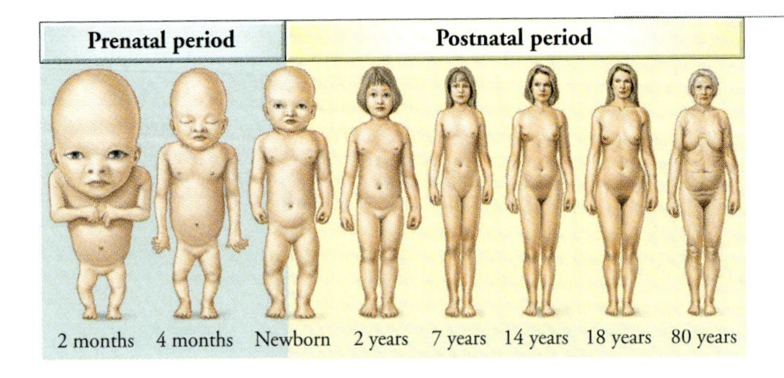

Fig. 18.11

NOTES

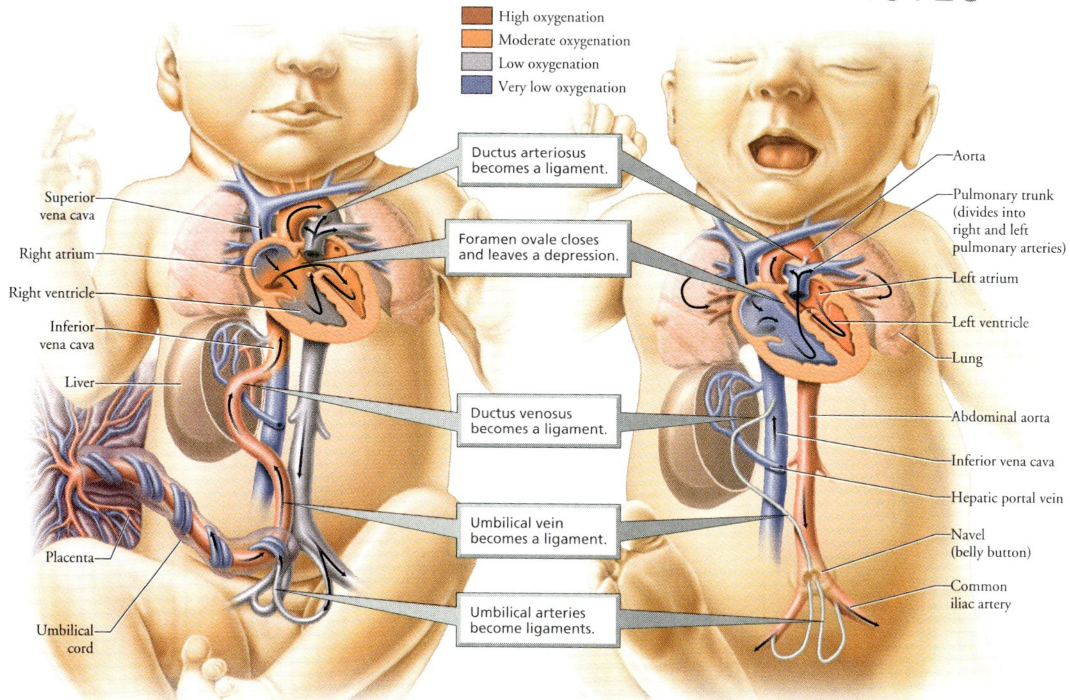

(a) Fetal circulation is characterized by a connection to the placenta (the umbilical cord) and several bypasses around organs that do not yet perform their postnatal functions.

(b) At birth, the umbilical cord is tied off and cut, leaving the navel (belly button). The bypasses close, allowing more blood to reach the now functional organs of the newborn.

Fig. 18.12

Major Events That Occur During Prenatal Development

Period	Major Events
Pre-embryonic (fertilization–week 2)	Fertilization Cleavage of the zygote Formation and implantation of the blastocyst Beginning of formation of extraembryonic membranes and placenta
Embryonic (week 3–week 8)	Gastrulation Formation of tissues, organs, and organ systems
Fetal (week 9–birth)	Continued differentiation and growth of tissues and organs Increase in crown-rump length Increase in weight

Table 18.1

NOTES

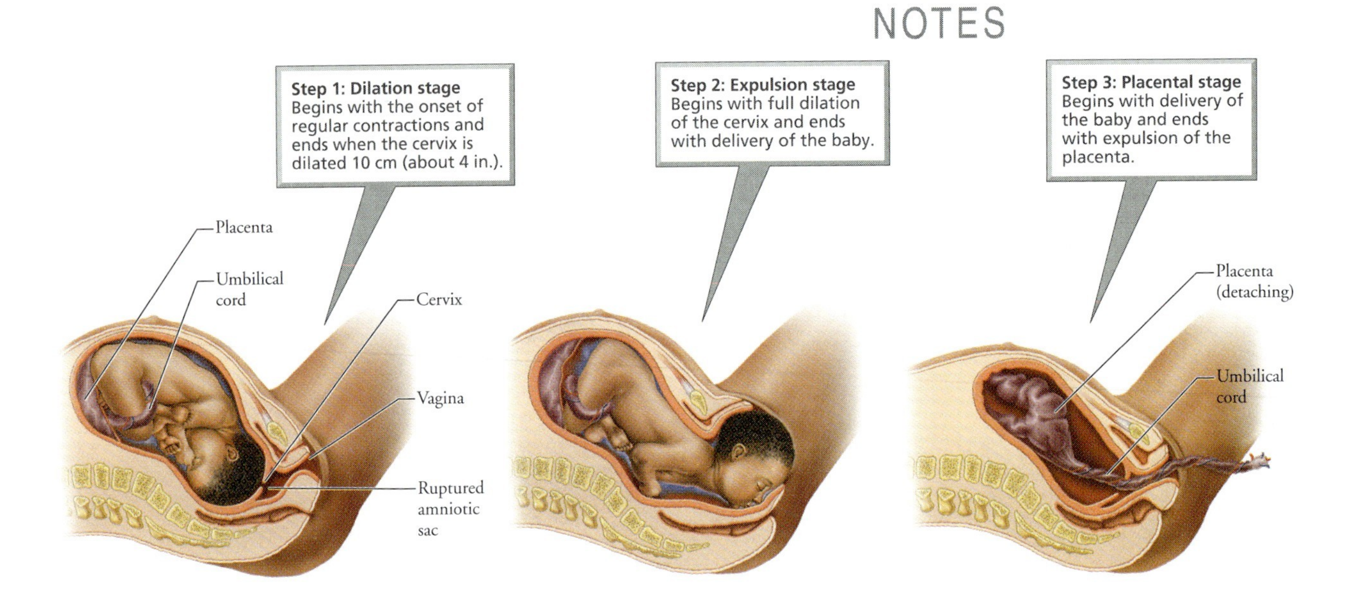

Fig. 18.14

NOTES

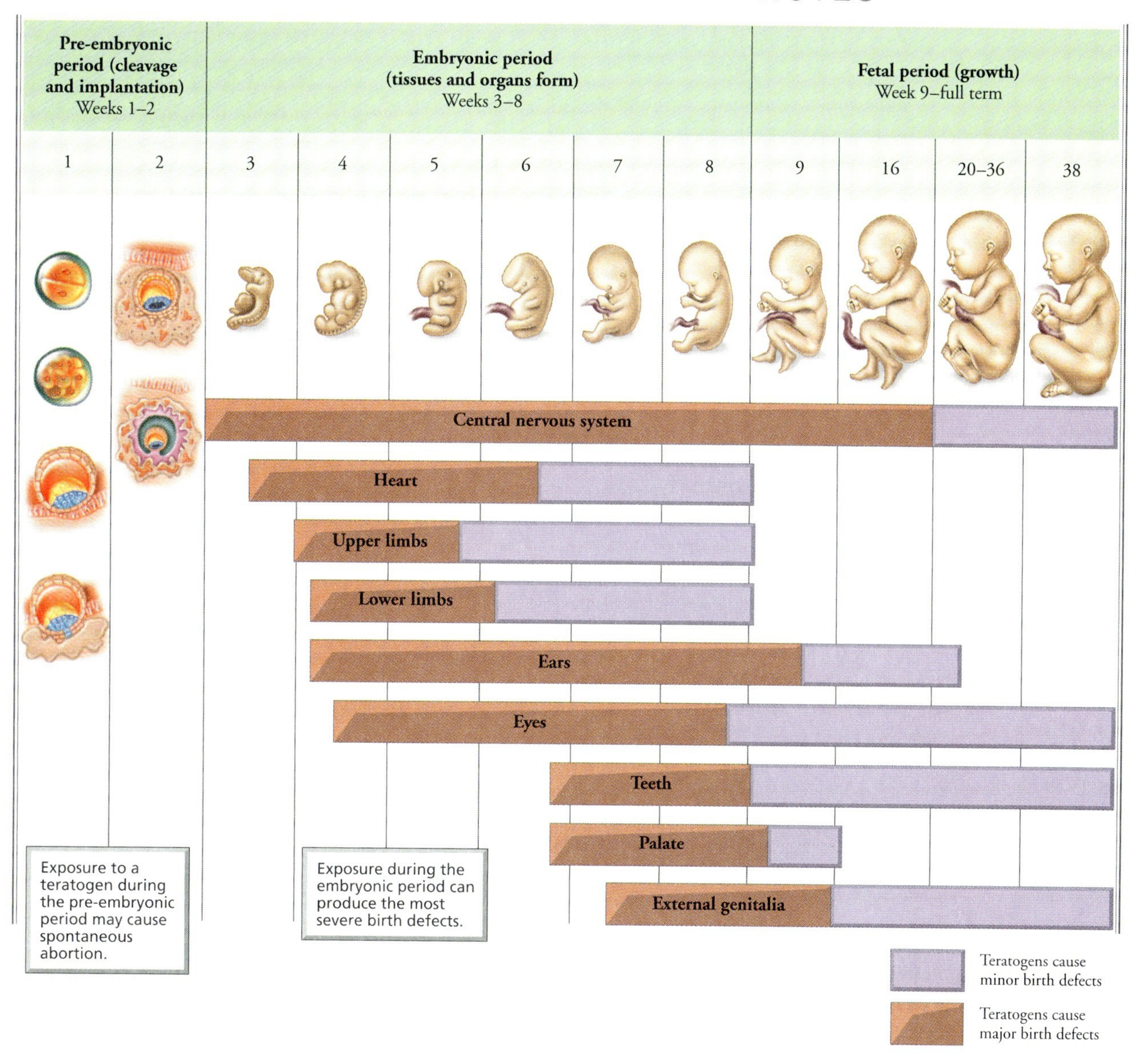

Fig. 18.16

NOTES

Changes in Body Systems Common in Old Age

ORGAN SYSTEM	SOME CHANGES THAT OCCUR WITH AGING
Integumentary	Wrinkles appear as skin becomes thinner and less elastic. Sweat glands decrease in number, making regulation of body temperature more challenging. Hair thins owing to death of hair follicles and turns gray as pigment-producing cells die.
Skeletal	Bones become lighter and more brittle, especially in women after menopause. About 7.6 cm (3 in.) of height are lost as the intervertebral disks deteriorate and the vertebrae move closer together. Joints become stiff and painful because of decreased production of synovial fluid.
Muscular	Muscle mass decreases owing to loss of muscle cells and decreased size of remaining muscle cells.
Nervous	Brain mass decreases. Movements and reflexes slow as conduction velocity of nerve fibers decreases and release of neurotransmitters slows. Hearing becomes less acute as hair cells in the inner ear stiffen. Ability of the eye to focus declines as the lens of the eye stiffens. Smell and taste become less acute.
Endocrine	In women, production of estrogen and progesterone decreases with menopause. In men, production of testosterone decreases. In both sexes, production of growth hormone decreases.
Circulatory	Cardiac output decreases as walls of the heart stiffen. Blood pressure rises as arteries become less elastic and clogged by fatty deposits.
Respiratory	Lung capacity decreases as alveoli break down and lung tissue becomes less elastic.
Digestive	Basal metabolic rate declines. Ability of the liver to detoxify substances declines.
Urinary	In both sexes, kidney mass declines, as does the rate of filtration of the blood by nephrons. Particularly in women, the external urethral sphincter weakens, causing incontinence. In men, the prostate gland enlarges, causing painful and frequent urination.
Reproductive	In men, fewer viable sperm are produced. In women, ovulation and menstruation cease at menopause.

Table 18.2

NOTES

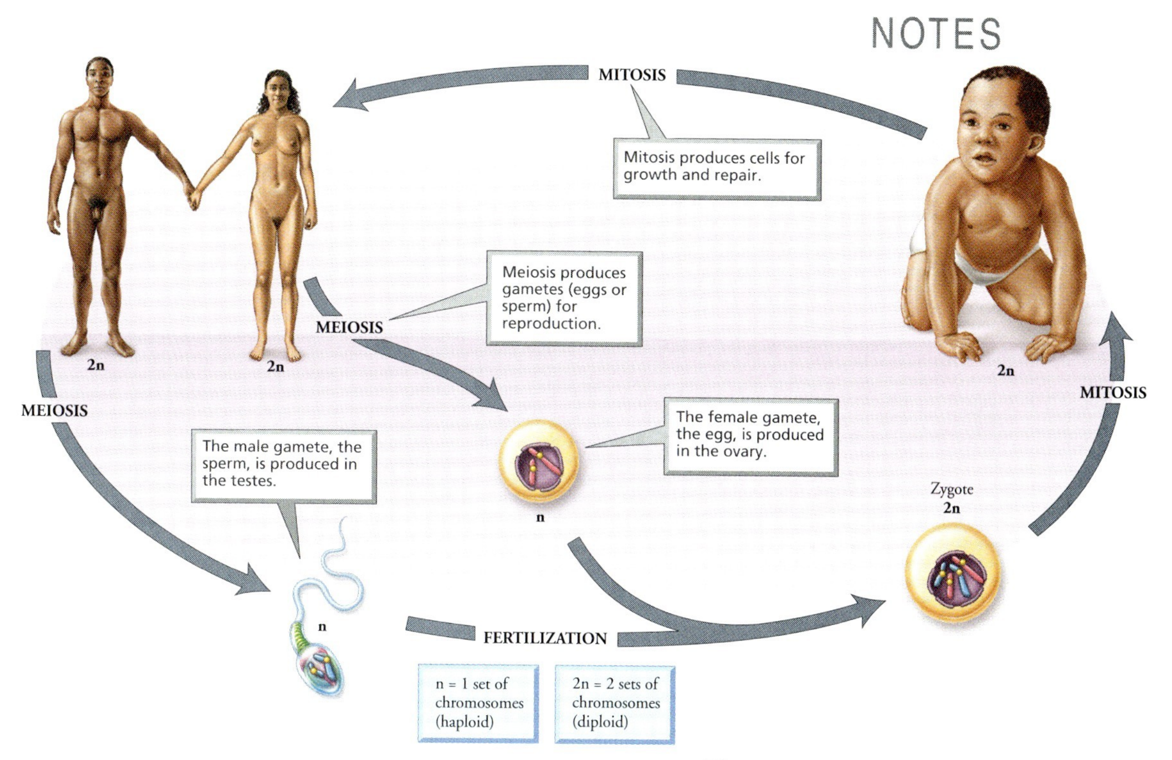

Fig. 19.1

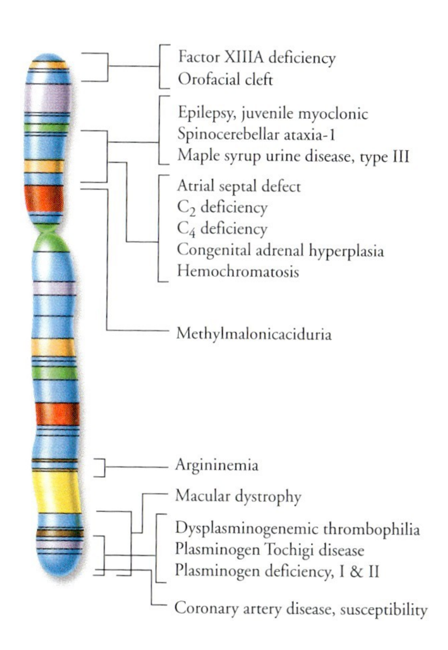

Fig. 19.2

NOTES

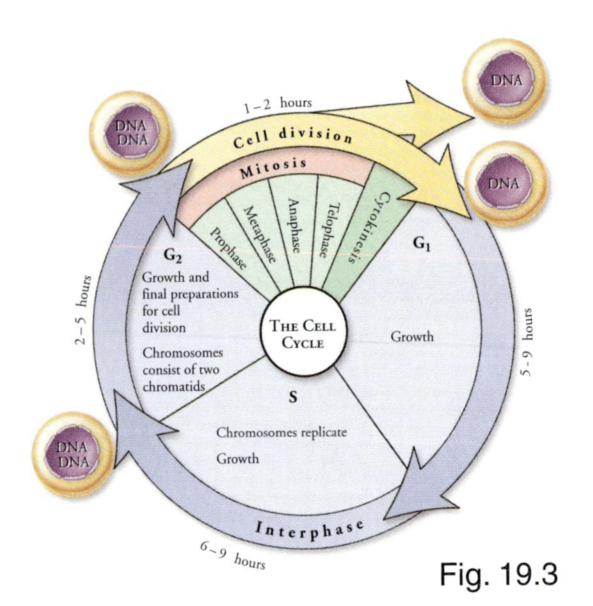

Fig. 19.3

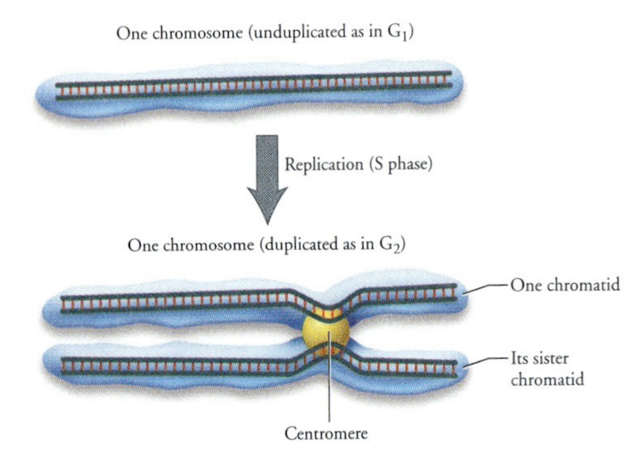

Fig. 19.4

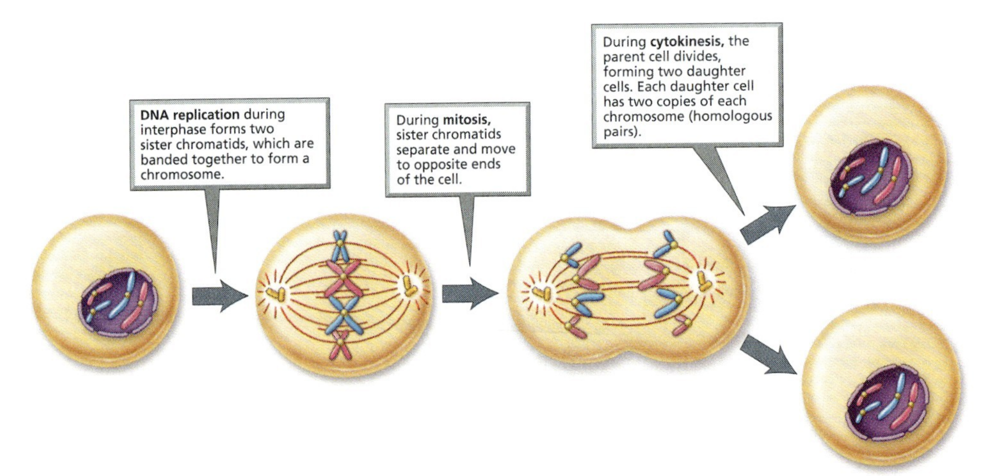

Fig. 19.5

NOTES

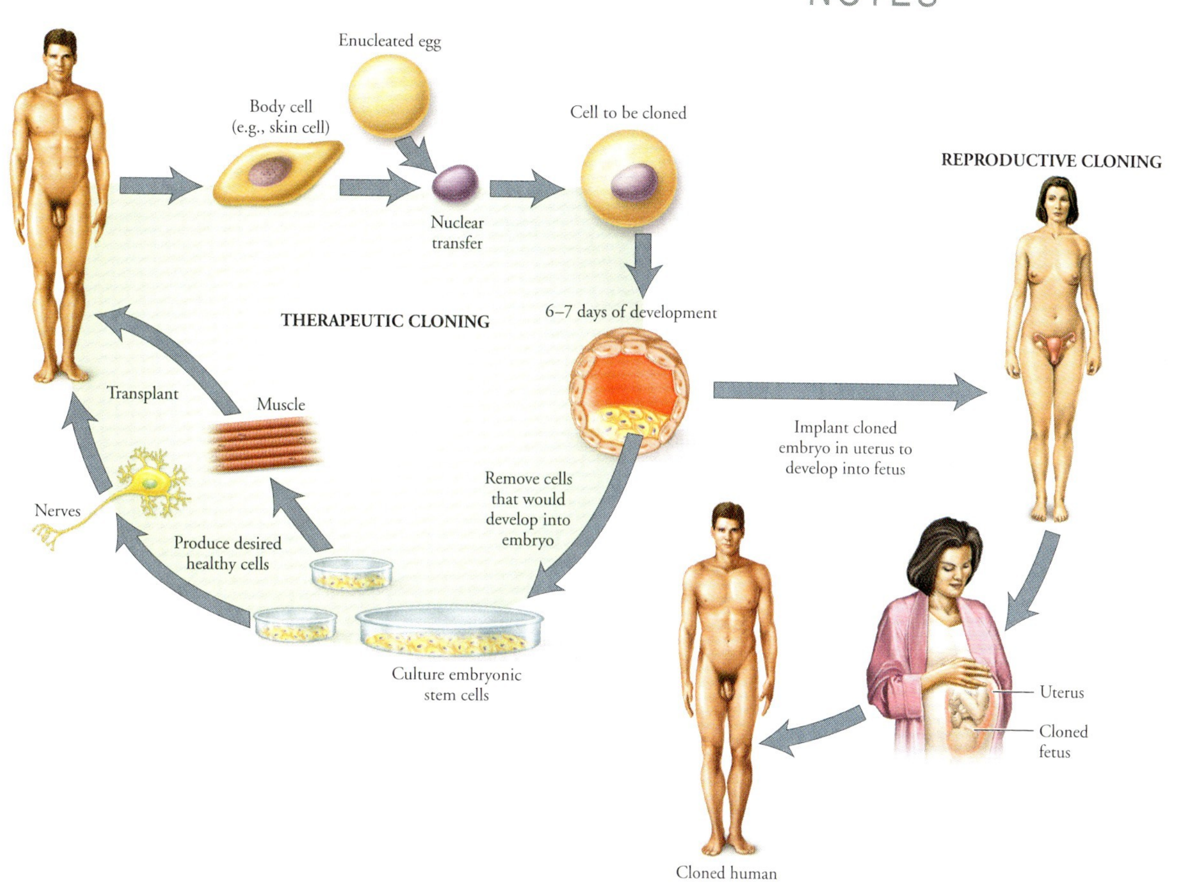

Fig. 19.A

NOTES

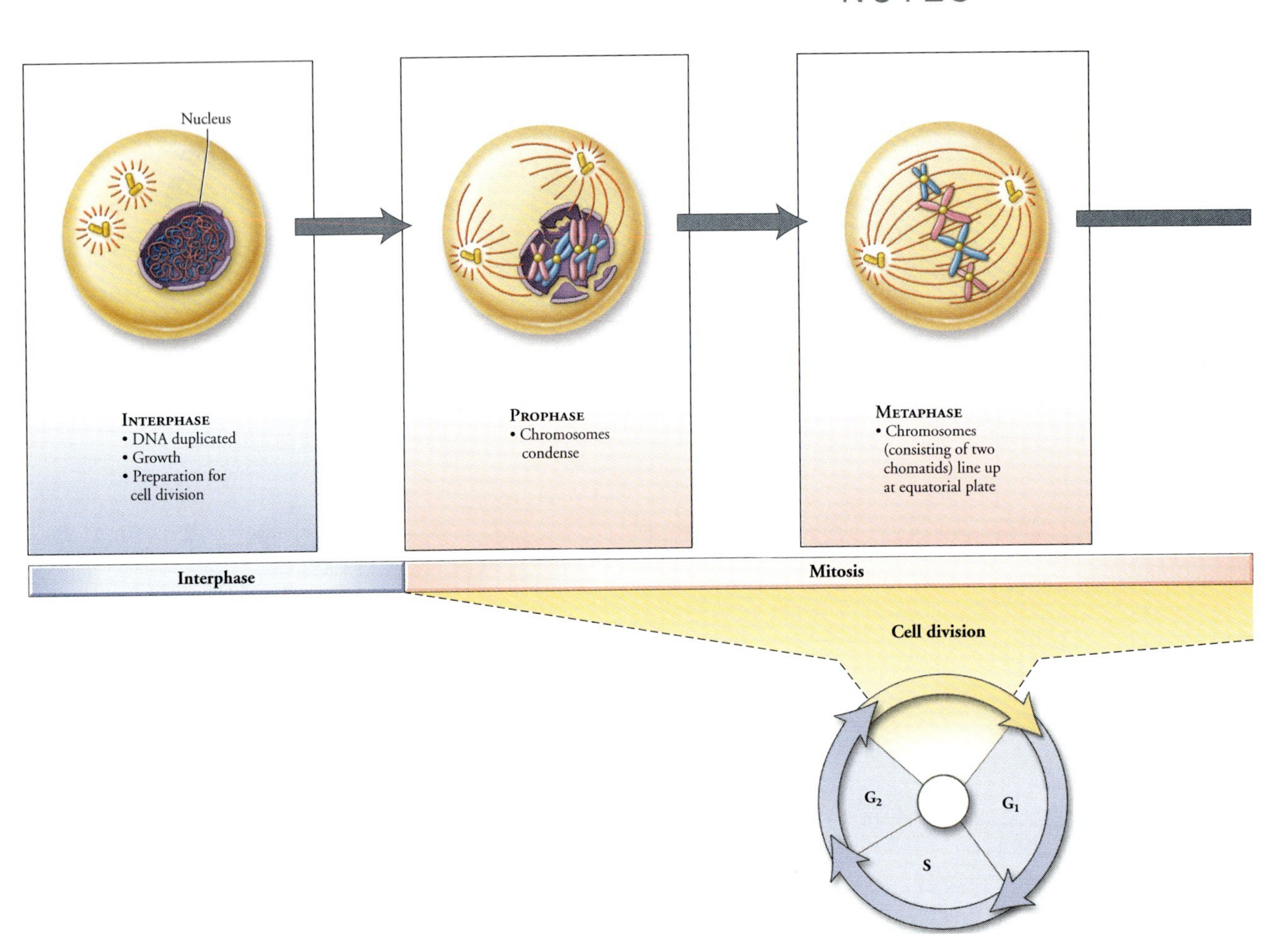
Nucleus
Interphase
• DNA duplicated
• Growth
• Preparation for cell division
Prophase
• Chromosomes condense
Metaphase
• Chromosomes (consisting of two chomatids) line up at equatorial plate
Interphase
Mitosis
Cell division
G2
G1
S

NOTES

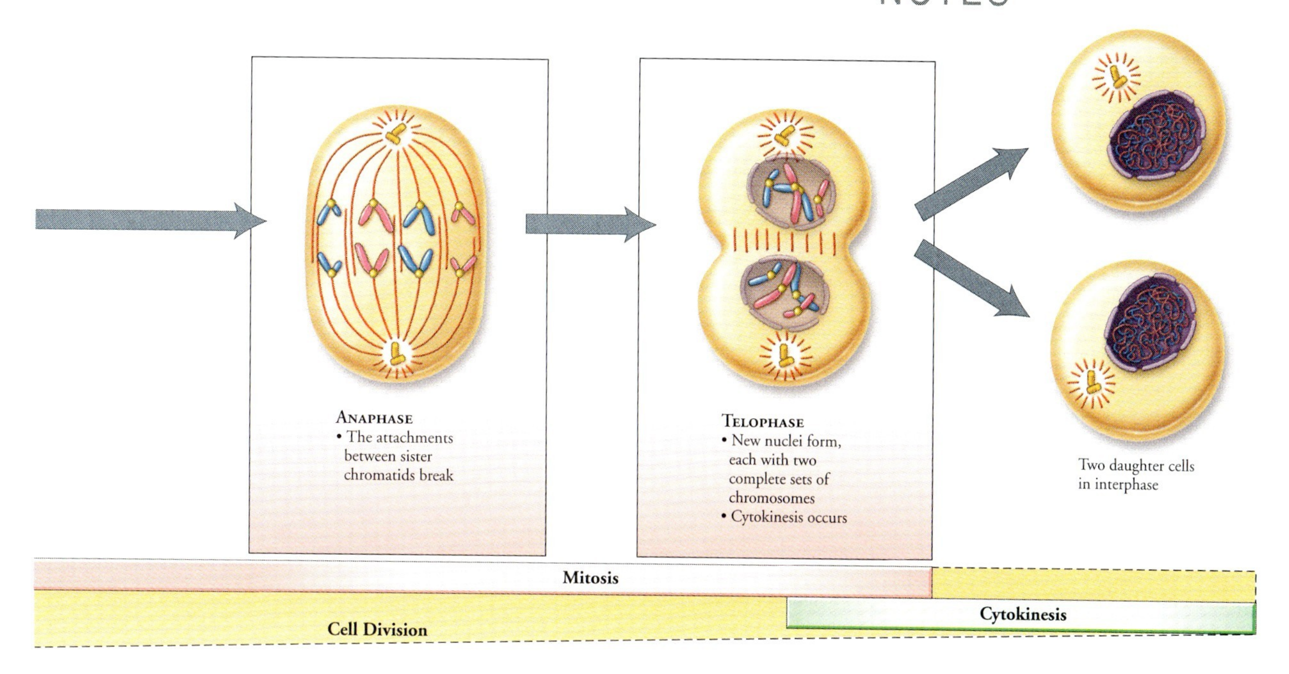

Fig. 19.6

NOTES

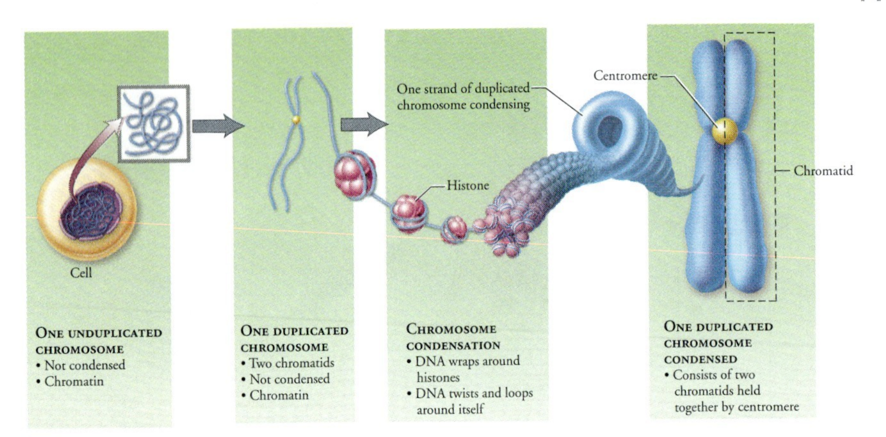

Fig. 19.7

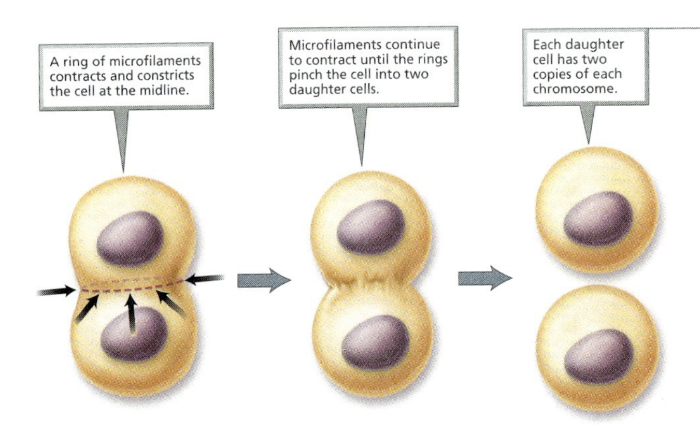

Fig. 19.9

PHASES OF THE CELL CYCLE

PHASE	MAJOR EVENTS
Interphase	Production of proteins and other macromolecules and organelles; duplication of centriole and DNA
Cell division	
Mitosis	
Prophase	Chromosomes condense; nucleoli disappear; nuclear envelope breaks down; centrioles divide and move to opposite ends of the cell; mitotic spindles form and attach to chromosomes
Metaphase	Chromosomes line up at the equatorial plate
Anaphase	Centromere splits; chromatids begin to separate, each becoming a chromosome
Telophase	Chromosomes reach opposite poles; nuclear envelopes form; mitotic spindle disassembles; nucleoli reappear; chromosomes uncoil
Cytokinesis	
	Cytoplasm divides, forming two daughter cells

Table 19.1

NOTES

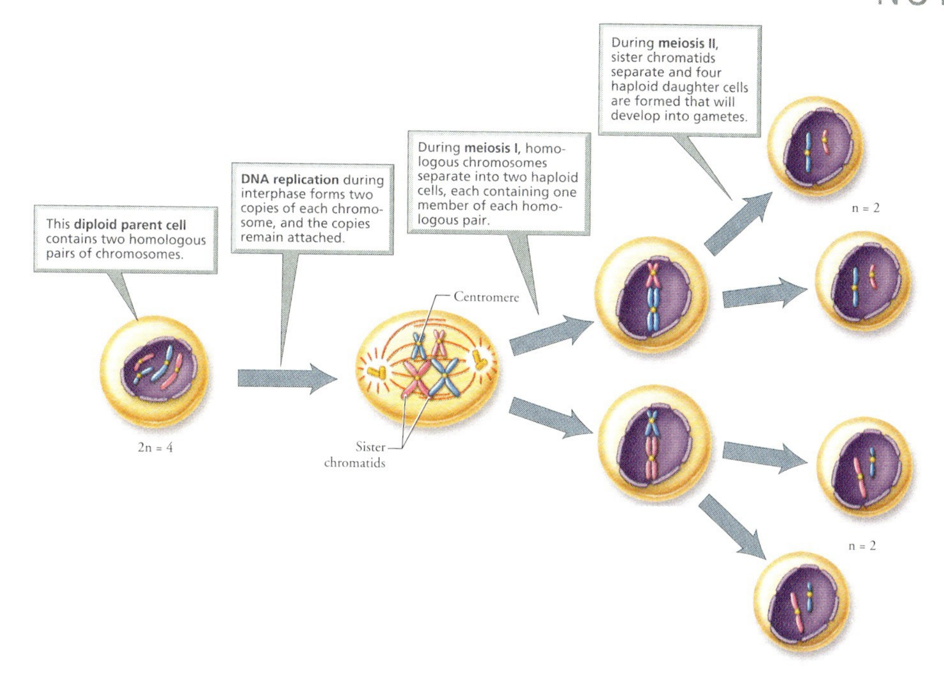

Fig. 19.10

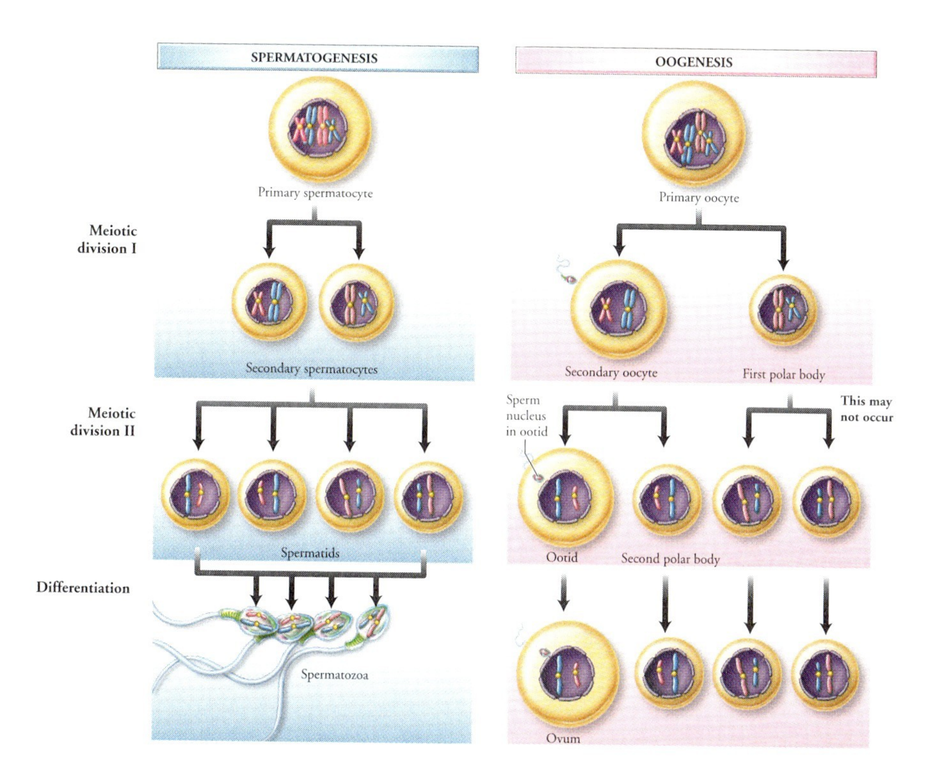

Fig. 19.11

NOTES

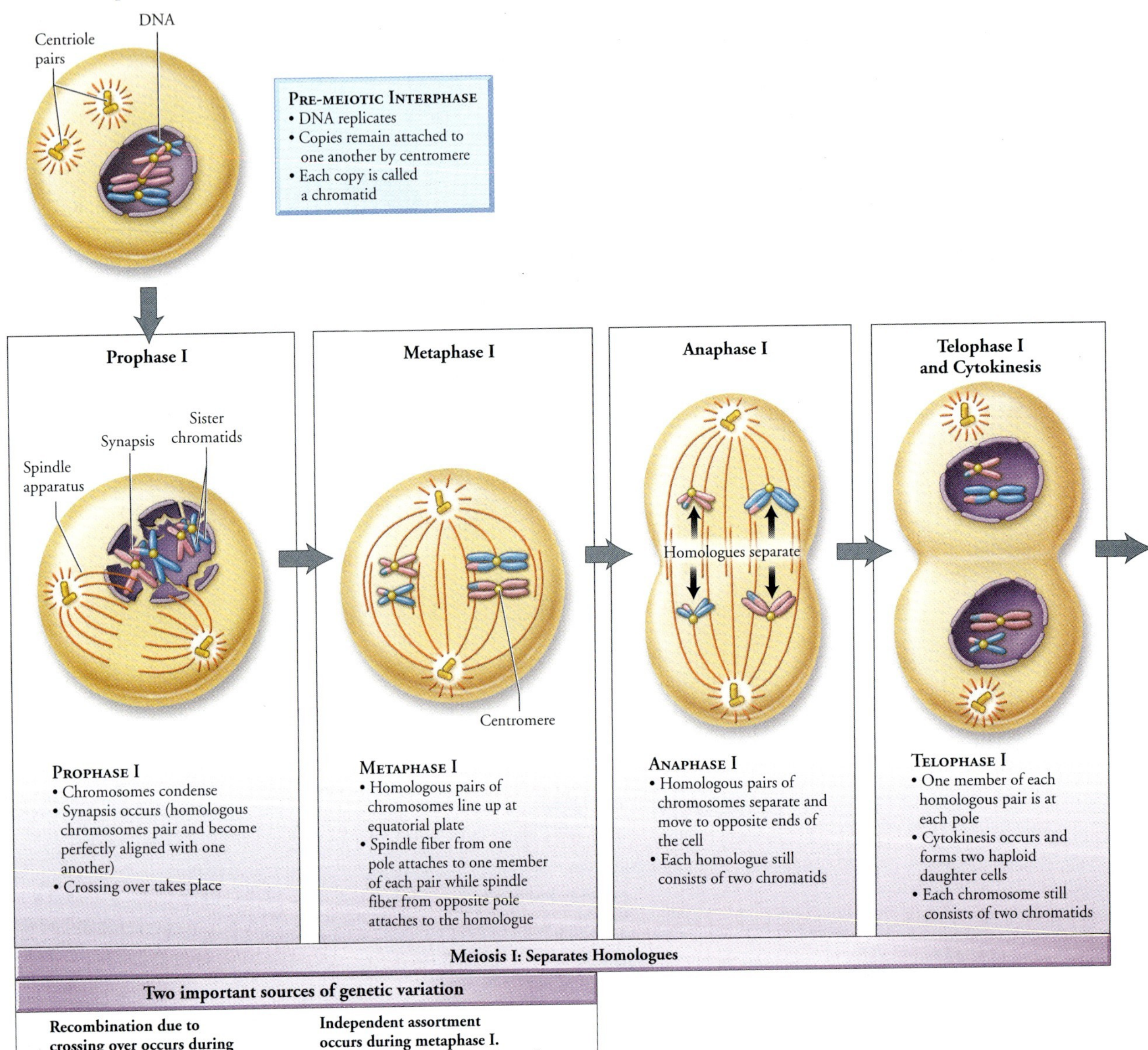
Interphase
DNA
Centriole pairs
PRE-MEIOTIC INTERPHASE
• DNA replicates
• Copies remain attached to one another by centromere
• Each copy is called a chromatid
Prophase I
Sister chromatids
Synapsis
Spindle apparatus
Metaphase I
Centromere
Anaphase I
Homologues separate
Telophase I and Cytokinesis
PROPHASE I
• Chromosomes condense
• Synapsis occurs (homologous chromosomes pair and become perfectly aligned with one another)
• Crossing over takes place
METAPHASE I
• Homologous pairs of chromosomes line up at equatorial plate
• Spindle fiber from one pole attaches to one member of each pair while spindle fiber from opposite pole attaches to the homologue
ANAPHASE I
• Homologous pairs of chromosomes separate and move to opposite ends of the cell
• Each homologue still consists of two chromatids
TELOPHASE I
• One member of each homologous pair is at each pole
• Cytokinesis occurs and forms two haploid daughter cells
• Each chromosome still consists of two chromatids
Meiosis I: Separates Homologues

Two important sources of genetic variation

Recombination due to crossing over occurs during prophase I.
Parts of nonsister chromatids are exchanged.

Independent assortment occurs during metaphase I.
Maternal and paternal members of homologous pairs align randomly at the equatorial plate, creating a random assortment of maternal and paternal chromosomes in the daughter cells.

NOTES

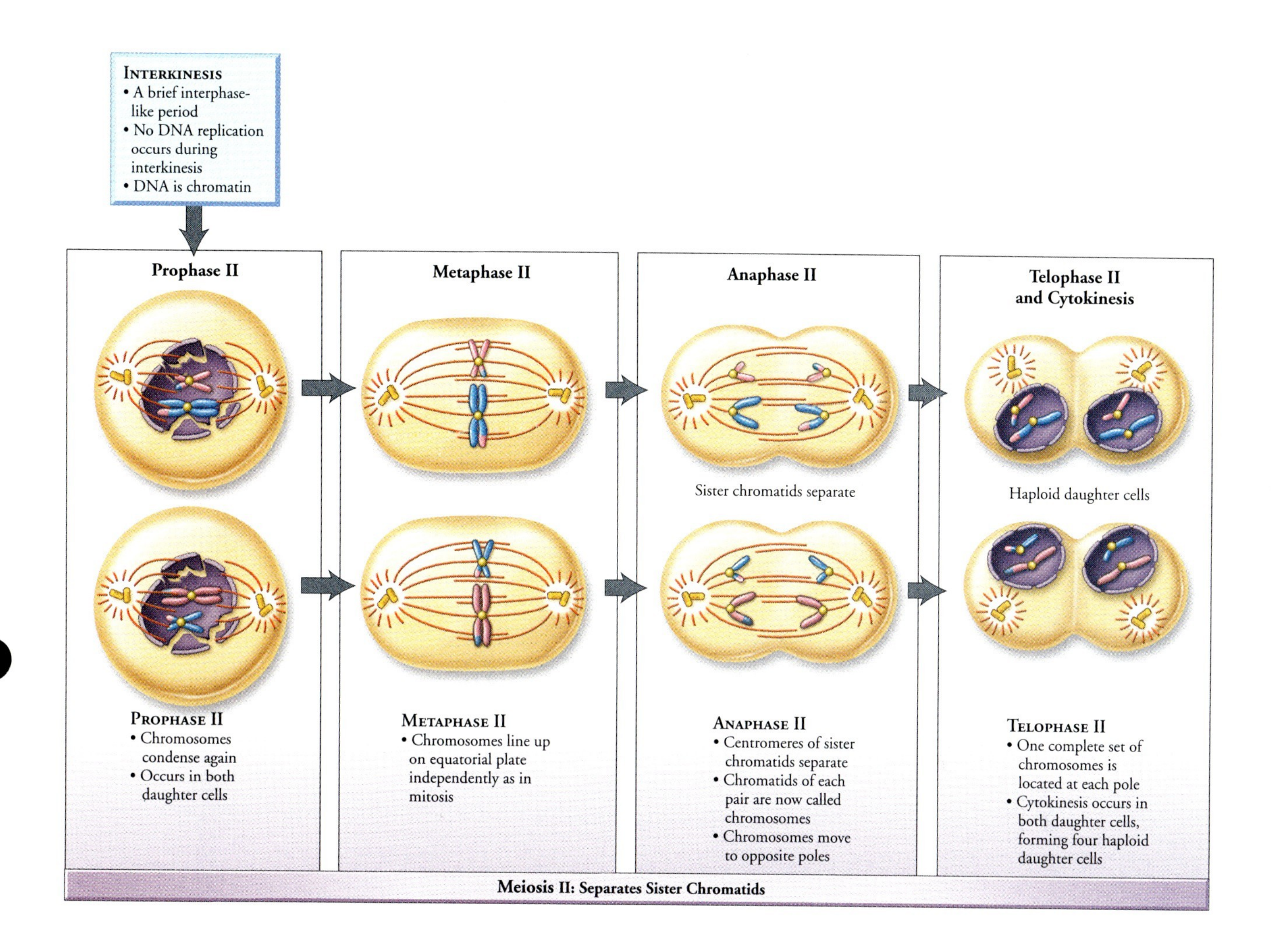

Fig. 19.12

NOTES

Mitosis and Meiosis Compared	
MITOSIS	**MEIOSIS**
Involves one cell division	Involves two cell divisions
Produces two diploid cells	Produces four haploid cells
Occurs in most somatic cells	Occurs only in ovary and testes during the formation of gametes (egg and sperm)
Results in growth and repair	Results in gamete (egg and sperm) production
No exchange of genetic material	Parts of chromosomes are exchanged in crossing over
Daughter cells are genetically similar	Daughter cells are genetically dissimilar

Table 19.2

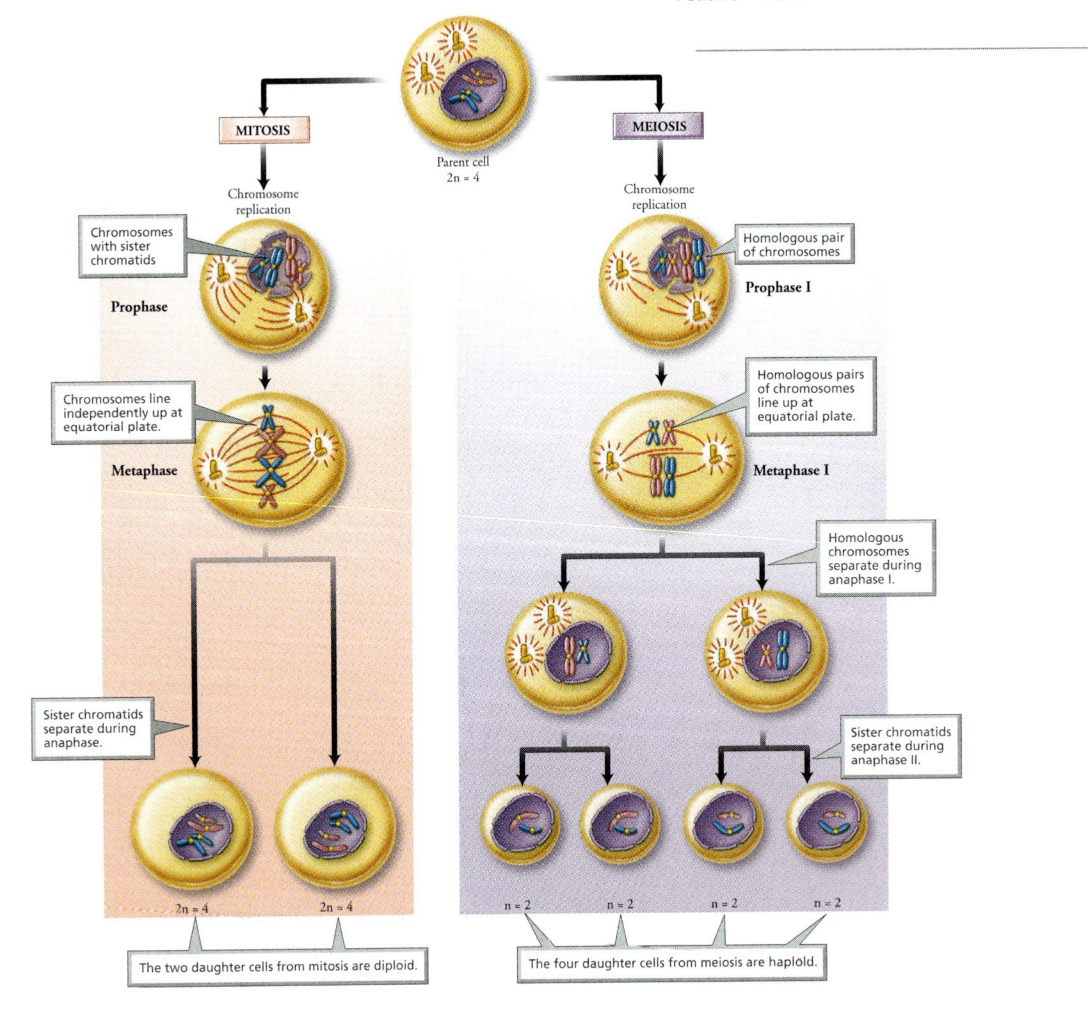

Fig. 19.13

NOTES

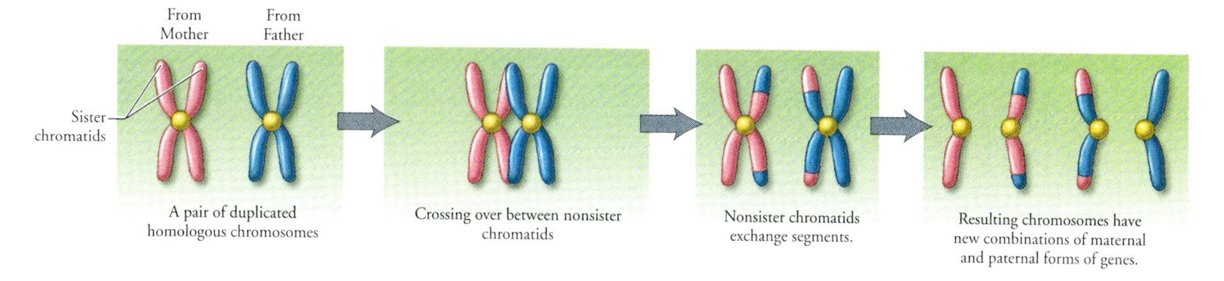

Fig 19.15

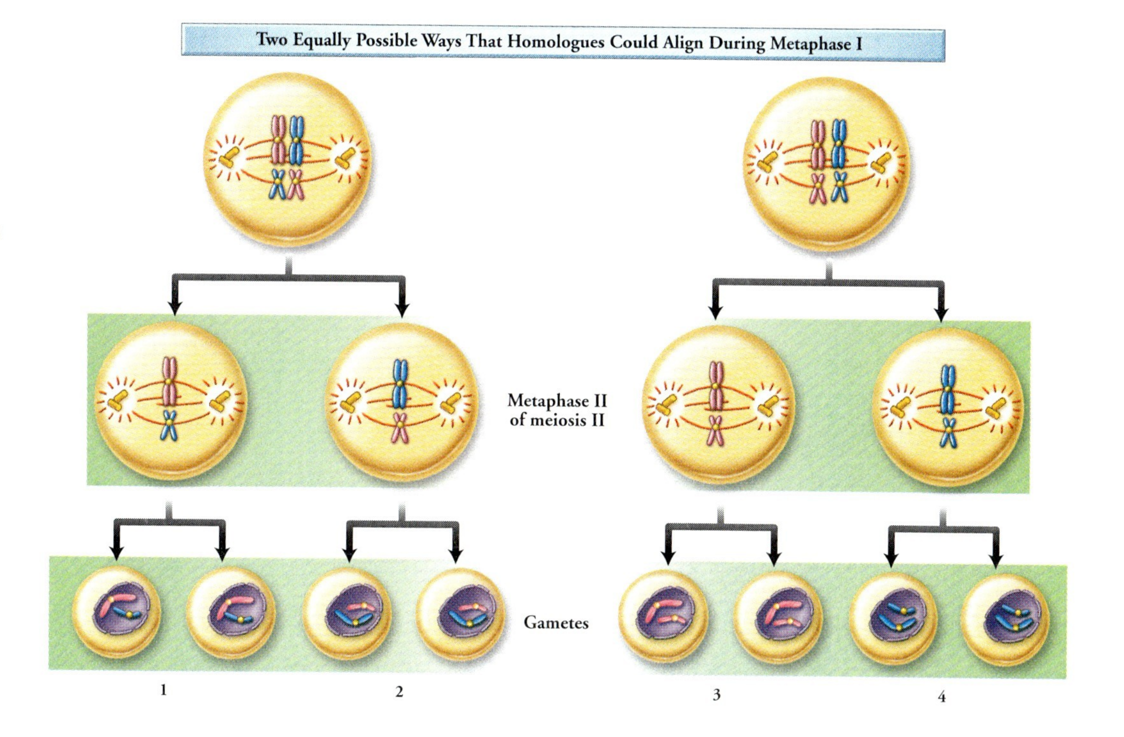

Fig 19.16

NOTES

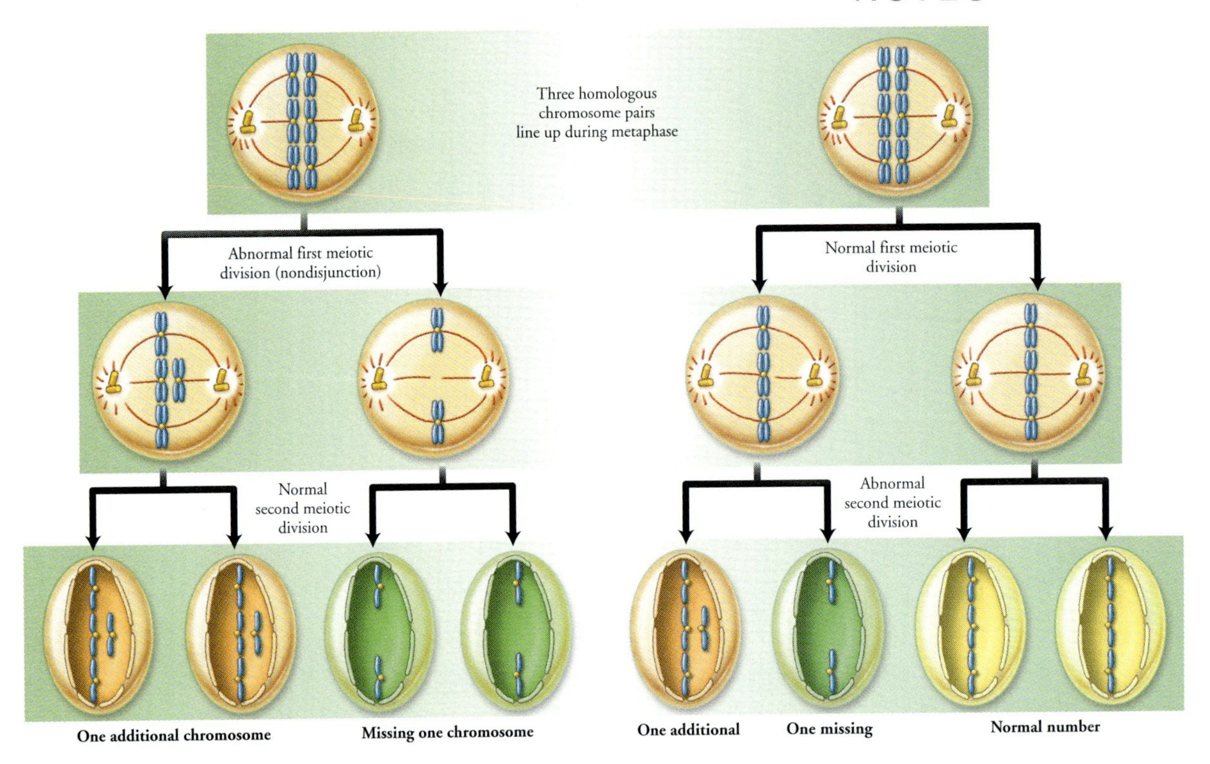

Fig. 19.17

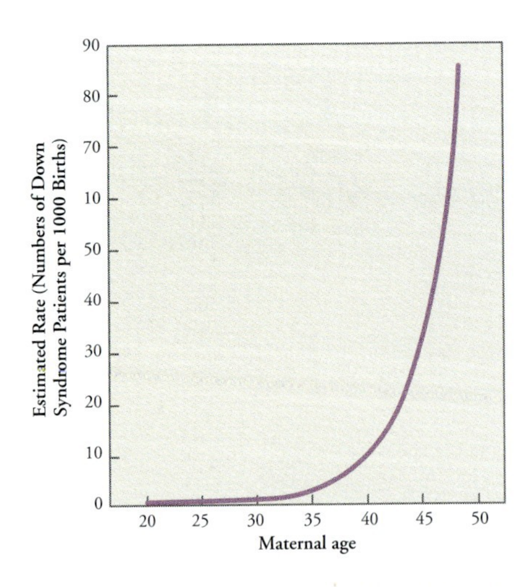

Fig. 19.18

NOTES

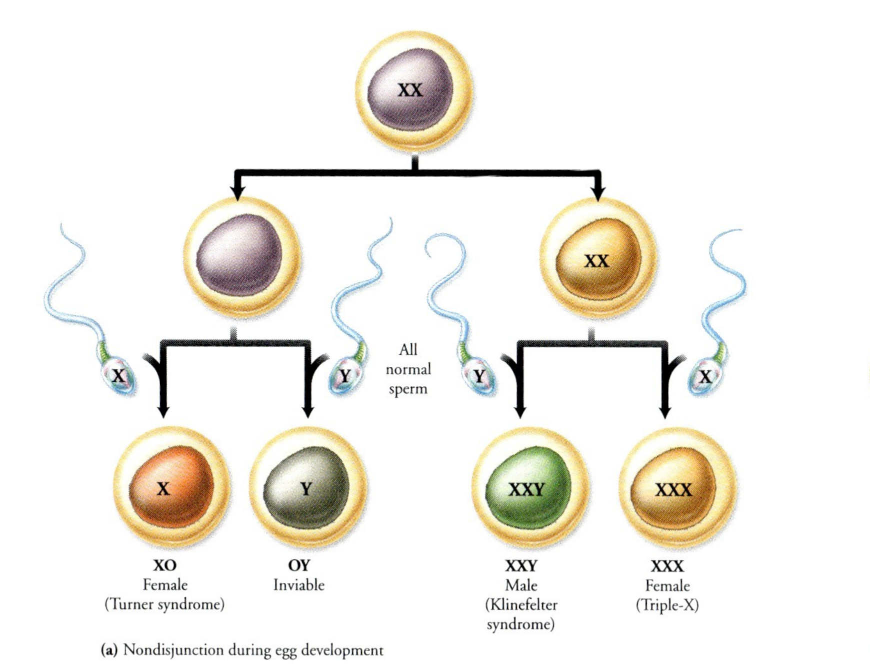

(a) Nondisjunction during egg development

XY
XY
X
Egg
X
Egg
XO
XXY
XO
(Turner syndrome)
XXY
(Klinefelter
syndrome)

(b) Nondisjunction during sperm development

Fig. 19.19

NOTES

NOTES

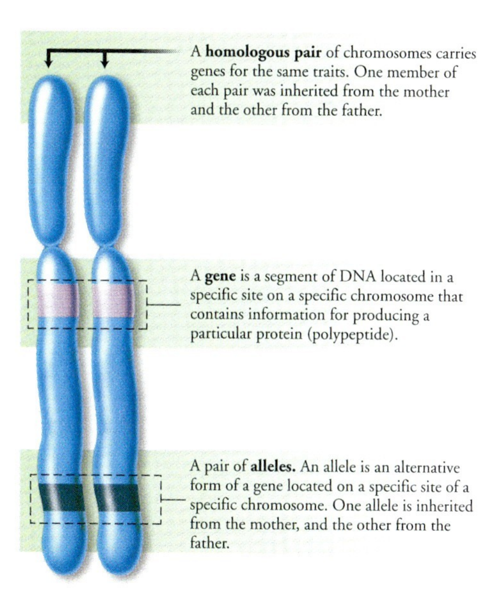

Fig. 20.1

COMMON TERMS IN GENETICS

GENOTYPE: THE ALLELES THAT ARE PRESENT	DESCRIPTION	PHENOTYPE: THE OBSERVABLE TRAIT
FF	Homozygous dominant: • Two dominant alleles present. • Dominant phenotype expressed.	Freckles
Ff	Heterozygous: • Different alleles present. • Dominant phenotype expressed.	Freckles
ff	Homozygous recessive • Two recessive alleles present. • Recessive phenotype expressed.	No freckles

Table 20.1

NOTES

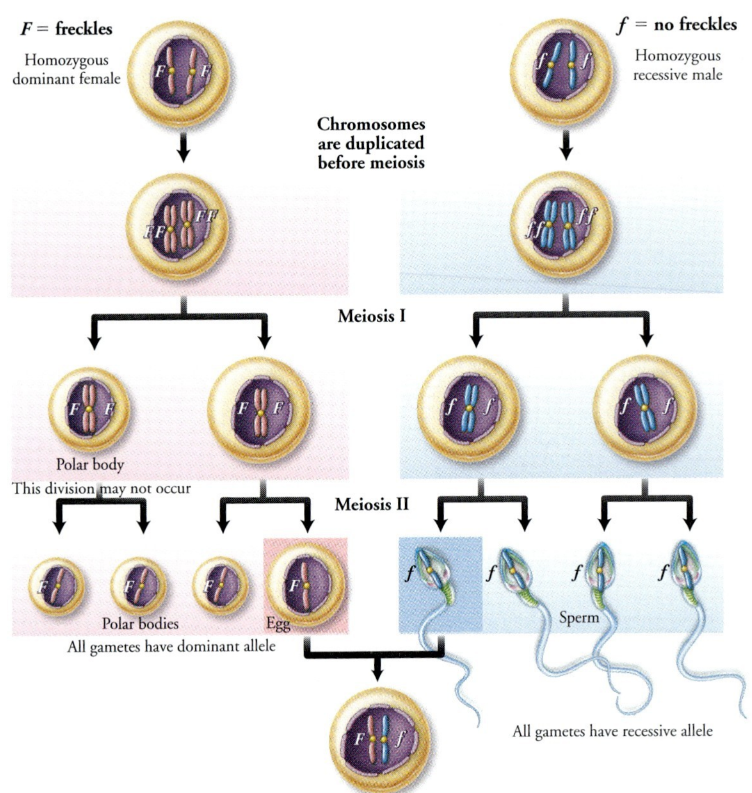

Fig. 20.3

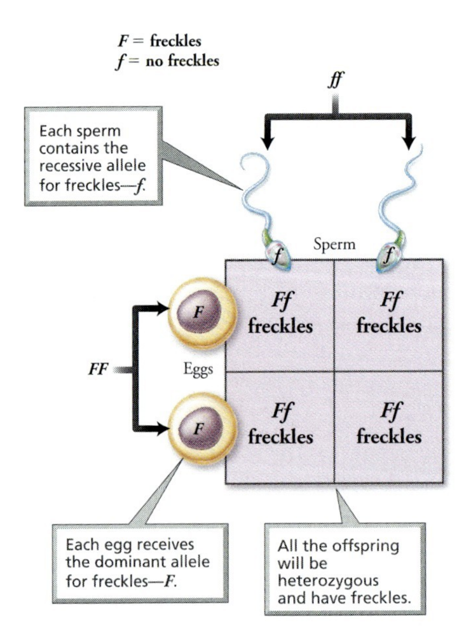

Fig. 20.4

NOTES

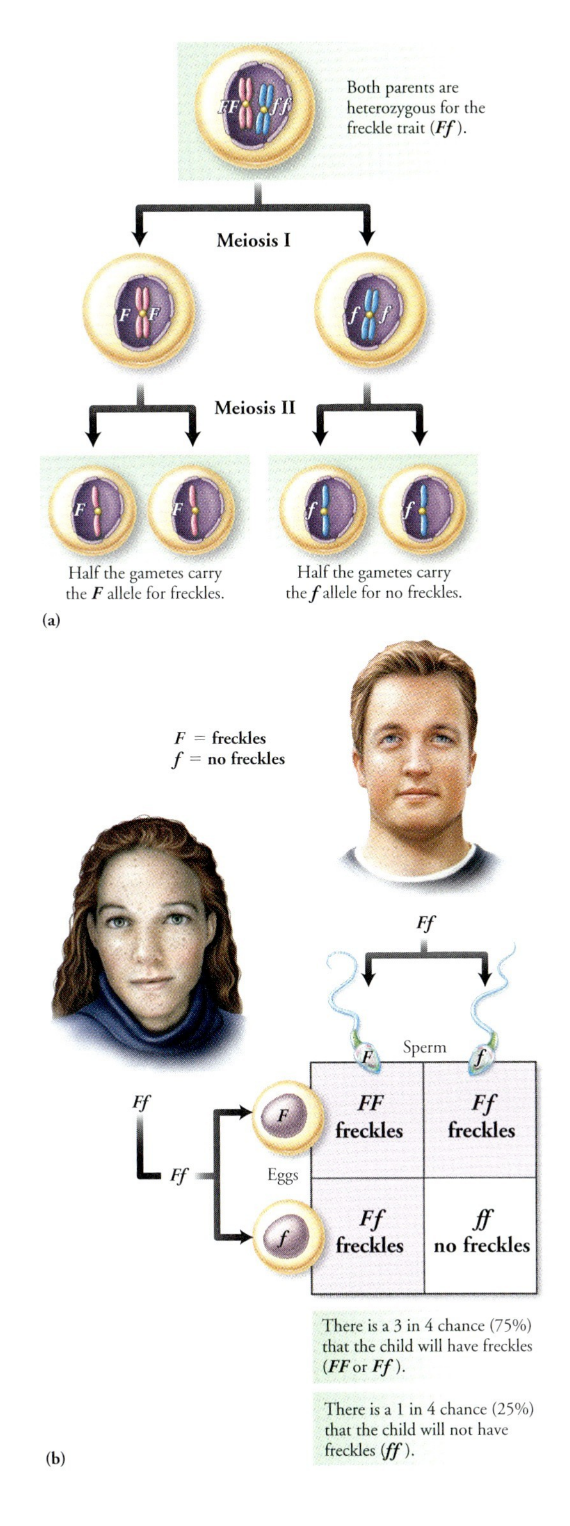

Fig. 20.5

NOTES

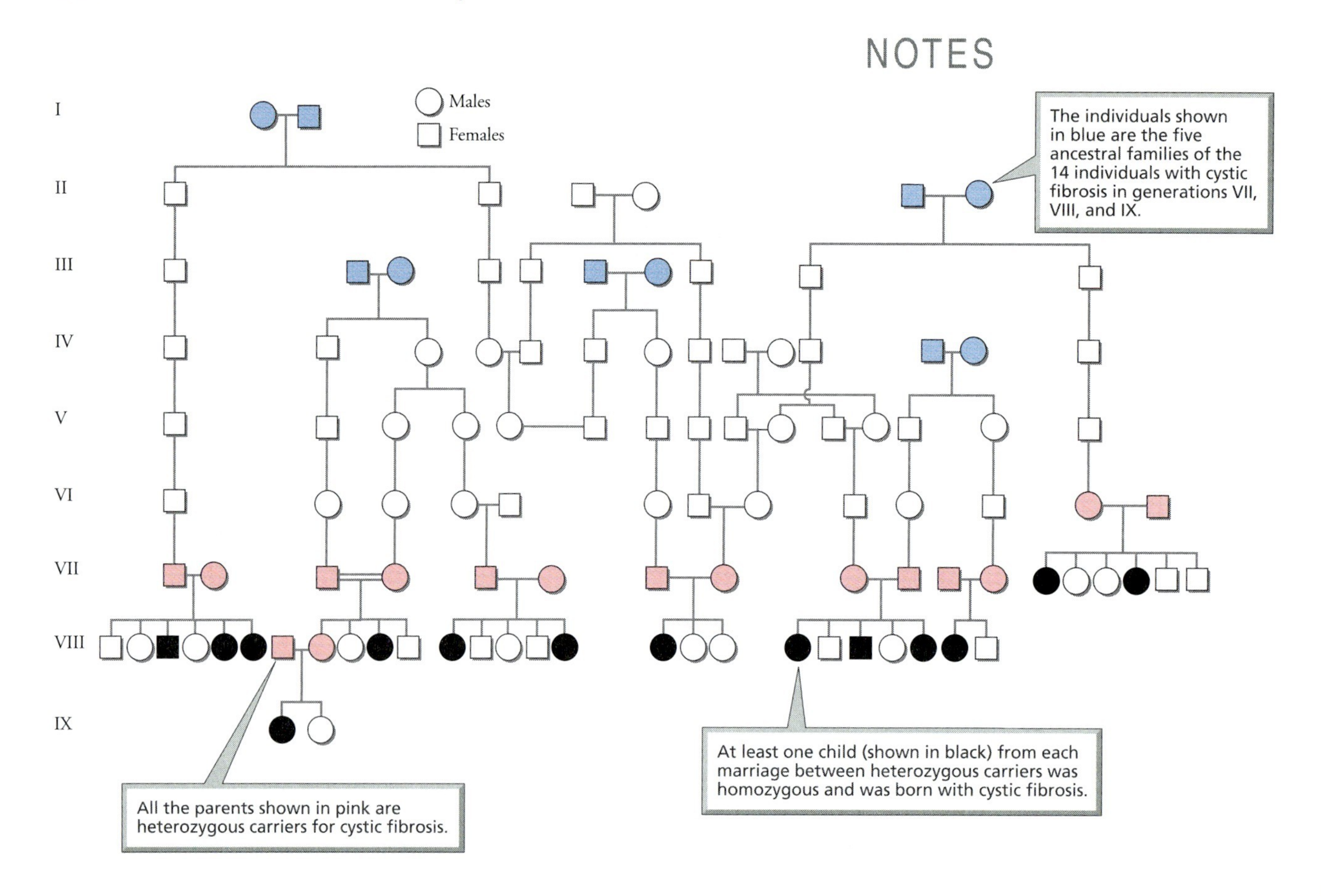

Fig. 20.7

NOTES

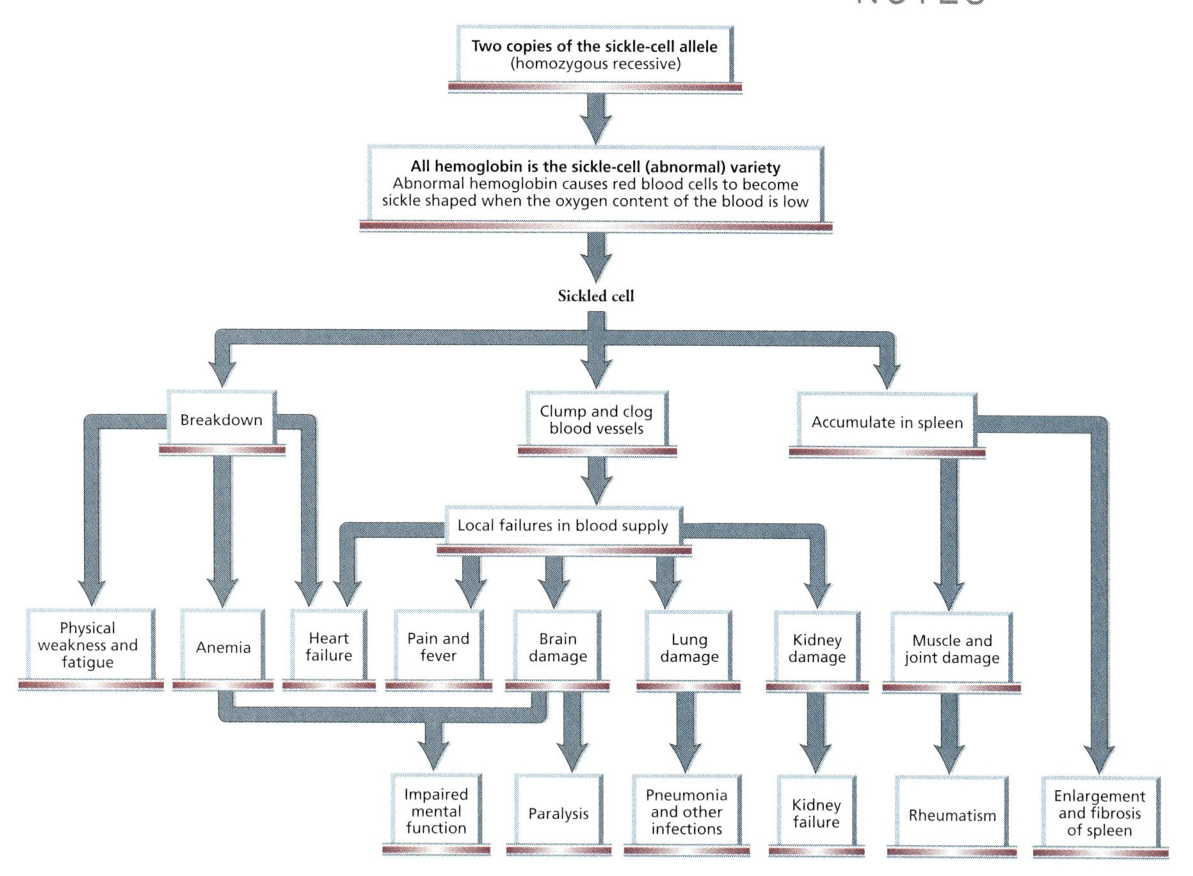

Fig. 20.9

THE RELATIONSHIP BETWEEN GENOTYPE AND ABO BLOOD GROUPS

GENOTYPE	BLOOD TYPE
I^AI^A, I^AI^O	A
I^BI^B, I^BI^O	B
I^AI^B	AB
I^OI^O	O

Table 20.2

NOTES

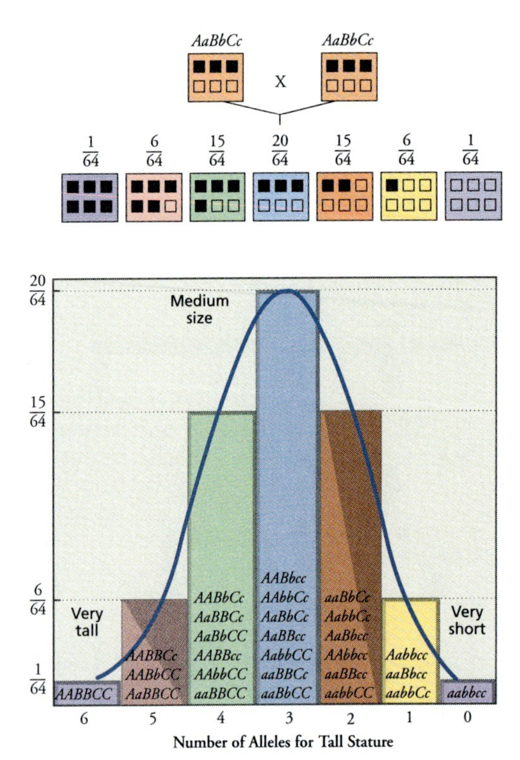

Fig. 20.10

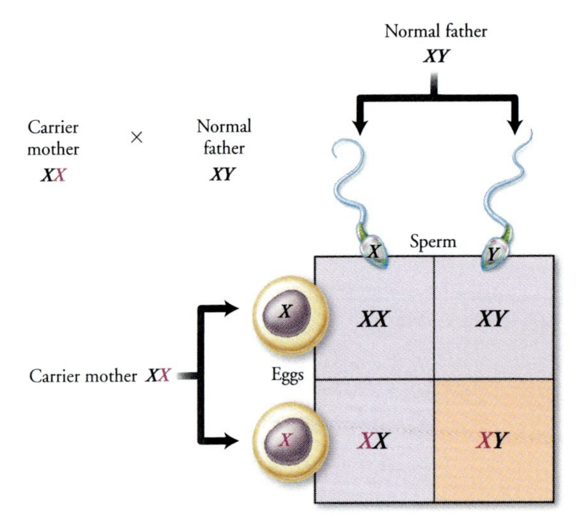

Fig. 20.11

NOTES

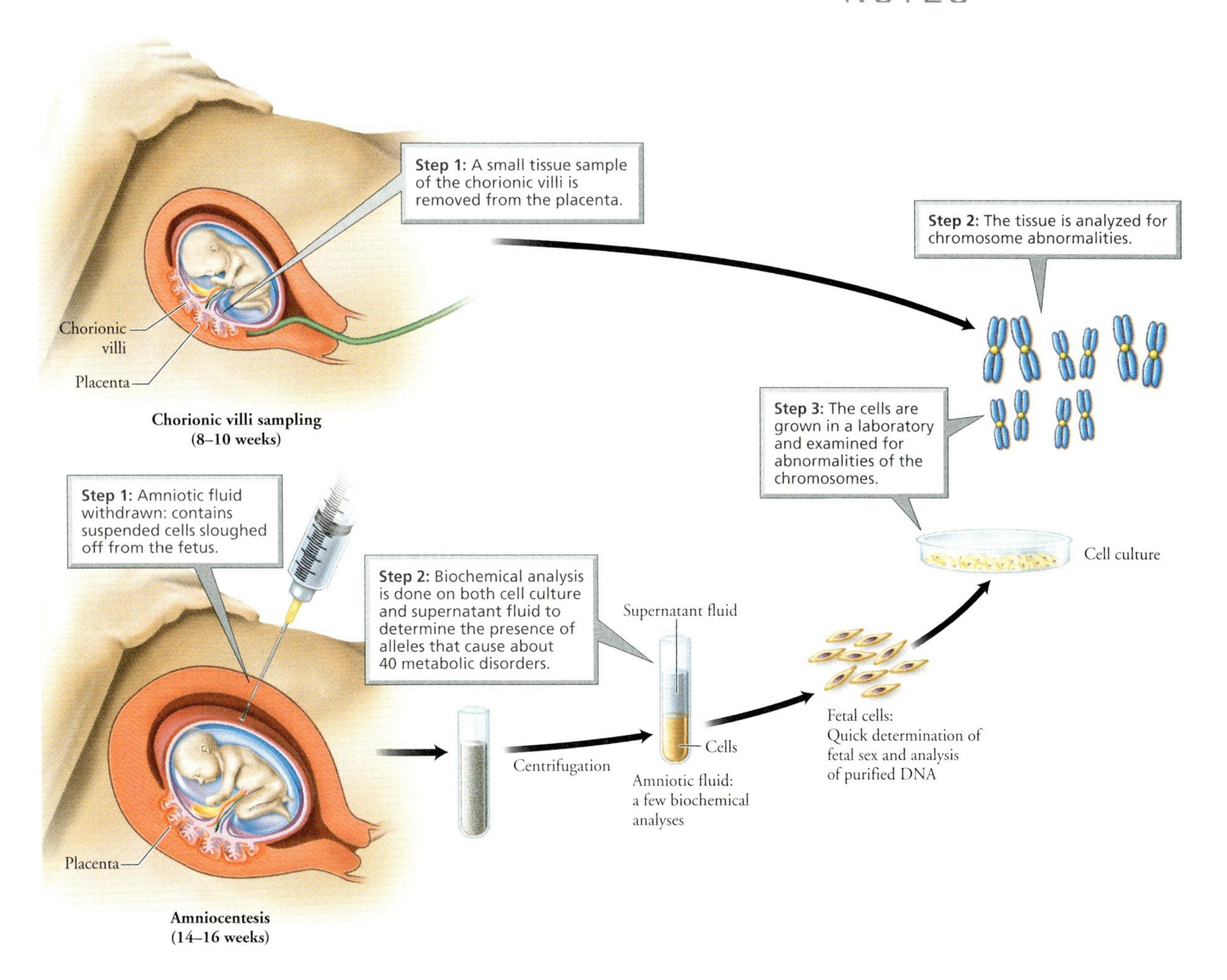

Fig. 20.14

NOTES

NOTES

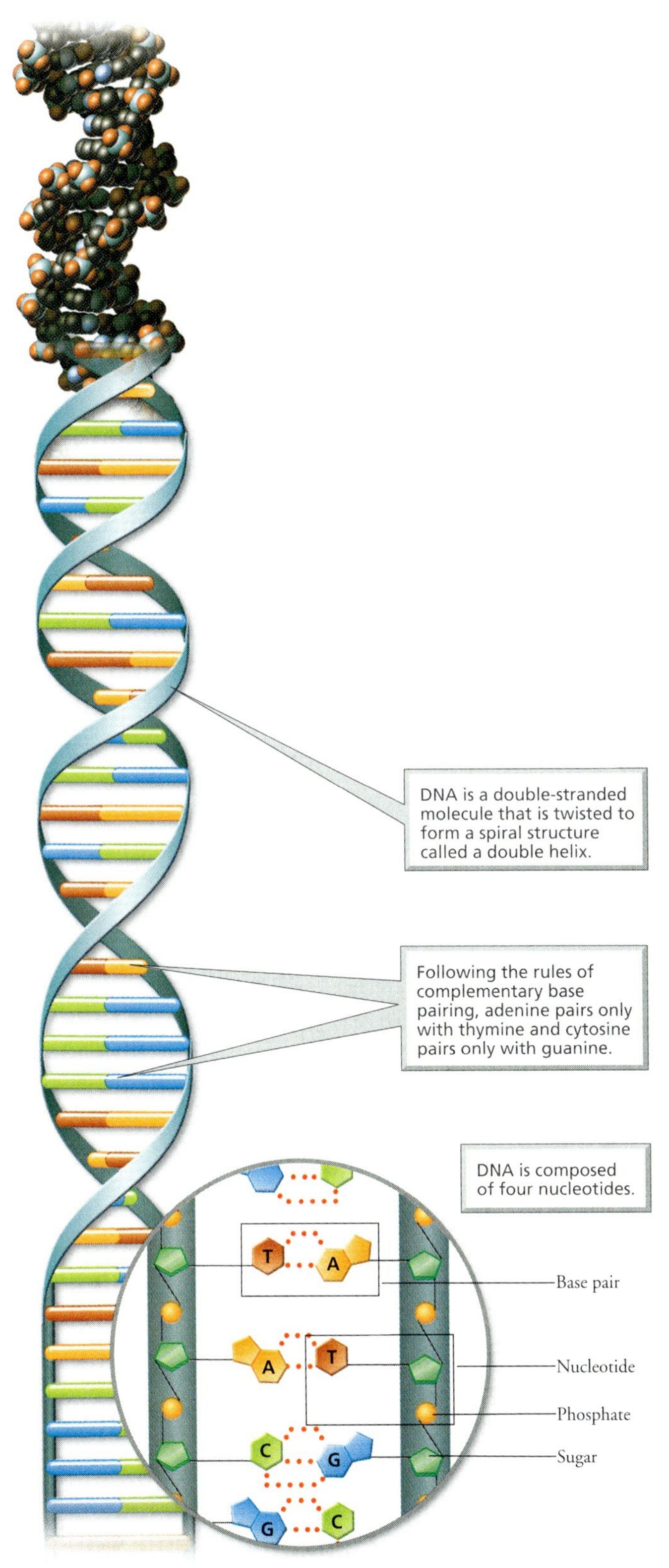

Fig. 21.1

NOTES

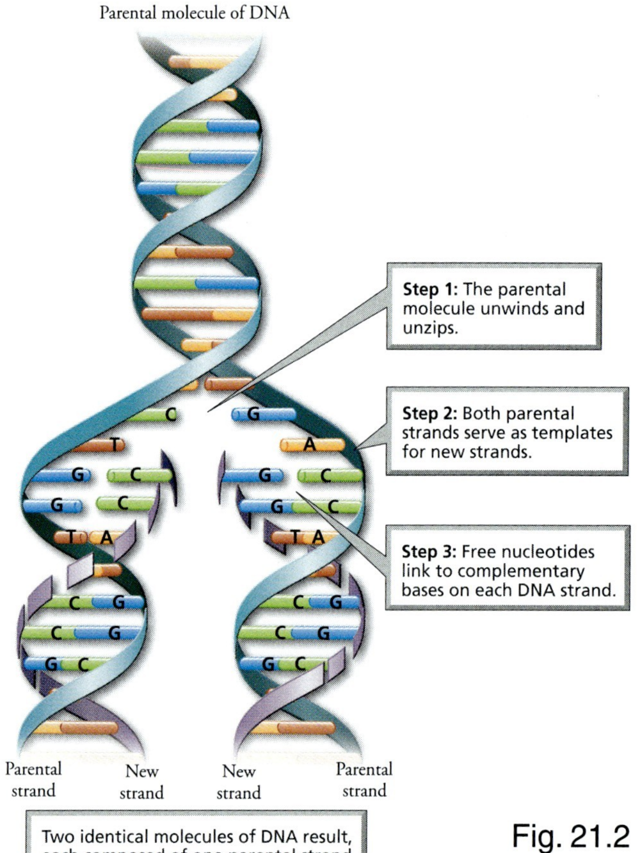

Fig. 21.2

Comparison of DNA and RNA		
	DNA	**RNA**
Similarities	Are nucleic acids	
	Are composed of linked nucleotides	
	Have a sugar-phosphate backbone	
	Have four types of bases	
Differences	Is a double-stranded molecule	Is a single-stranded molecule
	Has the sugar deoxyribose	Has the sugar ribose
	Contains the bases adenine, guanine, cytosine, and thymine	Contains the bases adenine, guanine, cytosine, and uracil (instead of thymine)
	Functions primarily in the nucleus	Functions primarily in the cytoplasm

Table 21.1

NOTES

FUNCTIONS OF RNA MOLECULES

MOLECULE	FUNCTIONS
Messenger RNA (mRNA)	Carries DNA's information in the sequence of its bases (codons) from the nucleus to the cytoplasm
Transfer RNA (tRNA)	Binds to a specific amino acid and transports it to the appropriate region of mRNA
Ribosomal RNA (rRNA)	Combines with protein to form ribosomes (structures on which polypeptides are synthesized)

Table 21.2

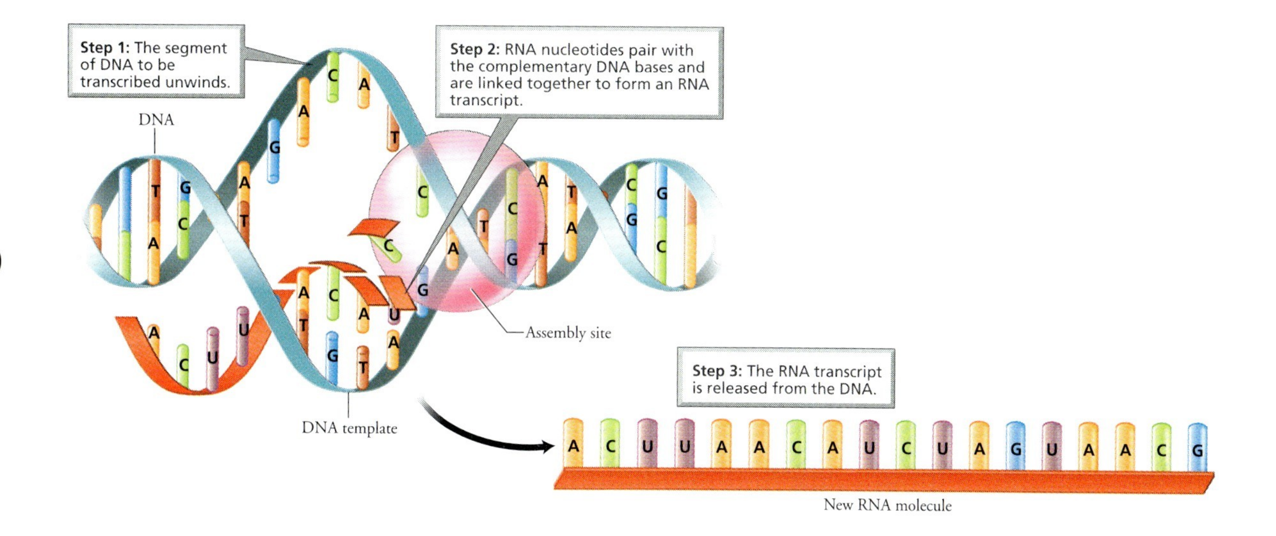

Fig. 21.3

NOTES

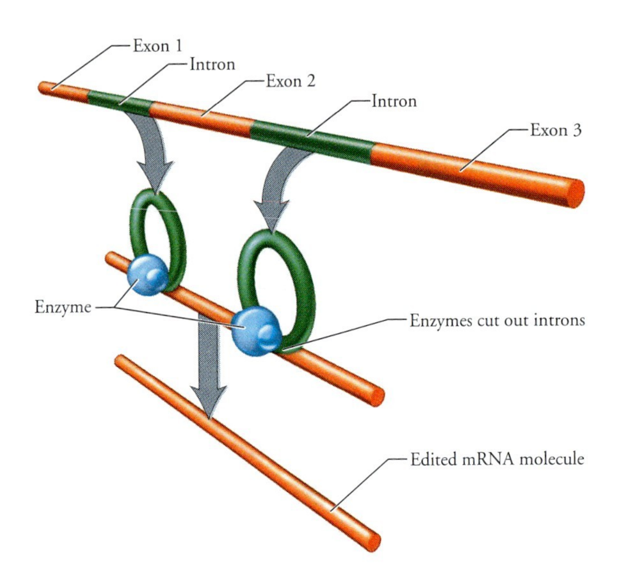

Fig. 21.4

The Genetic Code

First base	Second base: U		C		A		G		Third base
U	UUU	Phenylalanine	UCU	Serine	UAU	Tyrosine	UGC	Cysteine	U
	UUC		UCC		UAC		UGC		C
	UUA	Leucine	UCA		UAA	Stop	UGA	Stop	A
	UUG		UCG		UAG	Stop	UGG	Tryptophan	G
C	CUU	Leucine	CCU	Proline	CAU	Histidine	CGU	Arginine	U
	CUC		CCC		CAC		CGC		C
	CUA		CCA		CAA	Glutamine	CGA		A
	CUG		CCG		CAG		CGG		G
A	AUU	Isoleucine	ACU	Threonine	AAU	Asparagine	AGU	Serine	U
	AUC		ACC		AAC		AGC		C
	AUA		ACA		AAA	Lysine	AGA	Arginine	A
	AUG	Methionine Start	ACG		AAG		AGG		G
G	GUU	Valine	GCU	Alanine	GAU	Asparagine	GGU	Glycine	U
	GUC		GCC		GAC		GGC		C
	GUA		GCA		GAA	Glutamic Acid	GGA		A
	GUG		GCG		GAG		GGG		G

Table 21.3

NOTES

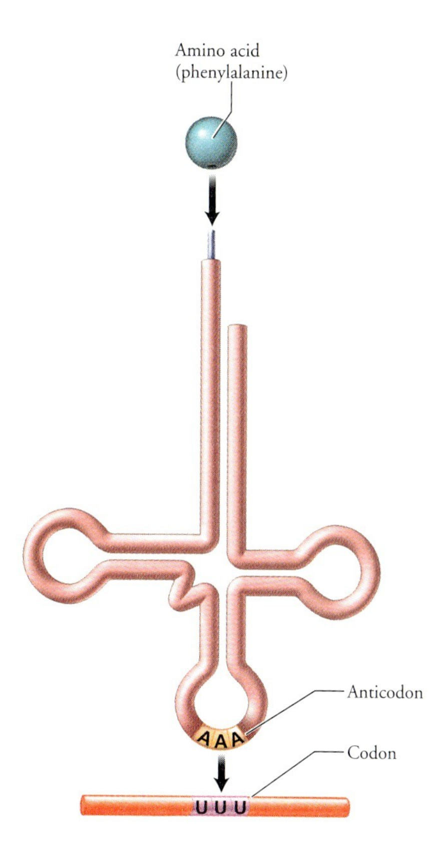

Fig. 21.5

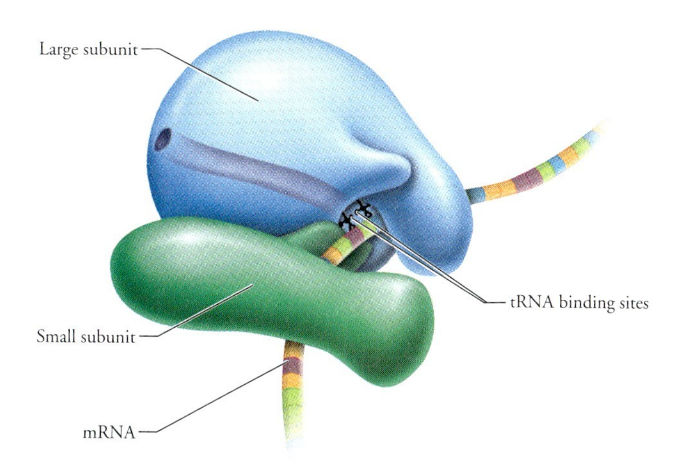

Fig. 21.6

NOTES

Initiation: Translation Begins

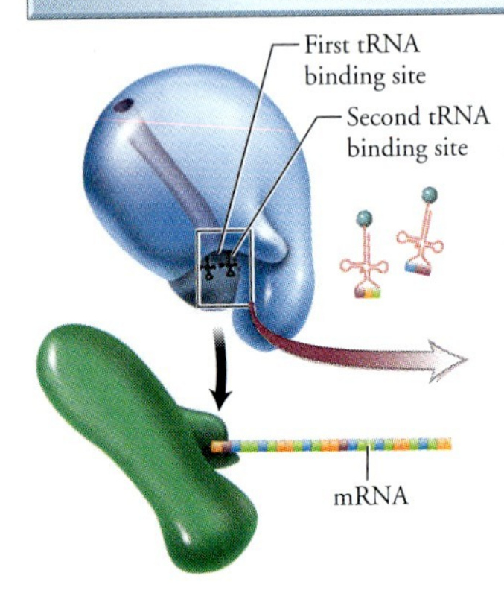

STEP 1
- The small ribosomal subunit joins to mRNA at the start codon, AUG

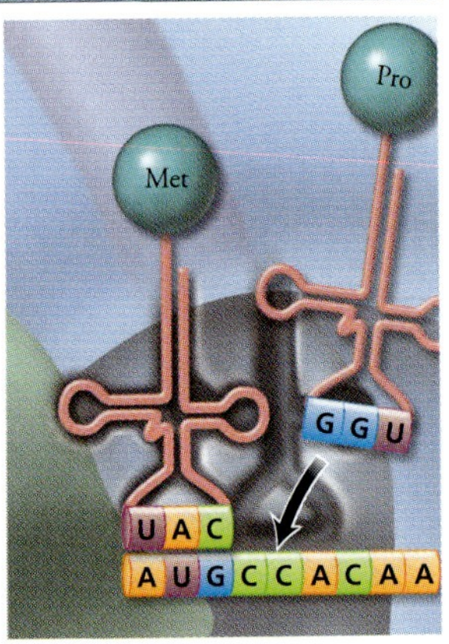

STEP 2
- A tRNA with a complementary anticodon pairs with the start codon
- Ribosomal subunits join to form a functional ribosome

Elongation: Amino Acids are Added to the Growing Chain One at a Time

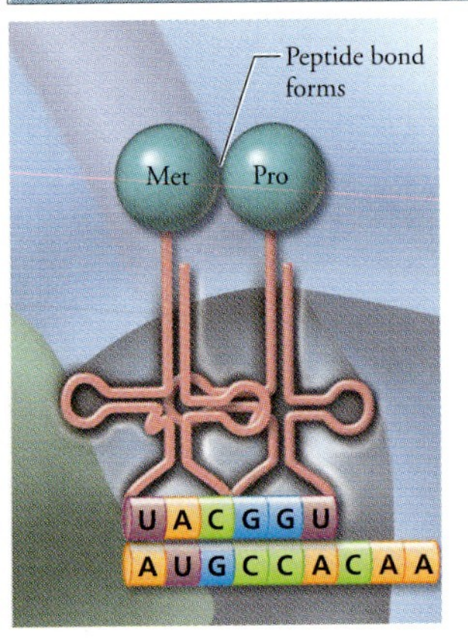

STEP 3
- A tRNA with the appropriate anticodon pairs with the next codon on mRNA
- Enzymes link the amino acids

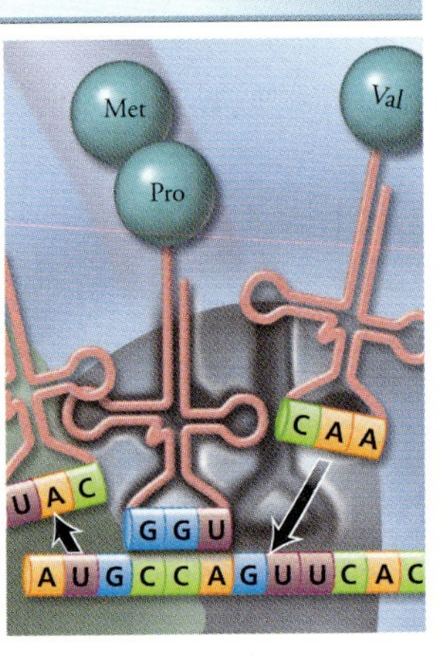

STEP 4
- The first tRNA leaves the ribosomes
- The ribosome moves along the mRNA, exposing the next codon
- Enzymes link the amino acids

NOTES

Termination: The Newly Synthesized Protein is Released

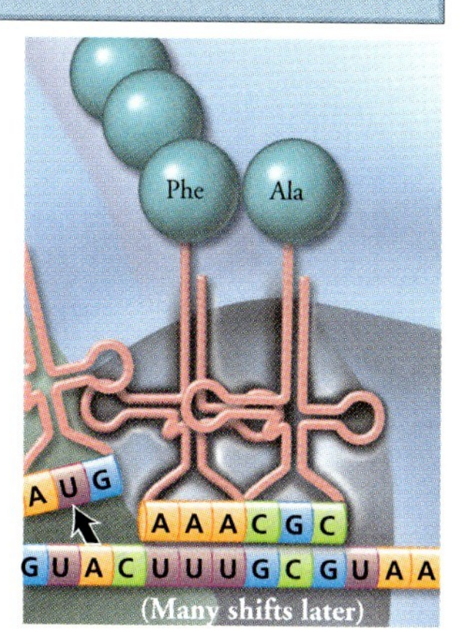

Step 5

- The tRNA in the first binding site leaves the ribosome
- The ribosome moves along the mRNA, exposing the next codon
- Enzymes link the amino acids
- The process is repeated many times

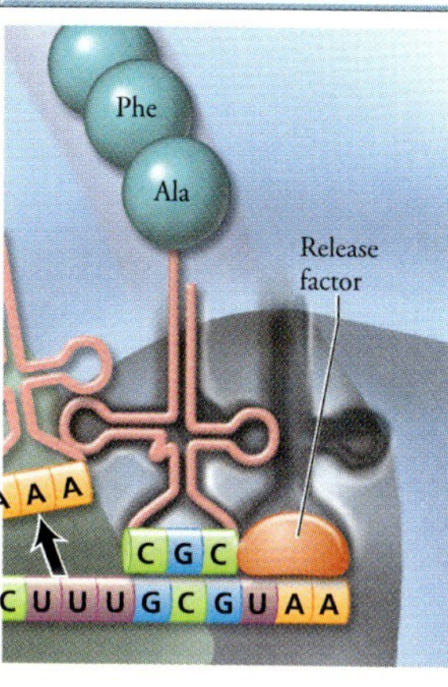

Step 6

- The stop codon moves into the ribosome

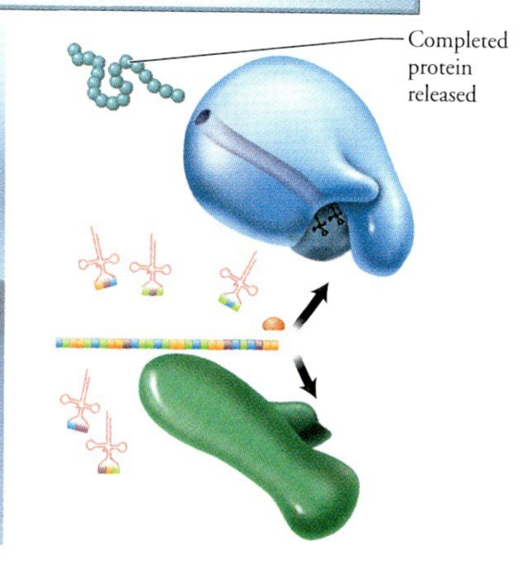

Step 7

- Release factors cause the release of the newly formed protein and the separation of the ribosomal subunits and the mRNA

Fig. 21.7

NOTES

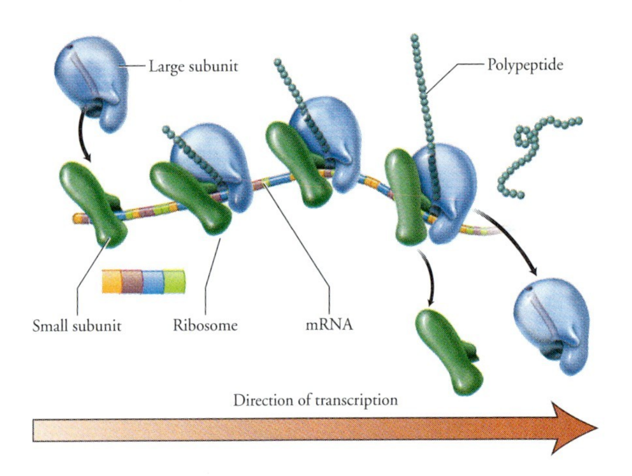

Fig. 21.8

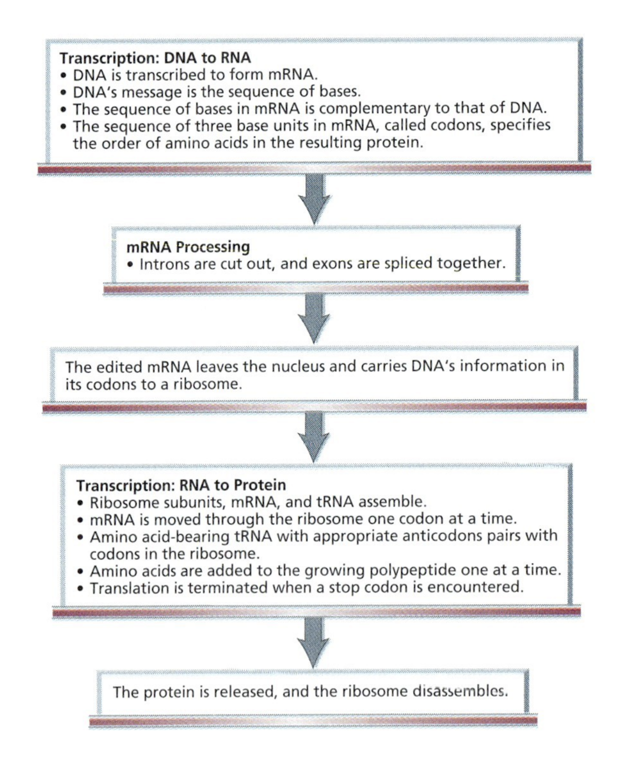

Fig. 21.9

NOTES

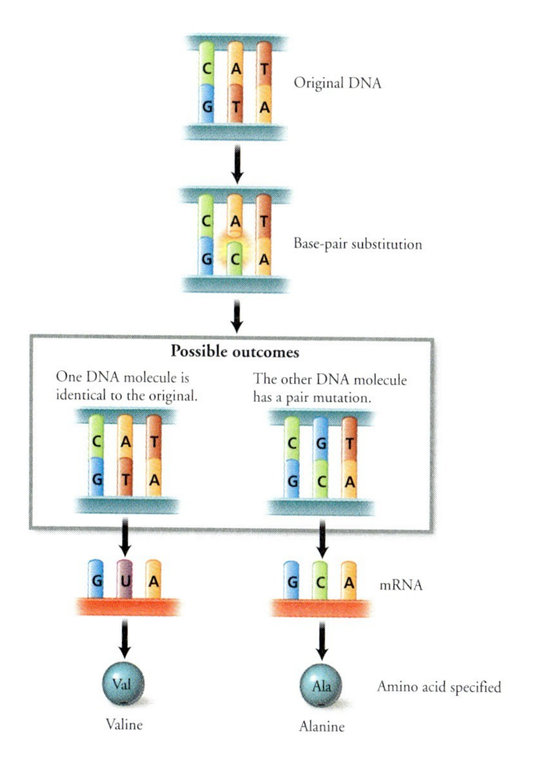

Fig. 21.10

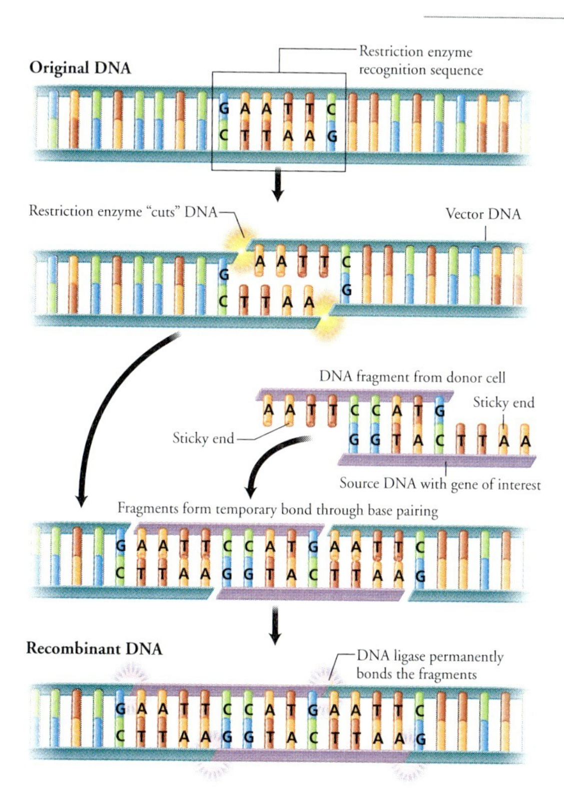

Fig.21.11

NOTES

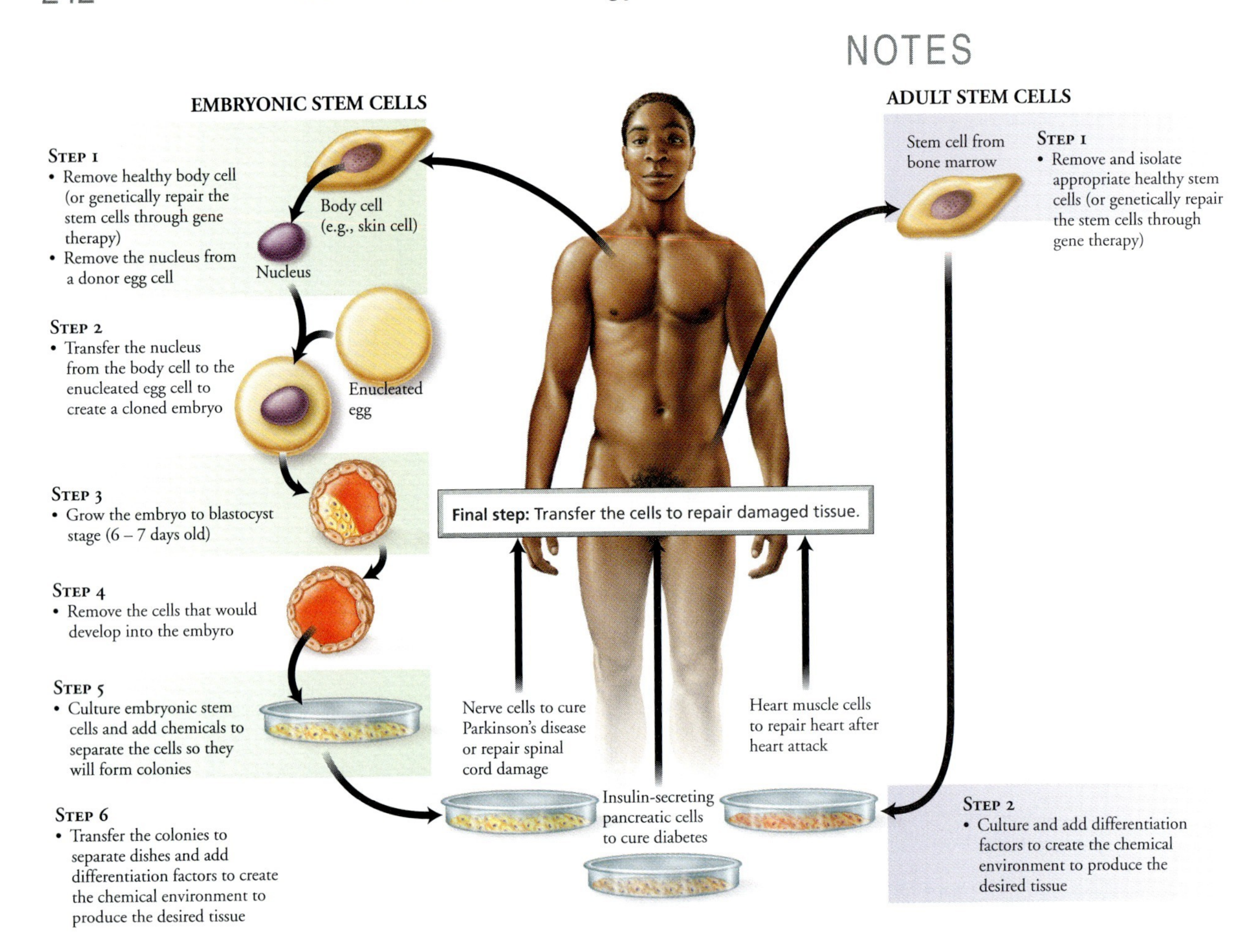

Fig. 21.A

NOTES

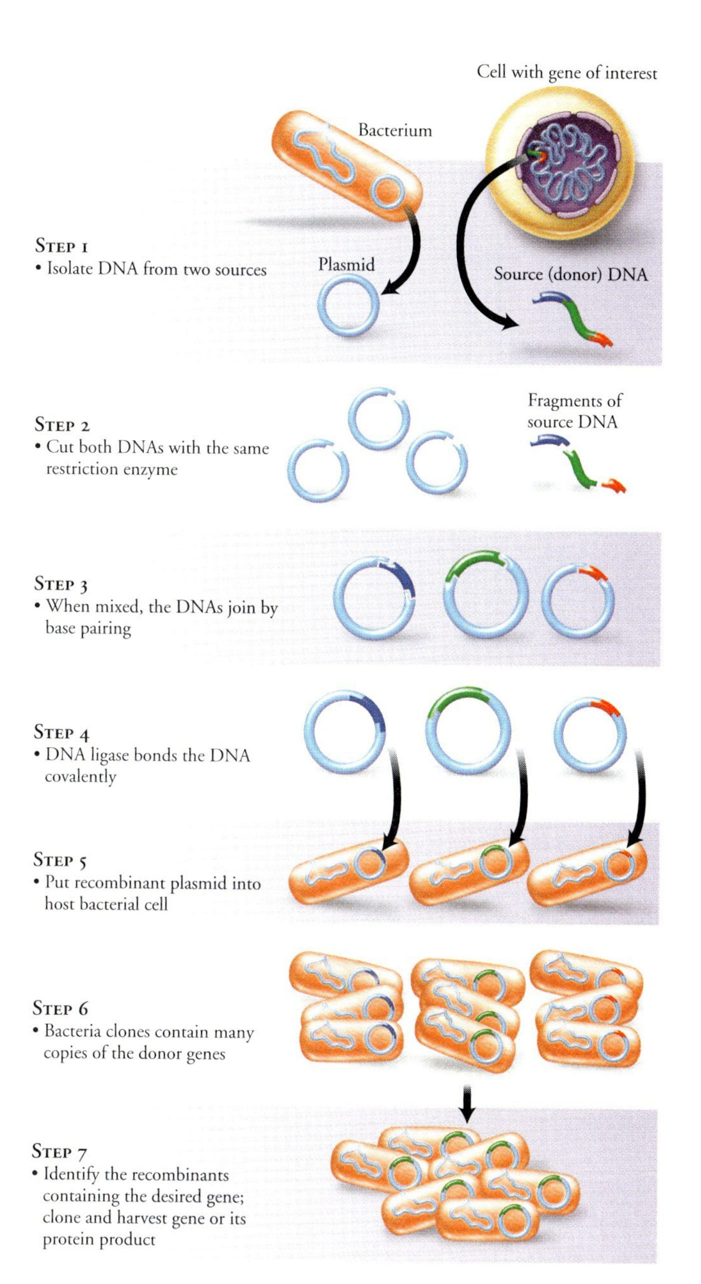

Fig. 21.12

NOTES

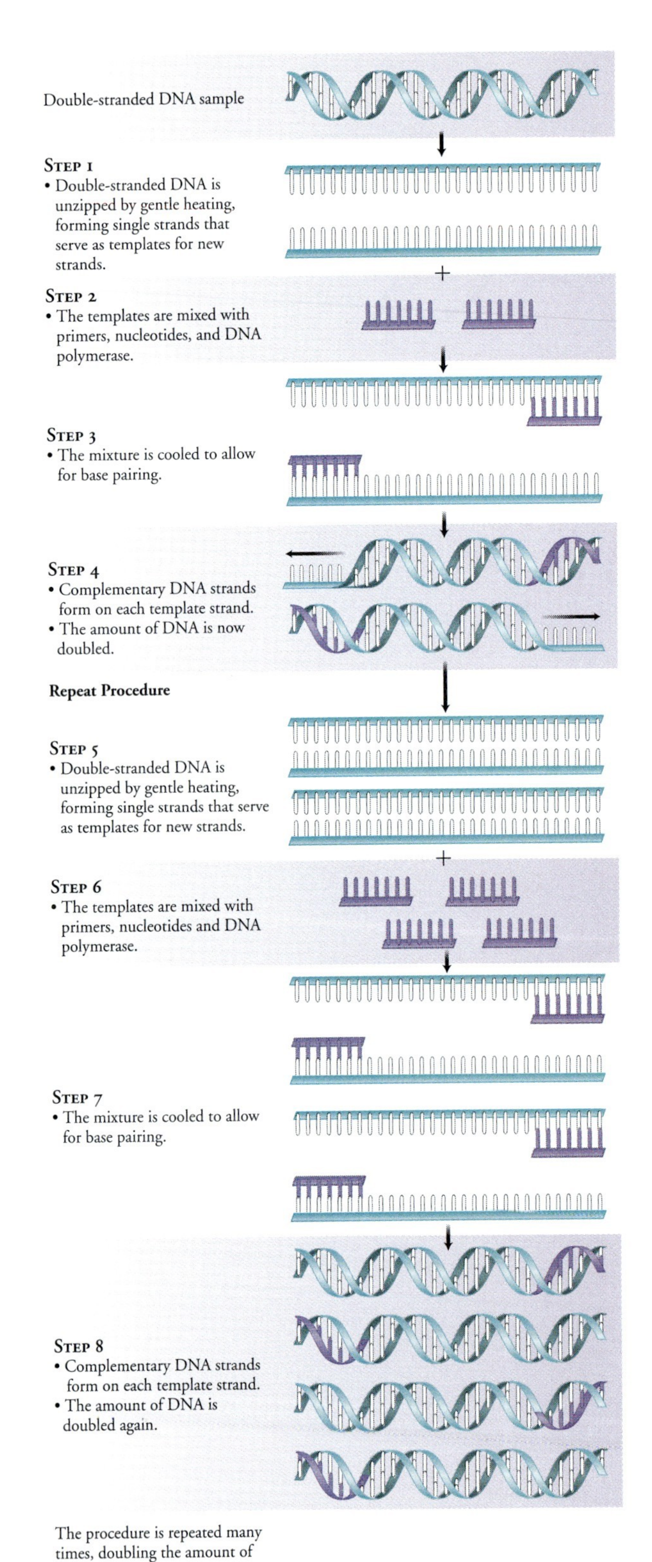

Fig. 21.13

NOTES

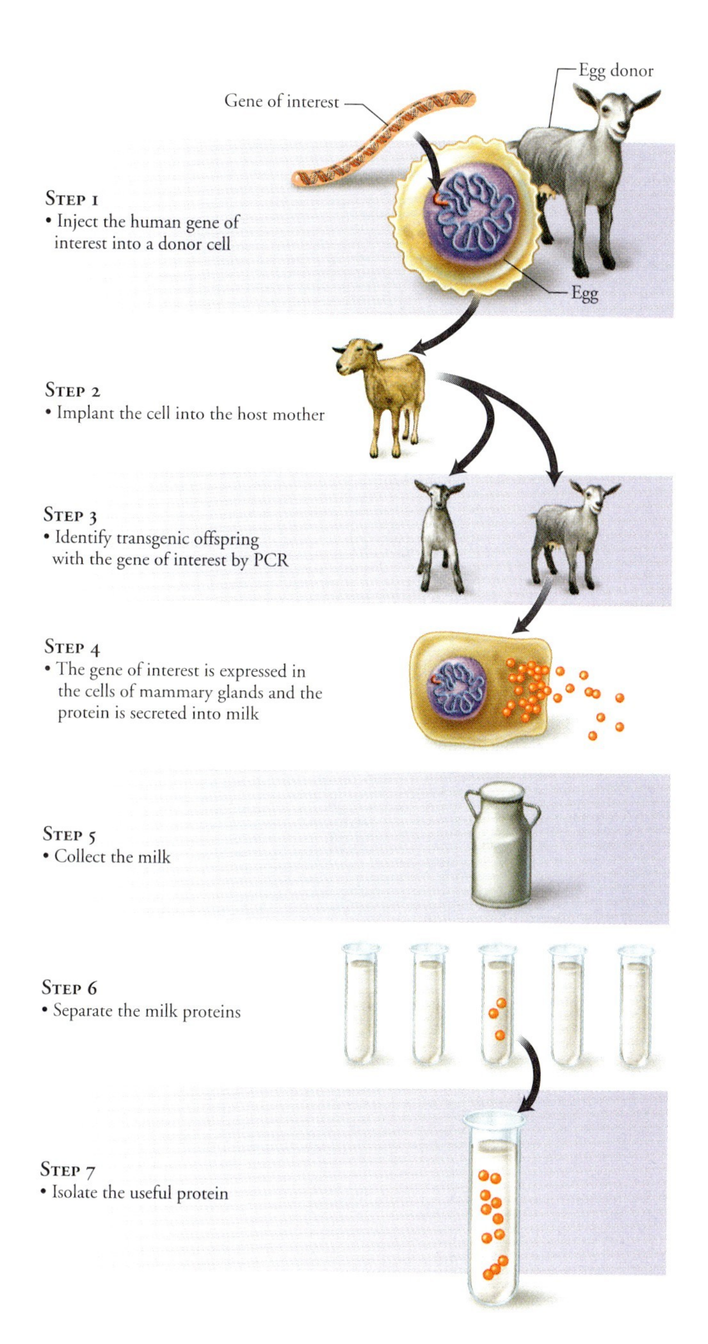

Fig. 21.15

NOTES

Concerns About Genetically Modified Food	
Health concerns	Is GM food safe for human consumption? Could GM crops increase bacterial resistance to antibiotics?
Environmental concerns	What effects will GM crops have on the level of use of pesticides? Can GM crops harm other organisms? Will GM crops become "superweeds"?
Social concerns	Can GM foods reduce world hunger? Who is responsible if GM foods prove to be harmful to humans or the environment?

Table 21.A

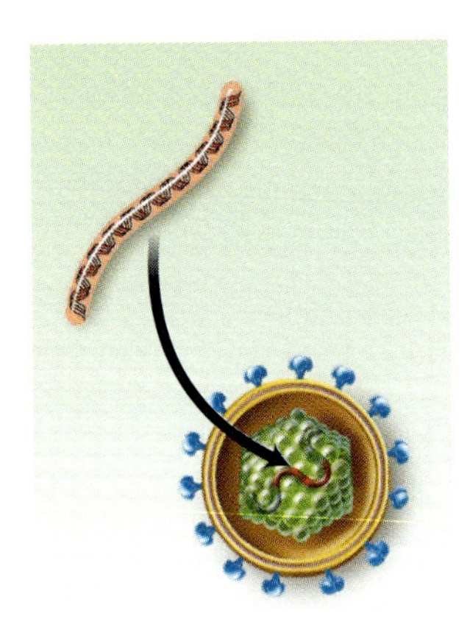

Step 1
Incorporate a healthy form of the gene into the virus.

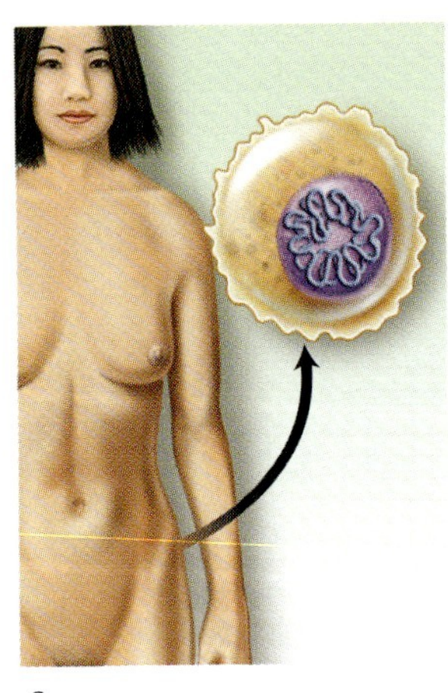

Step 2
Remove bone marrow stem cells from the patient.

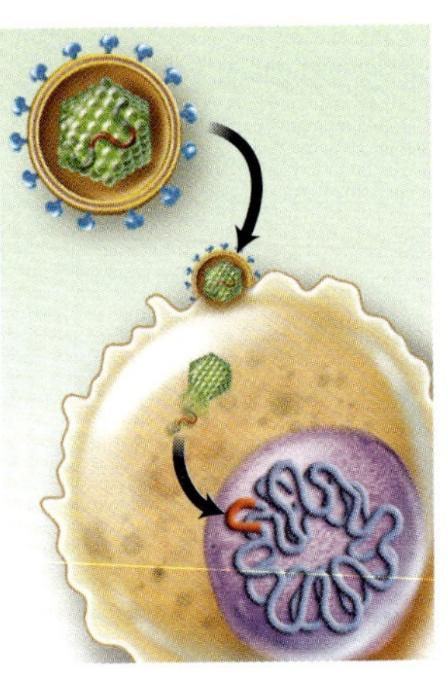

Step 3
Infect the patient's stem cells with the virus that is carrying the healthy form of the gene.

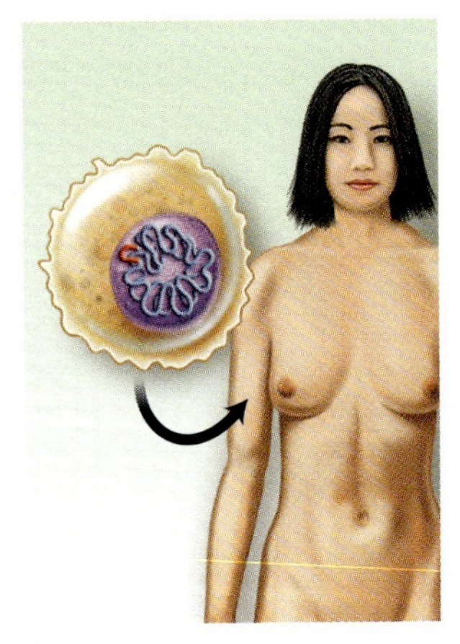

Step 4
Return the genetically engineered stem cells to the patient. The gene is expressed to produce the needed protein.

Fig. 21.18

NOTES

Leading sites of new cancer cases and deaths – 2004

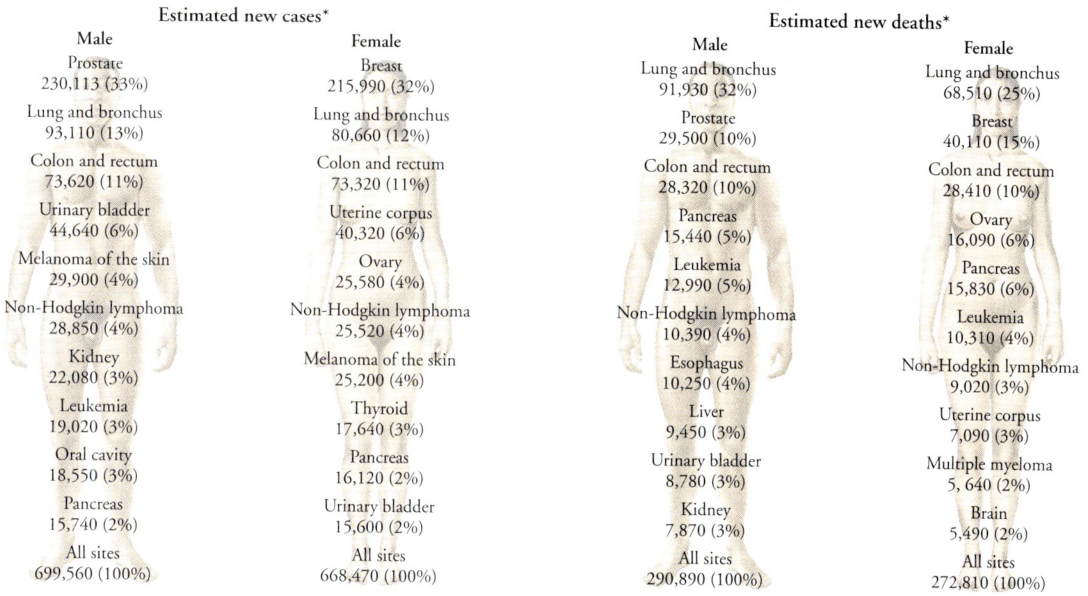

Estimated new cases* Male	Estimated new cases* Female	Estimated new deaths* Male	Estimated new deaths* Female
Prostate 230,113 (33%)	Breast 215,990 (32%)	Lung and bronchus 91,930 (32%)	Lung and bronchus 68,510 (25%)
Lung and bronchus 93,110 (13%)	Lung and bronchus 80,660 (12%)	Prostate 29,500 (10%)	Breast 40,110 (15%)
Colon and rectum 73,620 (11%)	Colon and rectum 73,320 (11%)	Colon and rectum 28,320 (10%)	Colon and rectum 28,410 (10%)
Urinary bladder 44,640 (6%)	Uterine corpus 40,320 (6%)	Pancreas 15,440 (5%)	Ovary 16,090 (6%)
Melanoma of the skin 29,900 (4%)	Ovary 25,580 (4%)	Leukemia 12,990 (5%)	Pancreas 15,830 (6%)
Non-Hodgkin lymphoma 28,850 (4%)	Non-Hodgkin lymphoma 25,520 (4%)	Non-Hodgkin lymphoma 10,390 (4%)	Leukemia 10,310 (4%)
Kidney 22,080 (3%)	Melanoma of the skin 25,200 (4%)	Esophagus 10,250 (4%)	Non-Hodgkin lymphoma 9,020 (3%)
Leukemia 19,020 (3%)	Thyroid 17,640 (3%)	Liver 9,450 (3%)	Uterine corpus 7,090 (3%)
Oral cavity 18,550 (3%)	Pancreas 16,120 (2%)	Urinary bladder 8,780 (3%)	Multiple myeloma 5, 640 (2%)
Pancreas 15,740 (2%)	Urinary bladder 15,600 (2%)	Kidney 7,870 (3%)	Brain 5,490 (2%)
All sites 699,560 (100%)	All sites 668,470 (100%)	All sites 290,890 (100%)	All sites 272,810 (100%)

* Excludes basal and squamous cell skin cancers and in situ carcinoma except urinary bladder.
Note: Percentages may not total 100% due to rounding.

Fig. 21a.1

Cancer Nomenclature by Tissue Type

TYPE OF CANCER	TISSUE TYPE
Carcinomas	Cancers of the epithelial tissues that infiltrate and spread (metastasize); include skin cancers, such as squamous cell, basal cell, and melanoma
Leukemias	Cancers of the bone marrow stem cells that produce the white blood cells
Sarcomas	Cancers of muscle, bone, cartilage, and connective tissues; may involve connective and muscle tissues of the bladder, kidneys, liver, lungs, parotid gland, and spleen
Lymphomas	Cancers of the lymph tissues such as lymph nodes; include Hodgkin lymphoma, non-Hodgkin lymphoma, and lymphosarcoma
Adenocarcinomas	Cancers of glandular epithelia, including liver, salivary glands, and breast

Table 21a.1

NOTES

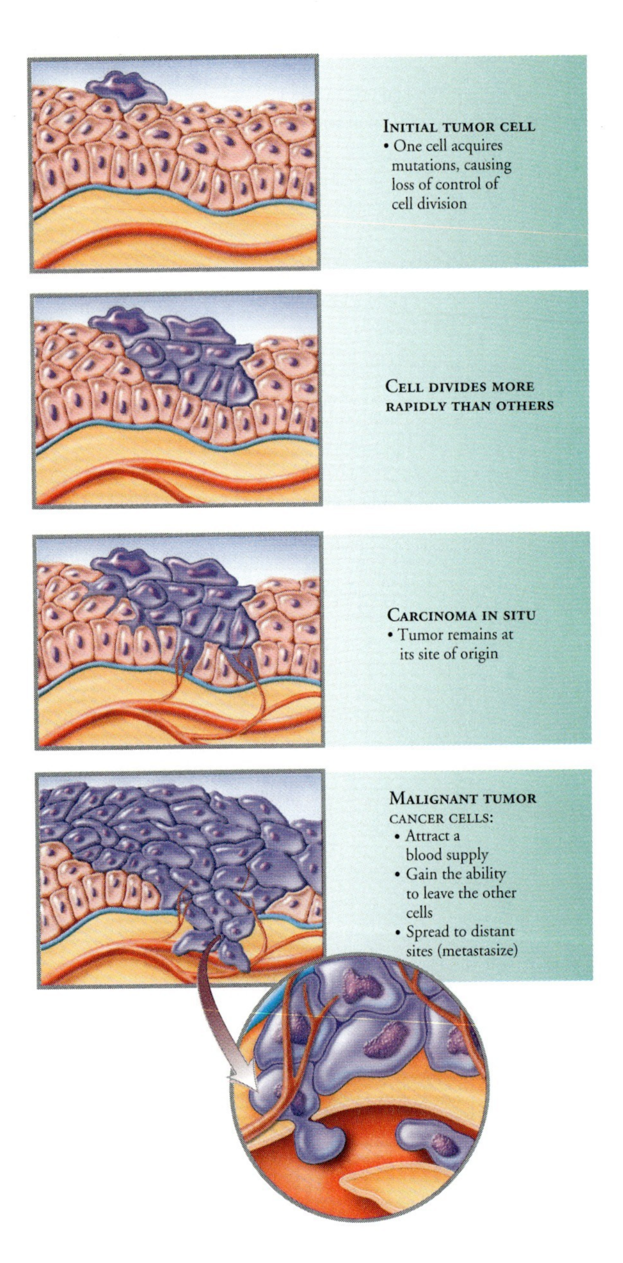

Fig. 21a.3

NOTES

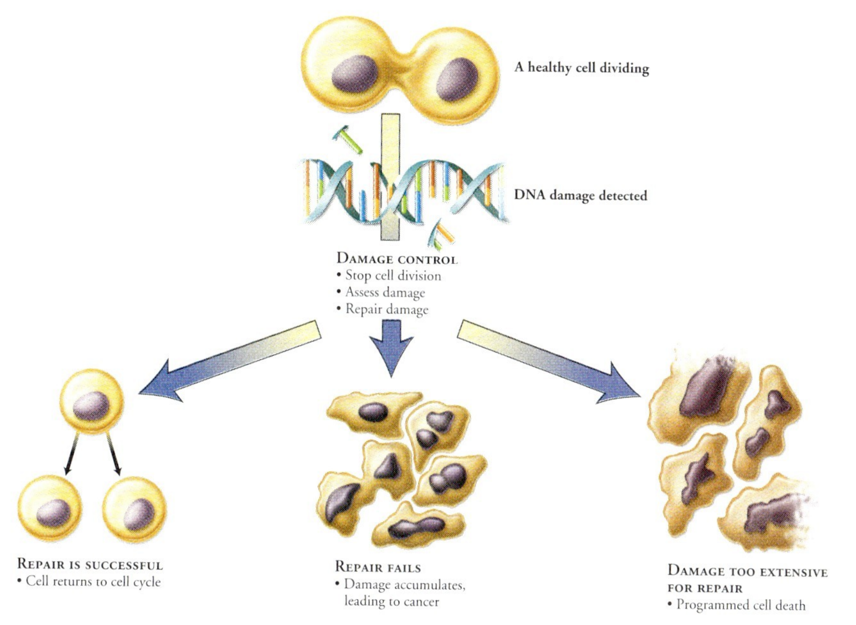

Fig. 21a.4

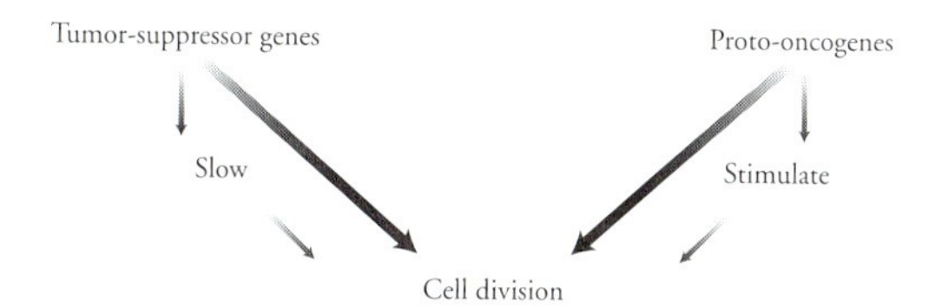

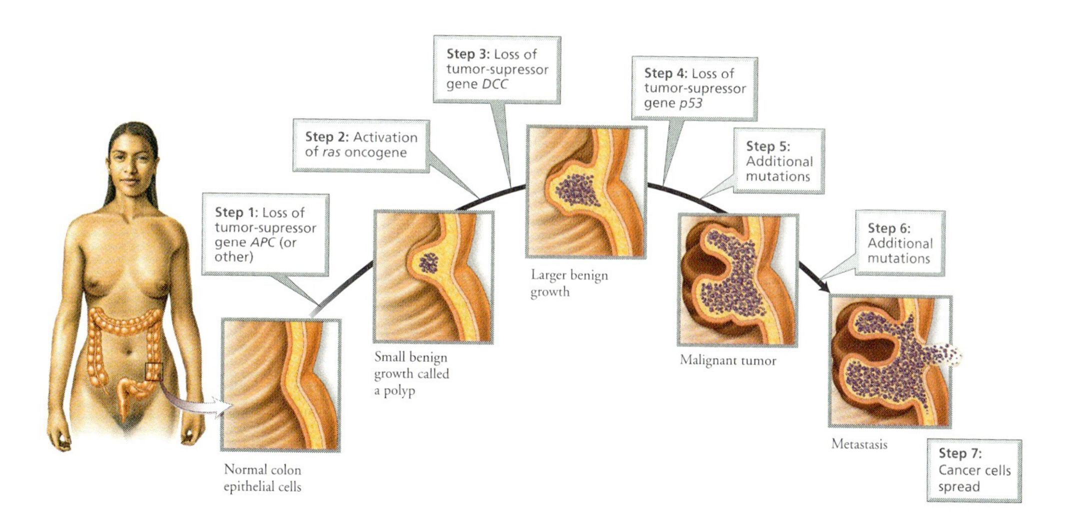

Fig. 21a.5

NOTES

Control Mechanisms and Methods and Prevention of Cancer Cell Evasion

MECHANISM THAT PROTECTS CELLS FROM CANCER	METHOD OF EVASION USED BY CANCER CELLS	TREATMENT AIMED AT PREVENTING EVASION TACTICS
Genetic controls on cell division		
Proto-oncogenes stimulate cell division through effects on growth factors	Oncogenes accelerate the rate of cell division	Gene therapy
Tumor-suppressor genes inhibit cell division	Mutations in tumor-suppressor genes take the "brakes" off cell division	Gene therapy
Programmed cell death		
A genetic program that initiates events that lead to the death of the cell when damaged DNA is detected or another signal is received	Mutations in tumor-suppressor genes: Mutant gene *p53* no longer triggers cell death when damaged DNA is detected	Gene therapy
Limitations on the number of times a cell can divide		
Telomeres protect the ends of chromosomes, but a fraction of each is shaved off each time the DNA is copied; when the telomeres are gone, DNA containing important genes is damaged when the DNA is copied, and the cell can no longer divide	Genes to produce telomerase, the enzyme that reconstructs telomeres, are turned on in cancer cells so telomere length is stabilized	Drugs to block telomerase activity or turn off the genes that produce it
Controls that prevent the formation of new blood vessels		
These controls are normally in effect except in a few instances such as wound healing	Cancer cells produce growth factors that attract new blood vessels and proteins that counter the normal proteins that inhibit blood vessel formation	Drugs to block blood vessel formation
Controls that keep normal cells in place		
Cellular adhesion molecules (CAMs) hold cells in place; unanchored cells stop dividing and self-destruct	Cancer cell oncogenes send a false message to the nucleus that the cell is properly anchored	Gene therapy to insert the gene that produces CAMs into tumor cells; using drugs to block certain CAMs on cancer cells so that they cannot grip blood vessel walls as they metastasize

Table 21a.2

NOTES

SOME VIRUSES LINKED TO HUMAN CANCERS	
VIRUS	**TYPES OF CANCER**
Human papilloma viruses (HPVs)	Cervical, penile, and other anogenital cancers in men and women
Hepatitis B and C	Liver cancer
Epstein–Barr virus	B cell lymphomas, especially Burkitt's lymphoma; nasopharyngeal carcinoma
Human T cell leukemia virus (HTLV–1)	Adult T cell leukemia
Cytomegalovirus (CMV)	Lymphomas and leukemias
Herpes viruses (HHV8)	Kaposi's sarcoma

Table 21a.3

TIPS FOR REDUCING YOUR CANCER RISK

1. Do not use tobacco. If you do, quit. Avoid exposure to secondhand smoke.
2. Reduce the amount of saturated fat in your diet, especially the fat from red meat.
3. Minimize your consumption of salt-cured, pickled, and smoked foods.
4. Eat at least five servings of fruits and vegetables every day.
5. Avoid excessive alcohol intake. If you consume alcohol, one or two drinks a day should be the maximum.
6. Watch your caloric intake, and keep your body weight proper for your height.
7. Avoid excessive exposure to sunlight. Wear protective clothing. Use sunscreen.
8. Avoid unnecessary medical x-rays.
9. Have the appropriate screening exams on a regular basis. Women should have PAP tests and mammograms. Men should have prostate tests. All adults should have tests for colorectal cancer.

Table 21a.4

CANCER'S SEVEN WARNING SIGNS

Change in bowel or bladder habit or function

A sore that does not heal

Unusual bleeding or bloody discharge

Thickening or lump in breast or elsewhere

Indigestion or difficulty swallowing

Obvious change in wart or mole

Nagging cough or hoarseness

Source: *American Cancer Society.*

Table 21a.5

NOTES

RECOMMENDED CANCER SCREENING TESTS

Guidelines suggested by the American Cancer Society for the early detection of cancer in people without symptoms, age 20 to 40

Cancer-related checkup every 3 years

Should include the procedures listed below plus health counseling (such as tips on quitting cigarette smoking) and examinations for cancers of the thyroid, testes, prostate, mouth, ovaries, skin, and lymph nodes. Some people are at higher than normal risk for certain cancers and may need to have tests more frequently.

Breast	• Exam by doctor every 3 years • Self-exam every month • One baseline breast x-ray ages 35–40 Higher risk for breast cancer: Personal or family history of breast cancer; never had children; had first child after 30
Uterus	• Pelvic exam every 3 years
Cervix	• Yearly PAP test beginning at age 18 or when sexual activity begins Higher risk for cervical cancer: Early age at first intercourse; multiple sex partners

Guidelines suggested by the American Cancer Society for the early detection of cancer in people without symptoms, age 40 and over

Cancer-related checkup every year

Should include the procedures listed below plus health counseling (such as tips on quitting cigarette smoking) and examinations for cancers of the thyroid, testes, prostate, mouth, ovaries, skin, and lymph nodes. Some people are at higher than normal risk for certain cancers and may need to have tests more frequently.

Breast	• Exam by doctor every year • Self-exam every month • Breast x-ray every year after 40 Higher risk for breast cancer: Personal or family history of breast cancer; never had children; had first child after 30
Uterus	• Pelvic exam every year
Cervix	• Yearly PAP test Higher risk for cervical cancer: Early age at first intercourse; multiple sex partners
Endometrium	• Endometrial tissue sample at menopause if at risk Higher risk for endometrial cancer: Infertility, obesity, failure of ovulation, abnormal uterine bleeding, estrogen therapy
Prostate	• Yearly prostate-specific antigen (PSA) blood test and digital rectal exam after age 50
Colon and rectum	• Fecal occult blood test every year after age 50 • Flexible sigmoidoscopy beginning at age 50 and every 5 years thereafter Higher risk for colorectal cancer: Personal or family history of colon or rectal cancer; personal or family history of polyps in the colon or rectum; ulcerative colitis

Table 21a.6

NOTES

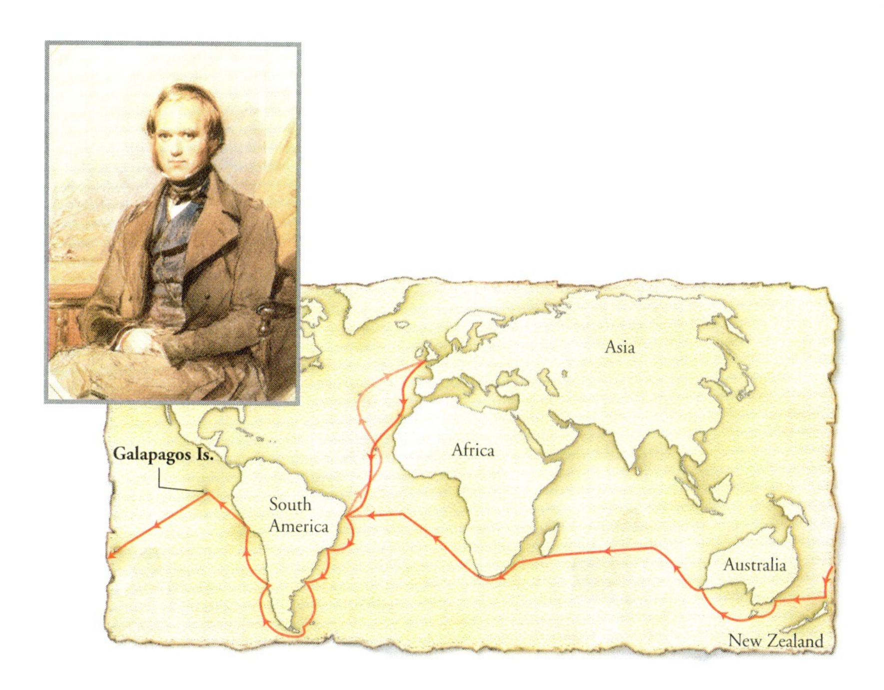

Fig. 22.1

Fig. 22.2

NOTES

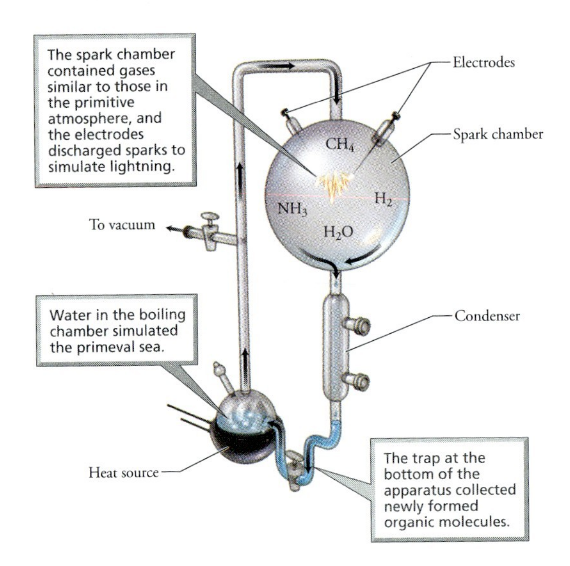

Fig. 22.3

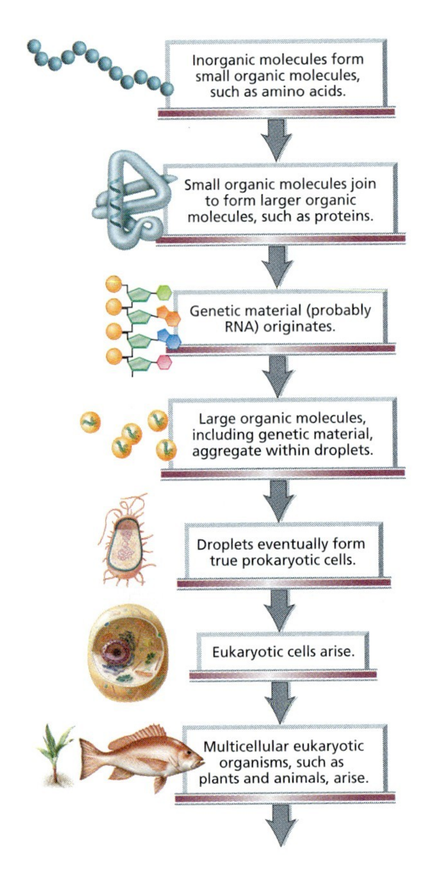

Fig. 22.4

NOTES

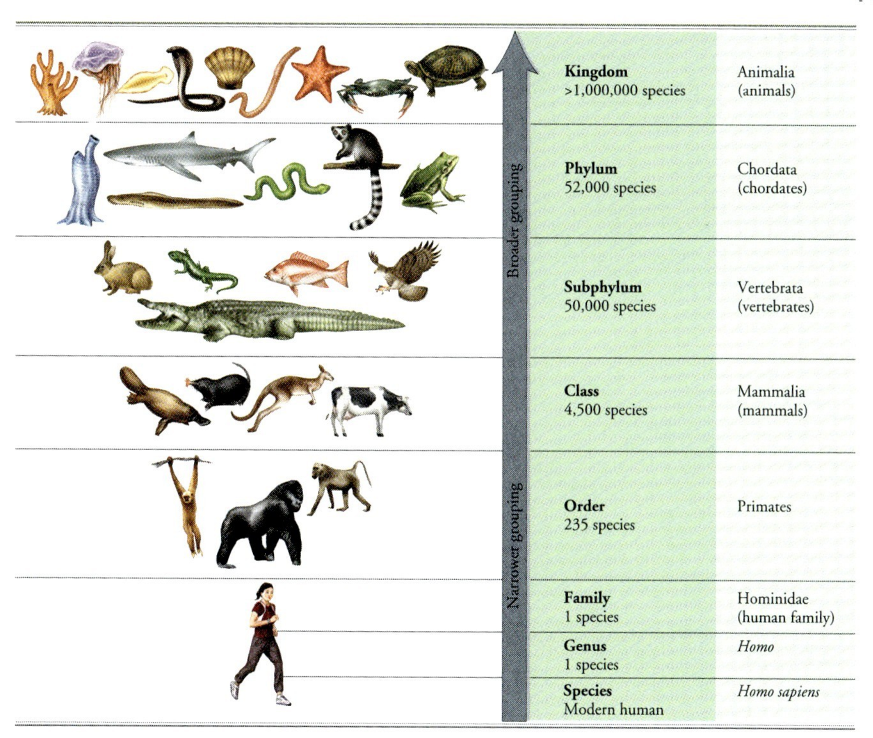

Fig. 22.8

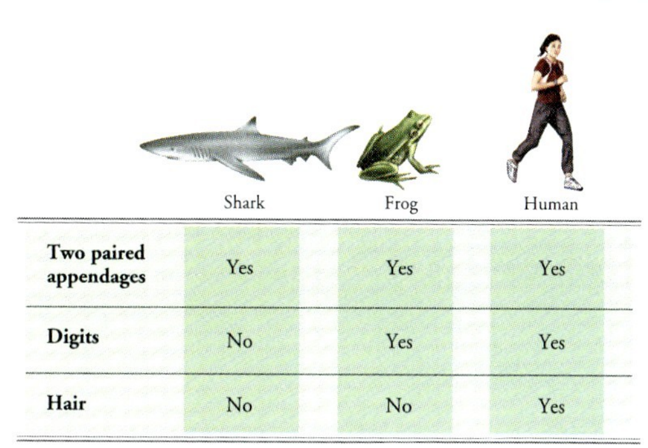

(a) A character matrix is used to construct a phylogenetic tree.

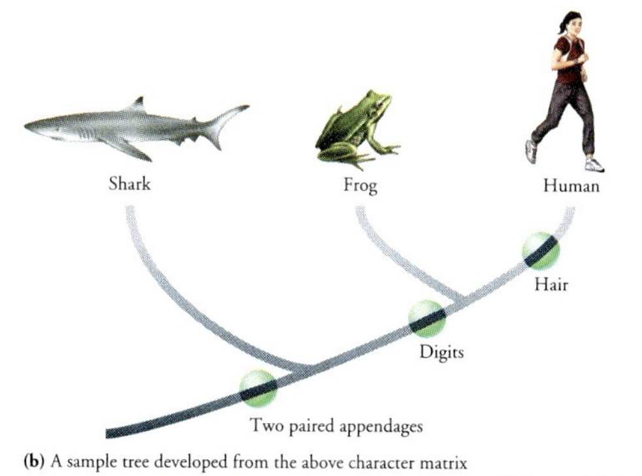

(b) A sample tree developed from the above character matrix

Fig. 22.9

NOTES

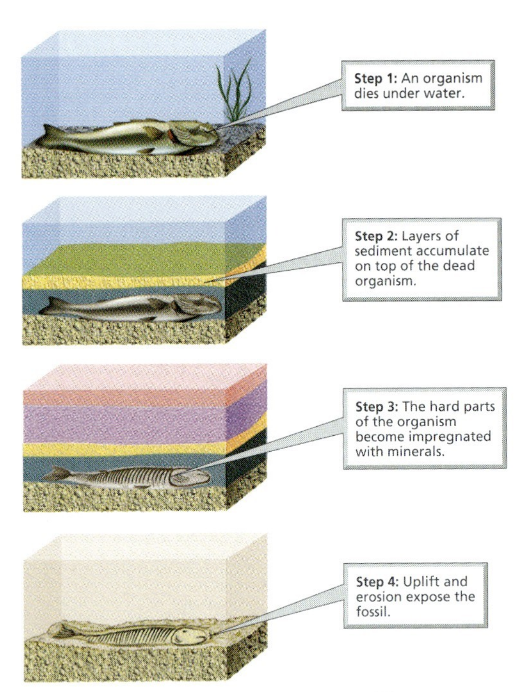

Fig. 22.11

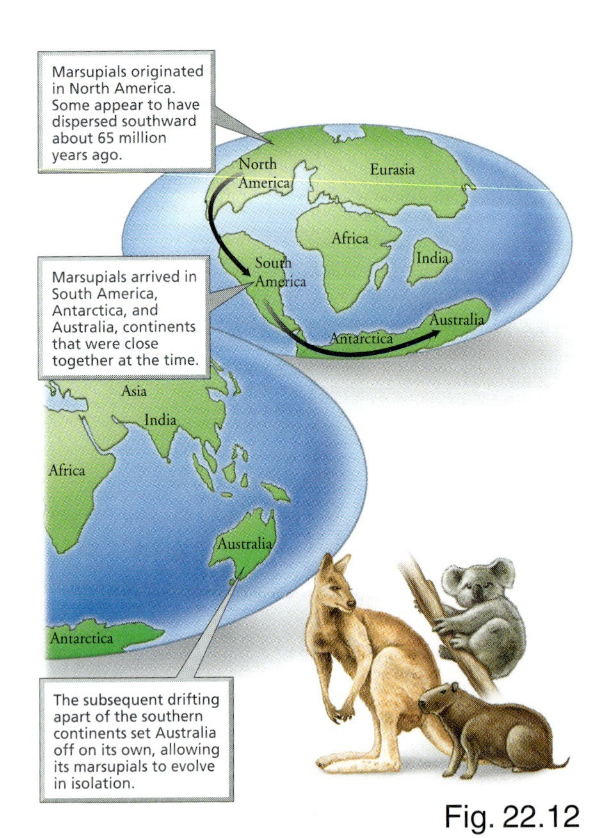

Fig. 22.12

NOTES

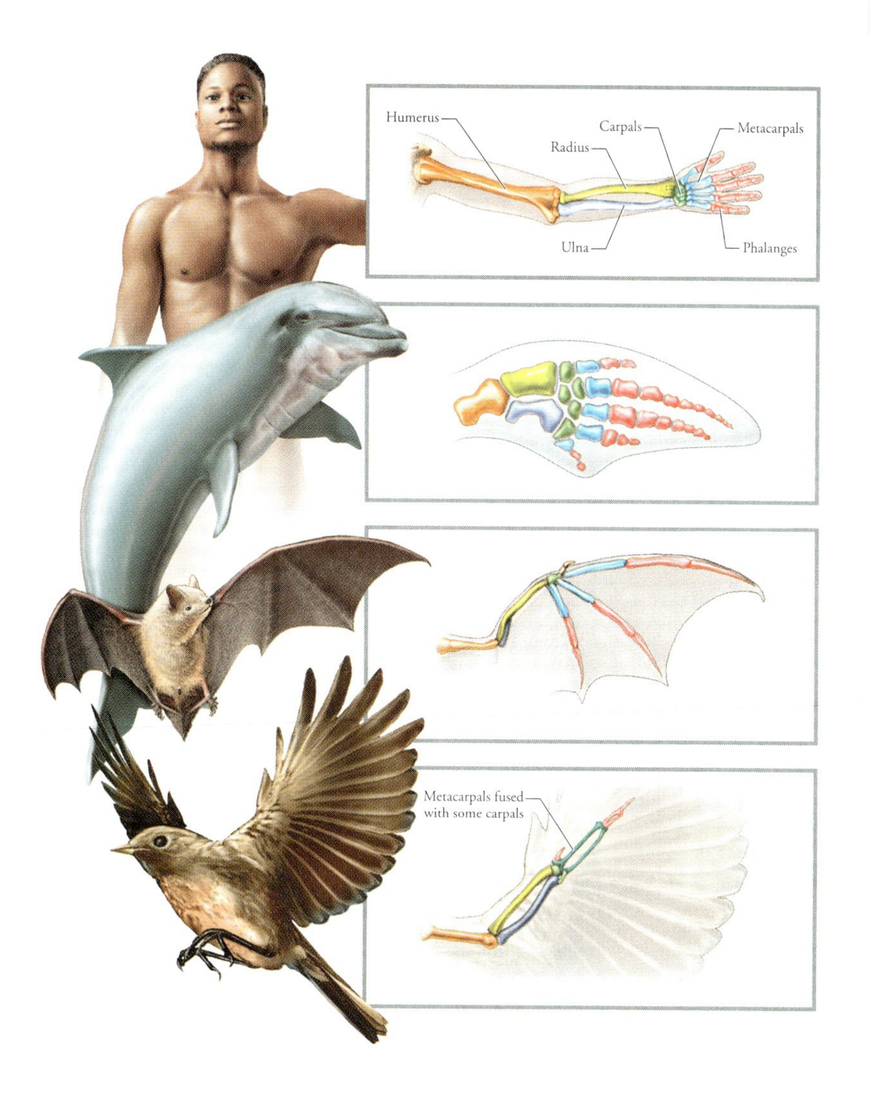

Fig. 22.13

NOTES

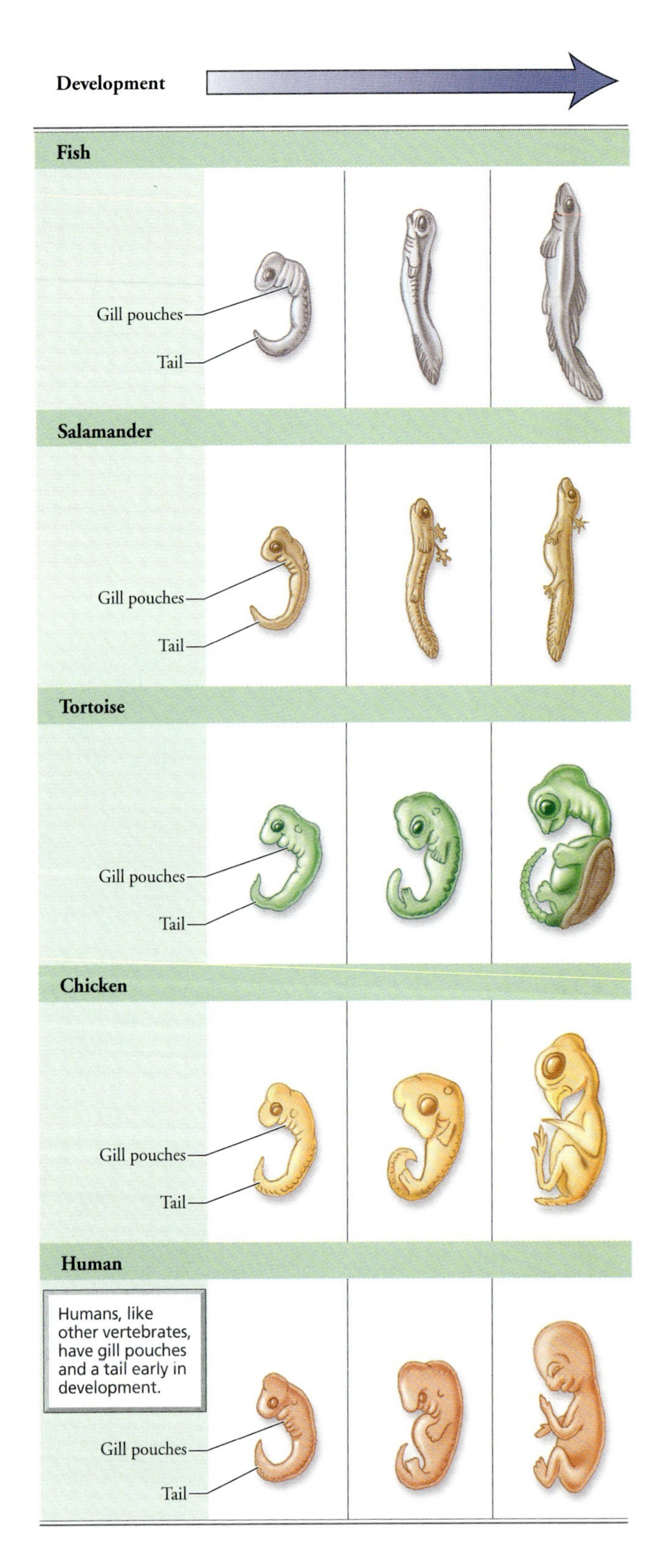

Fig. 22.14

NOTES

Differences in DNA Sequences Indicating Phylogenetic Distances between Pairs of Primate Species	
SPECIES PAIRS	PERCENTAGE DIFFERENCE IN NUCLEOTIDE SEQUENCES
Human–chimpanzee	2.5
Human–gibbon	5.1
Human–Old World monkey	9.0
Human–New World monkey	15.8
Human–lemur (prosimian)	42.0

Data from Stebbins, G. L. 1982. *Darwin to DNA, Molecules to Humanity.* San Francisco: W. H. Freeman.

Table 22 1

Characteristics of Primates
Flexible shoulder joints
Grasping forelimbs; some species with fully opposable thumbs
Grasping feet, with big toe separated from other toes (except humans)
Sensitive digits with flattened nails instead of claws
Forward-facing eyes
Small litter size
Complex social behavior, including extensive parental care
Reliance on learned behavior

Table 22 .2

NOTES

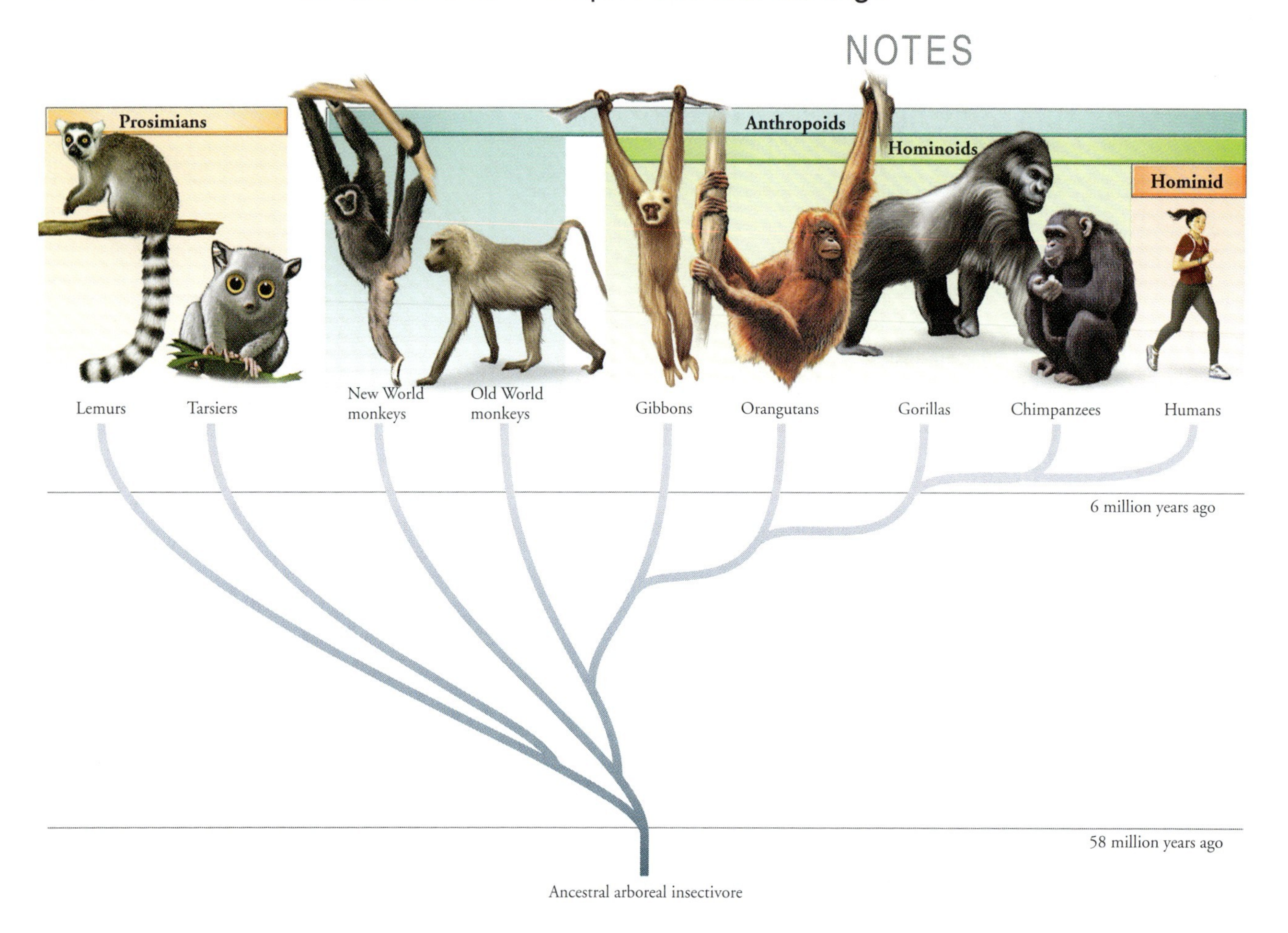

Fig. 22.19

MAJOR TRENDS IN HOMINID EVOLUTION
• Bipedalism
• Increasing brain size and associated cultural trends such as tool use, language, and behavioral complexity
• Shortening of the jaw and flattening of the face
• Reduced difference in size between males and females

Table. 22.3

NOTES

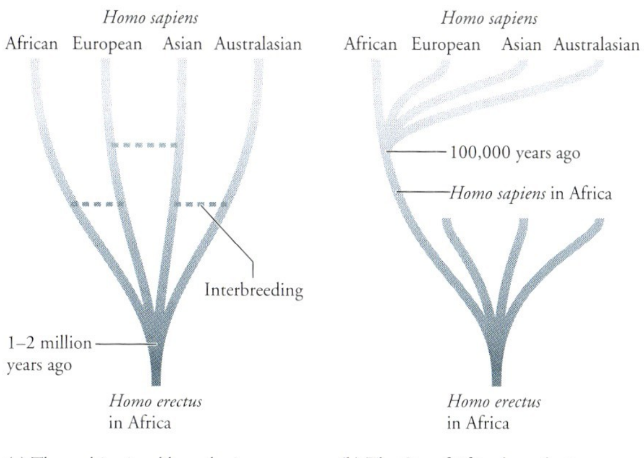

(a) The multiregional hypothesis suggests that modern humans evolved independently in several areas of the world from distinctive populations of *H. erectus*.

(b) The Out of Africa hypothesis suggests that modern humans evolved from *H. erectus* in Africa and later migrated to other parts of the world, replacing other descendants of *H. erectus*.

Fig. 22.23

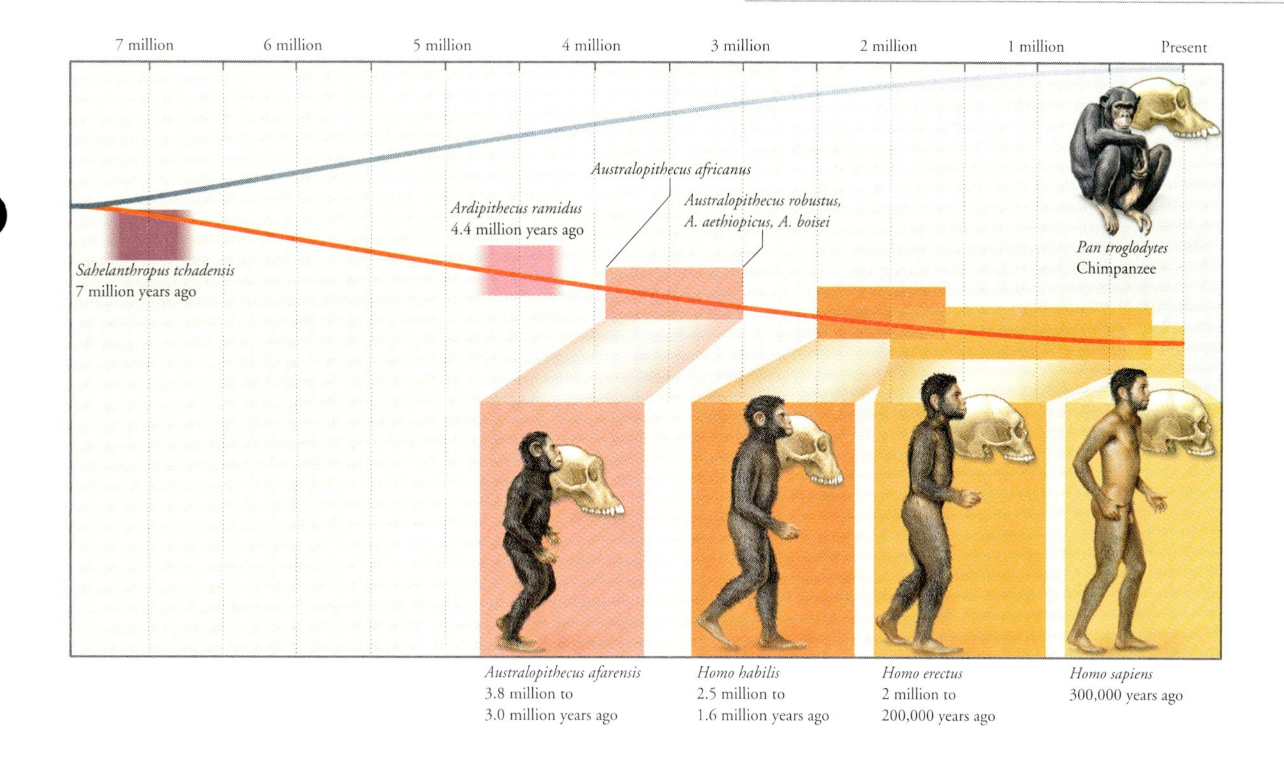

Fig. 22.24

NOTES

NOTES

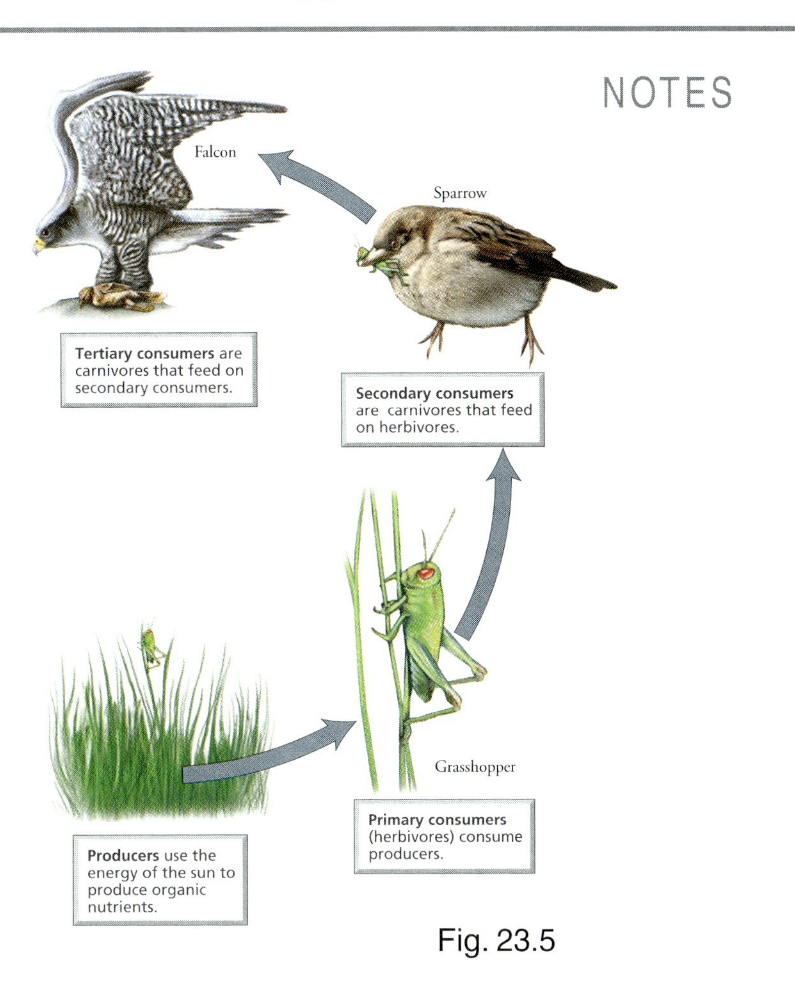

Fig. 23.5

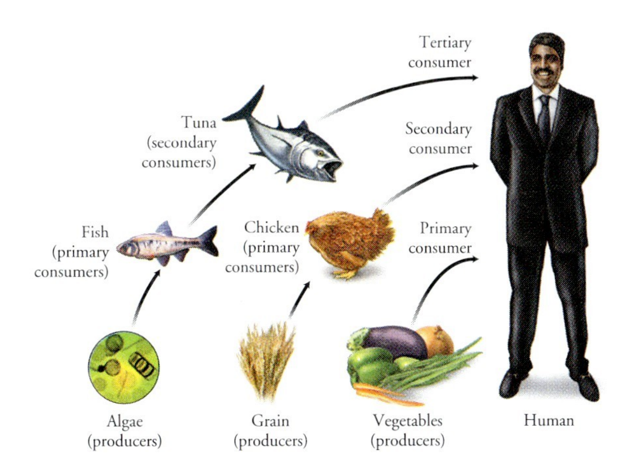

Fig. 23.6

NOTES

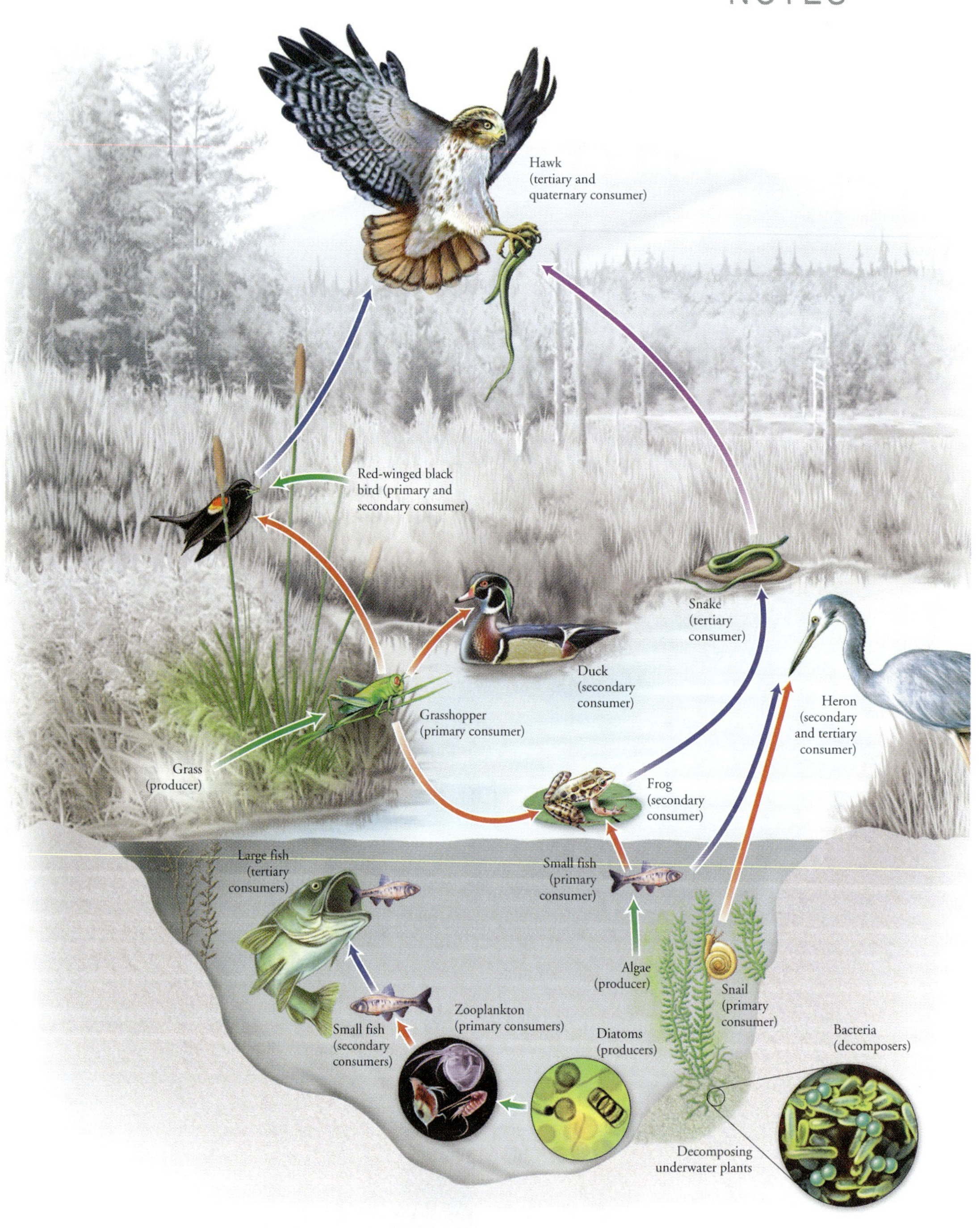

Fig. 23.7

NOTES

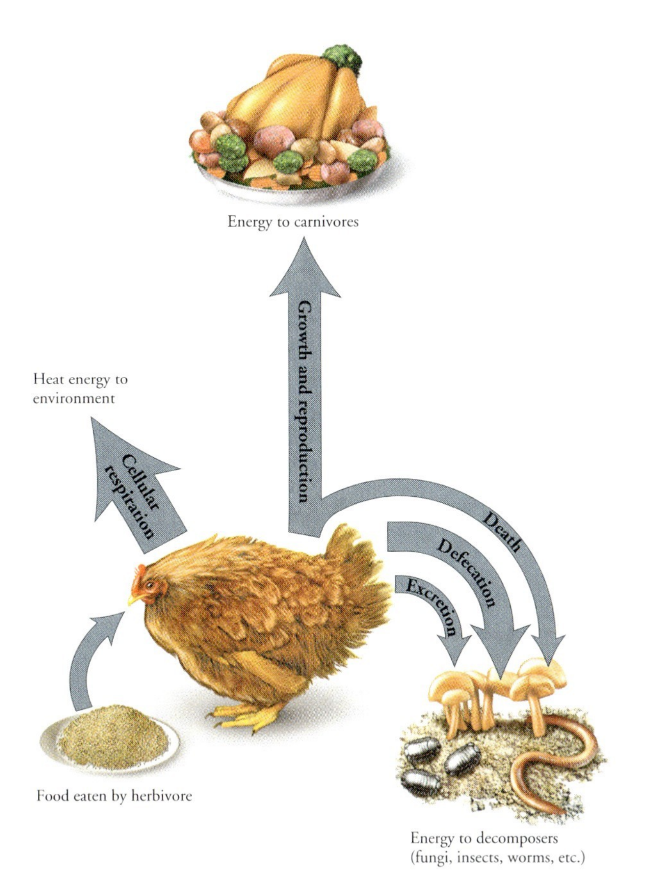

Fig. 23.8

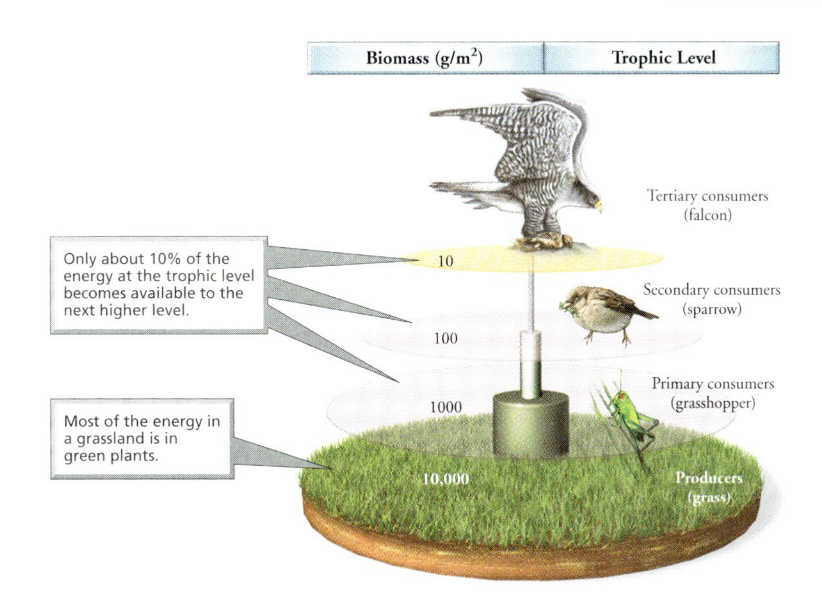

Fig. 23.9

NOTES

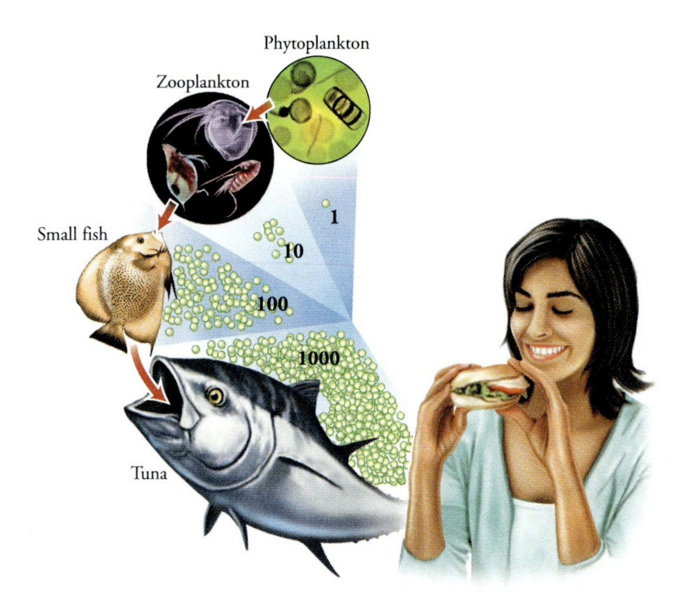

Fig. 23.10

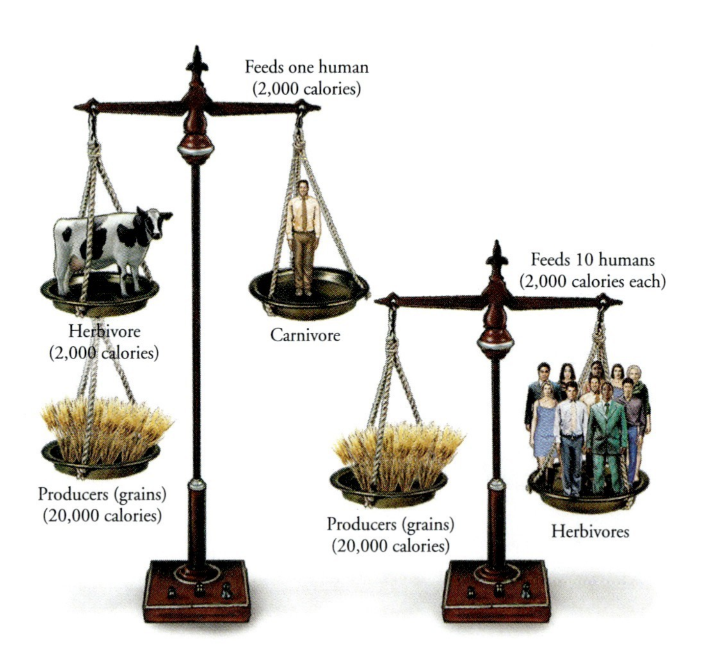

Fig. 23.11

NOTES

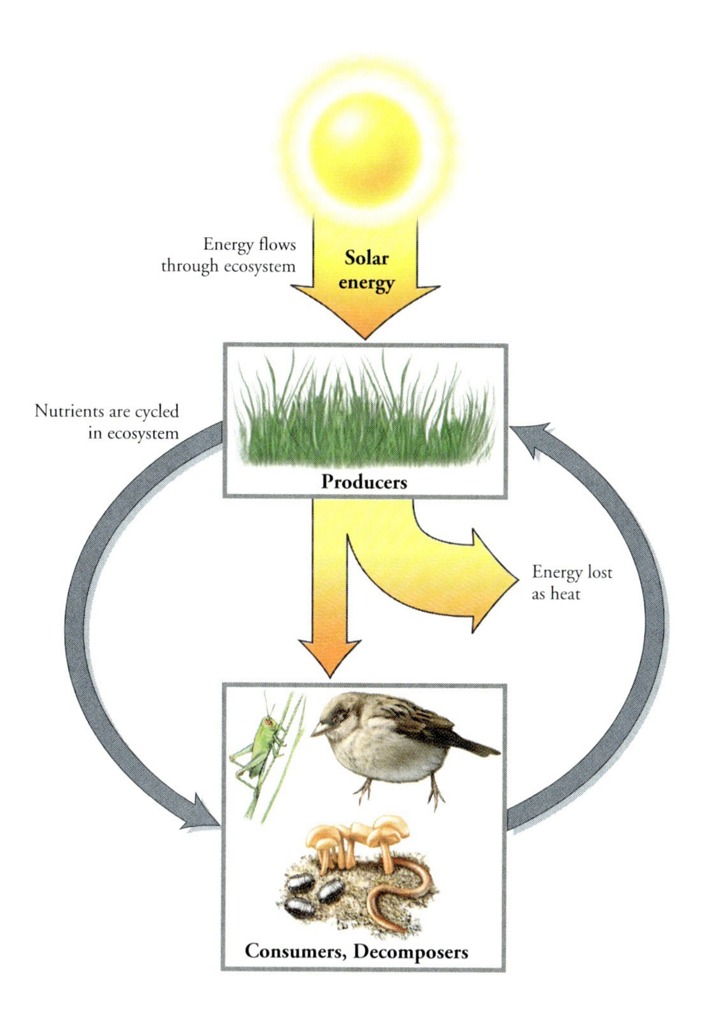

Fig. 23.12

NOTES

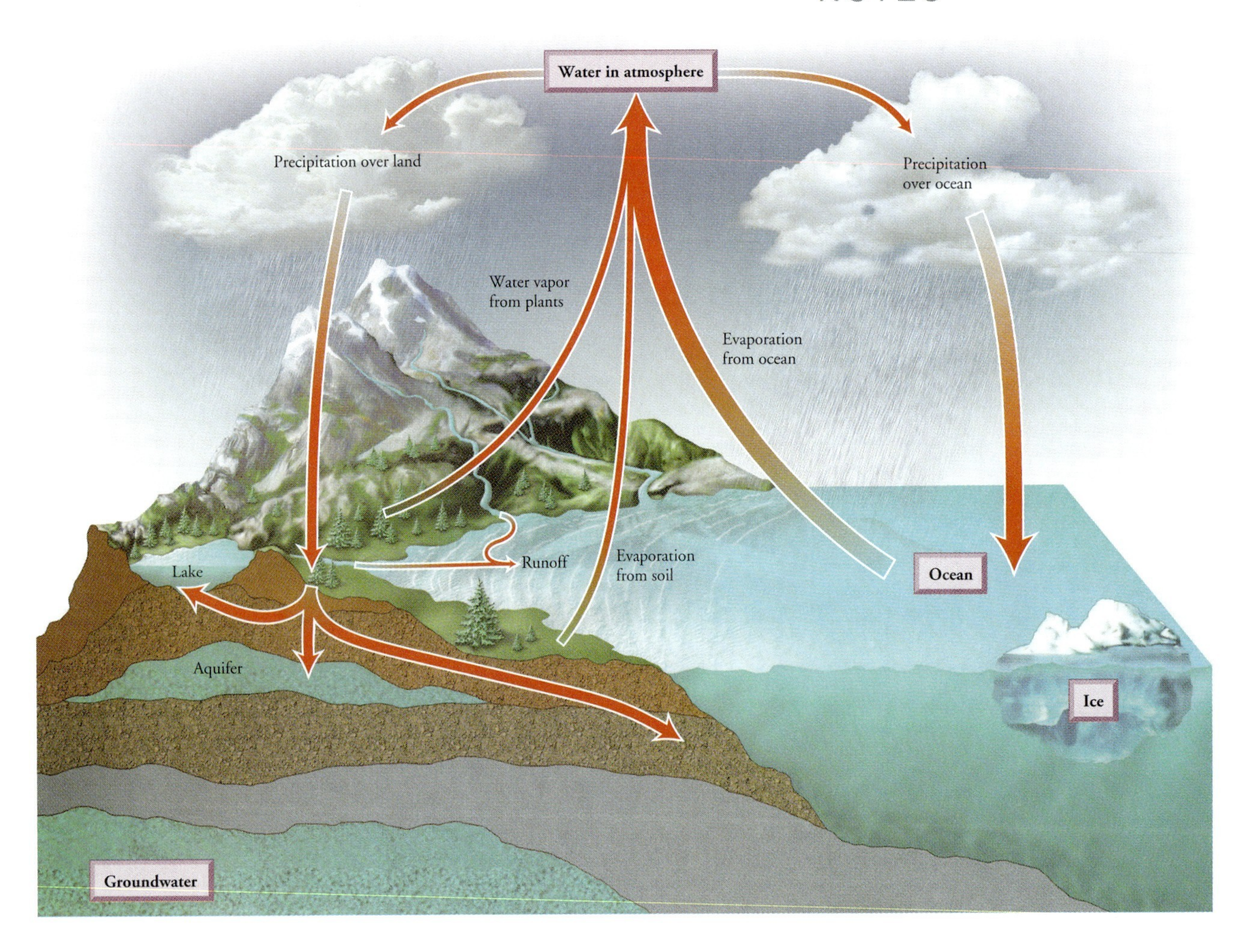

Fig. 23.13

NOTES

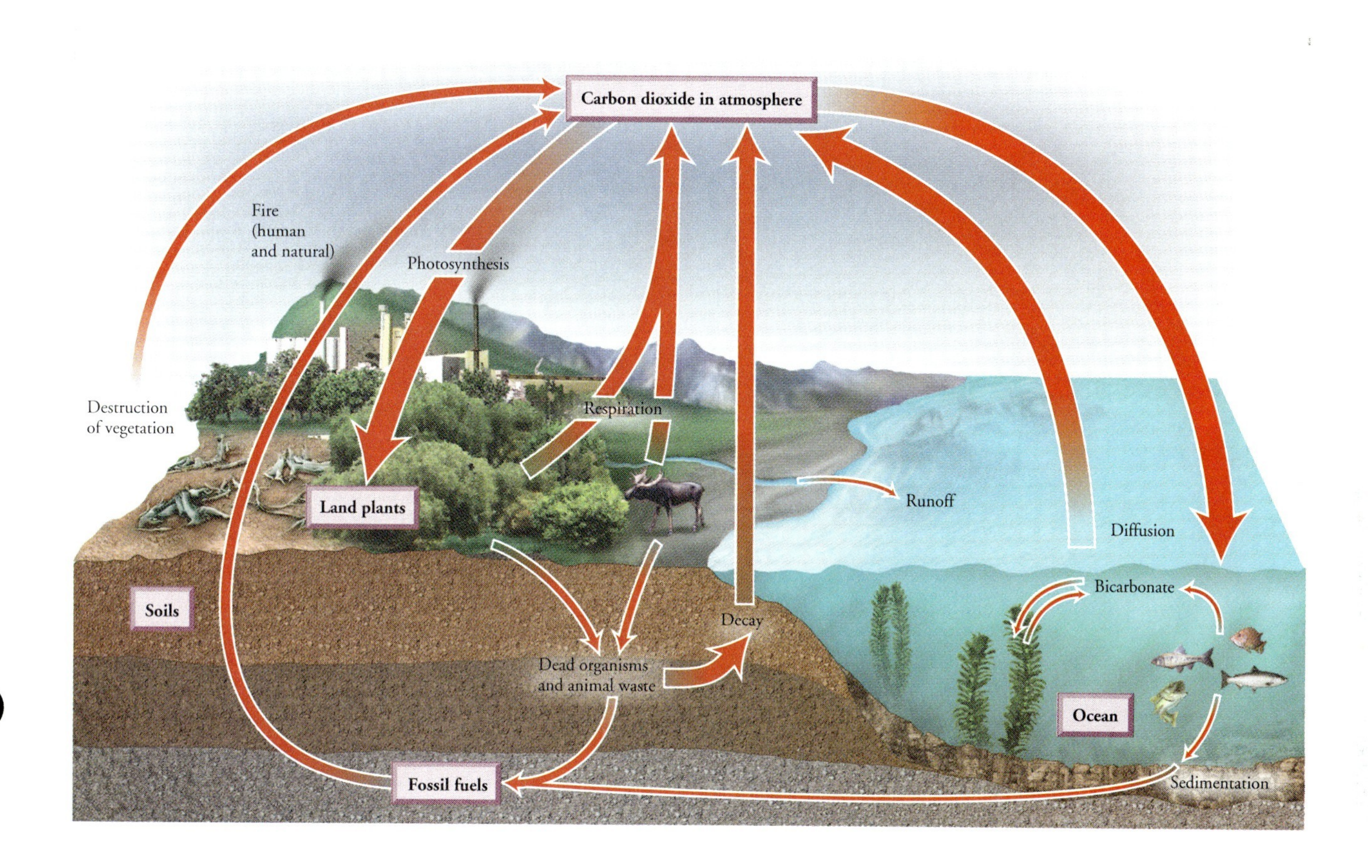

Fig. 23.14

NOTES

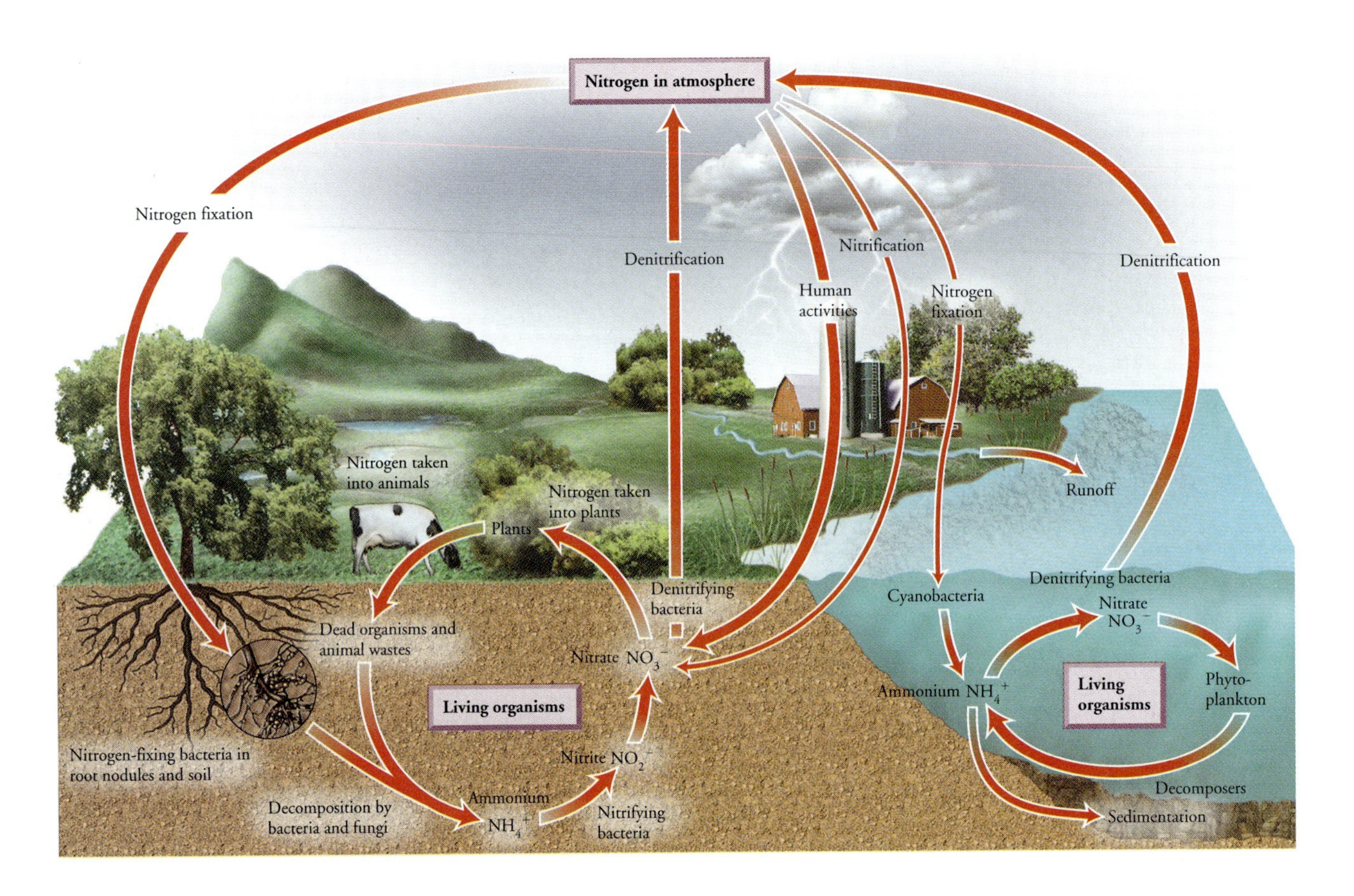

Fig. 23.15

NOTES

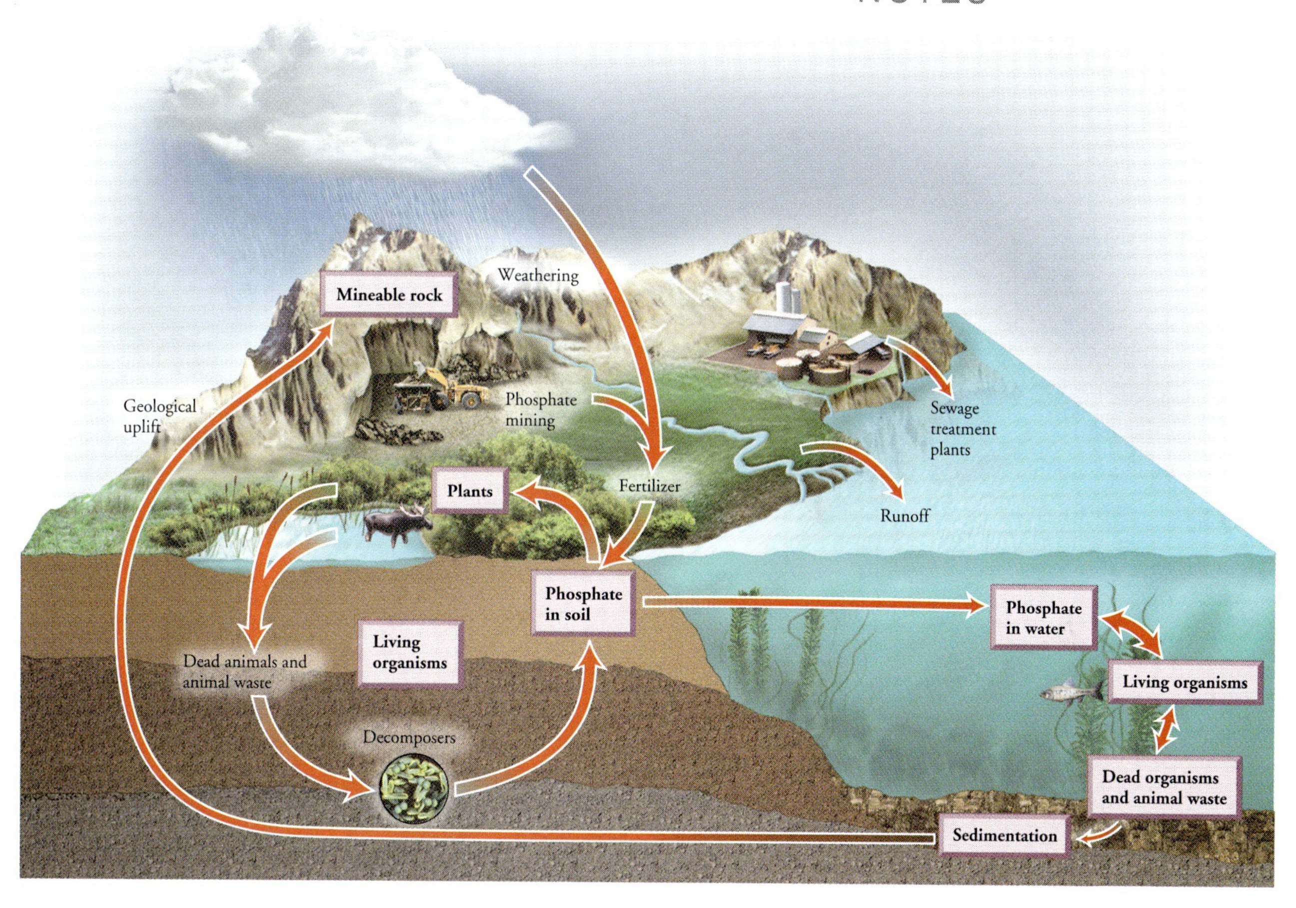

Fig. 23.16

NOTES

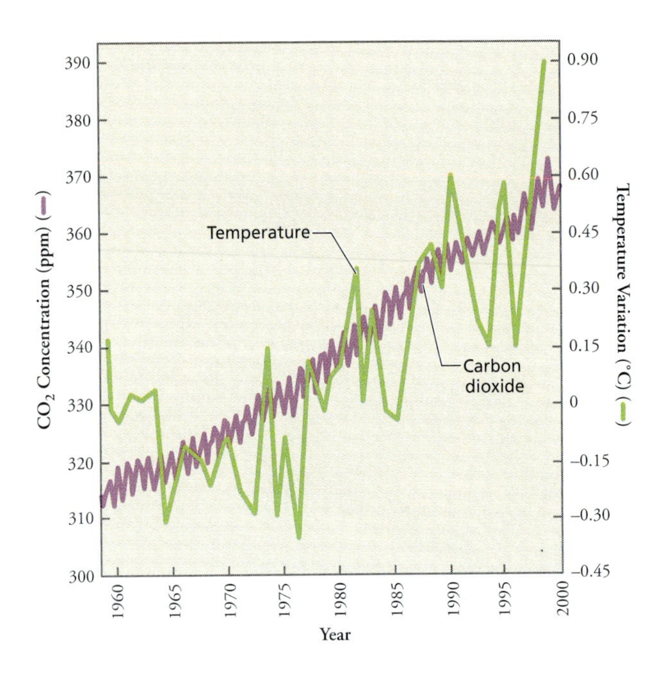

Fig. 23.17

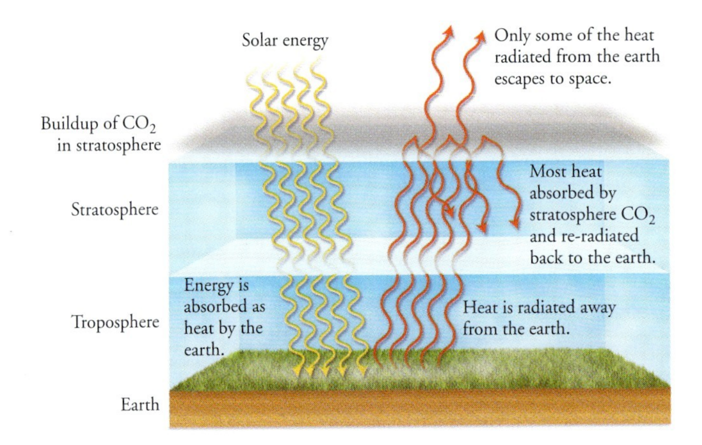

Fig. 23.18

NOTES

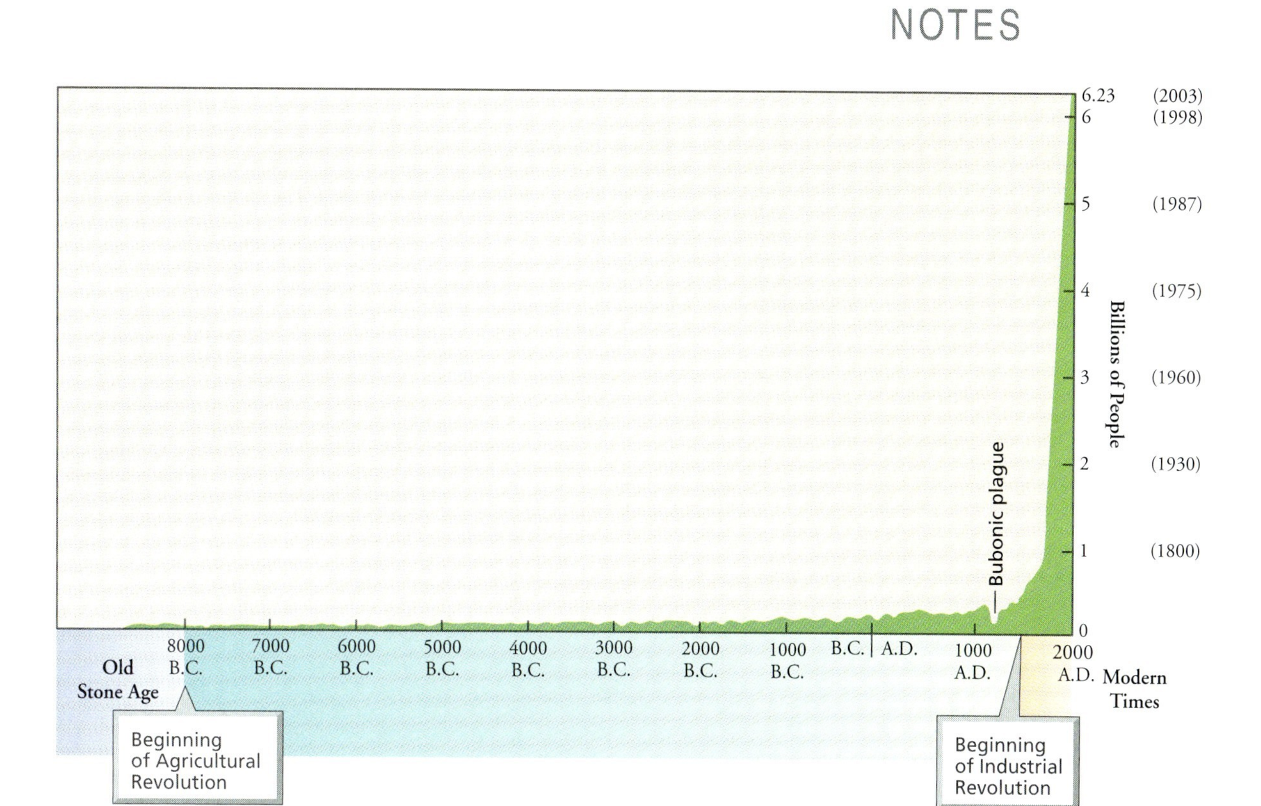

Fig. 24.1

NOTES

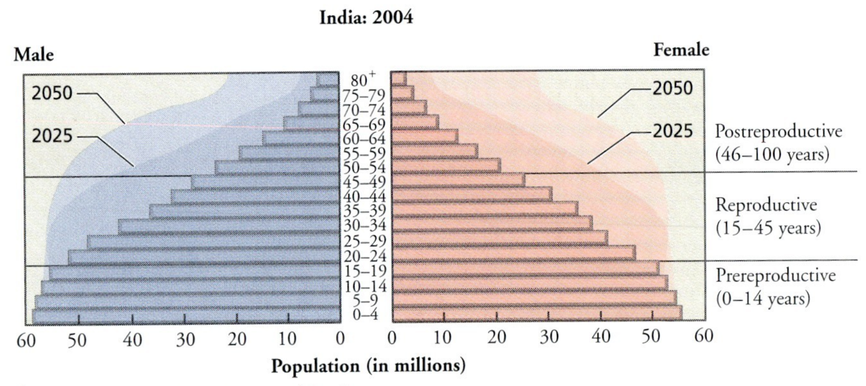

SOURCE: U.S. Census Bureau, International Data Base.

Expanding

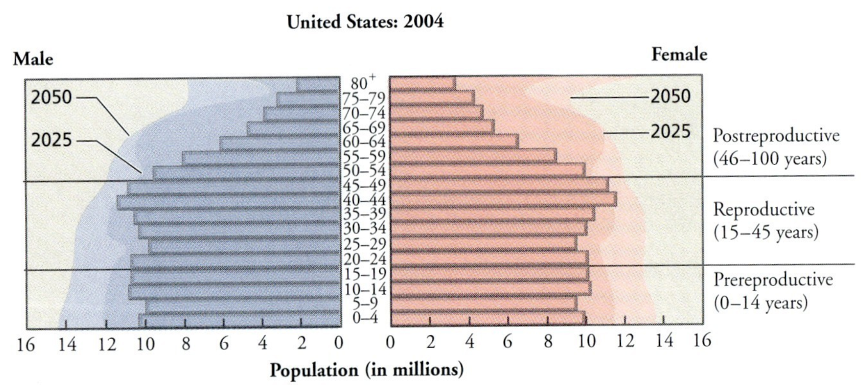

SOURCE: U.S. Census Bureau, International Data Base.

Moderately stable

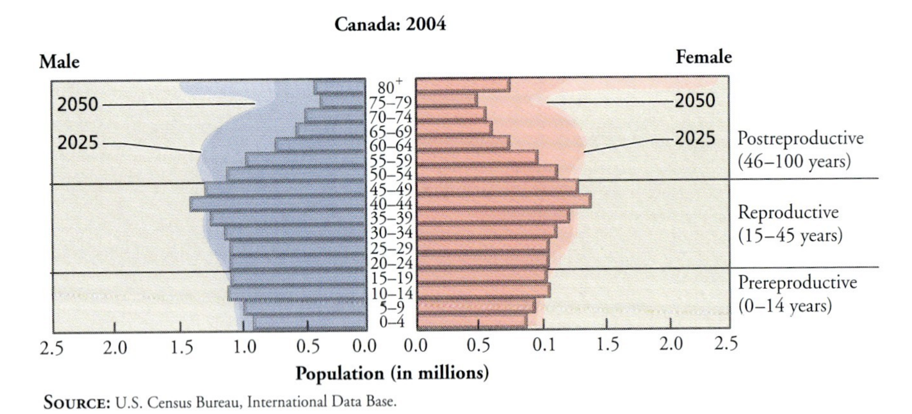

SOURCE: U.S. Census Bureau, International Data Base.

Declining

Fig. 24.2

NOTES

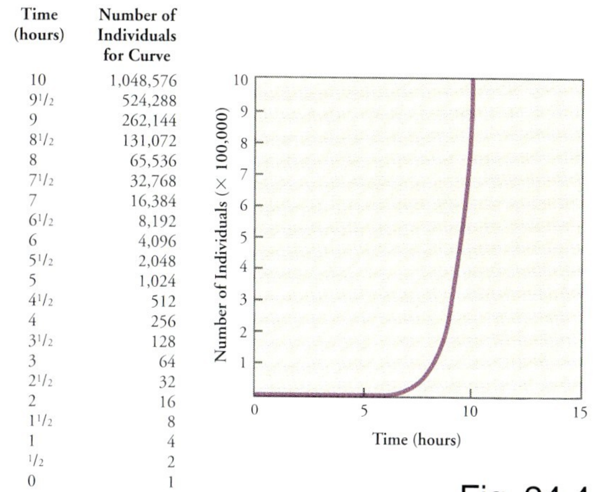

Time (hours)	Number of Individuals for Curve
10	1,048,576
$9\frac{1}{2}$	524,288
9	262,144
$8\frac{1}{2}$	131,072
8	65,536
$7\frac{1}{2}$	32,768
7	16,384
$6\frac{1}{2}$	8,192
6	4,096
$5\frac{1}{2}$	2,048
5	1,024
$4\frac{1}{2}$	512
4	256
$3\frac{1}{2}$	128
3	64
$2\frac{1}{2}$	32
2	16
$1\frac{1}{2}$	8
1	4
$\frac{1}{2}$	2
0	1

Fig. 24.4

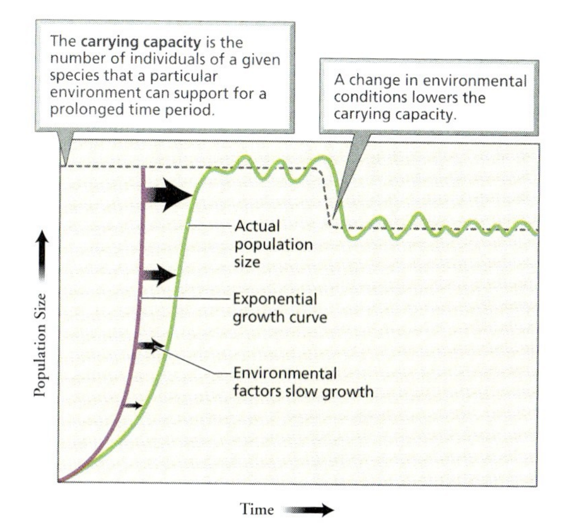

Fig. 24.5

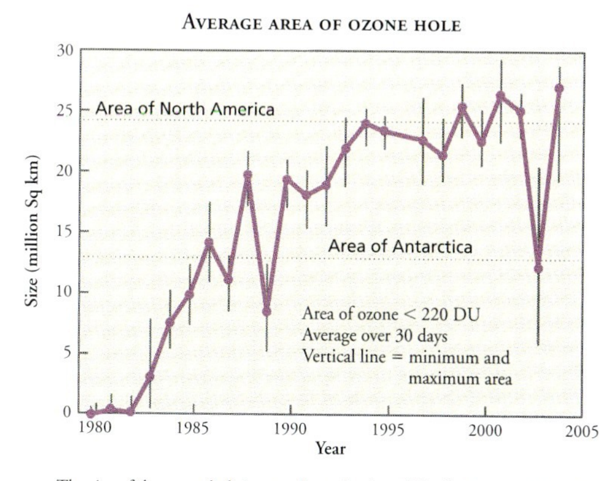

The size of the ozone hole is now about the size of North America. Although the size of the hole shrank in 2002, it increased again in 2003.

Fig. 24.7

NOTES

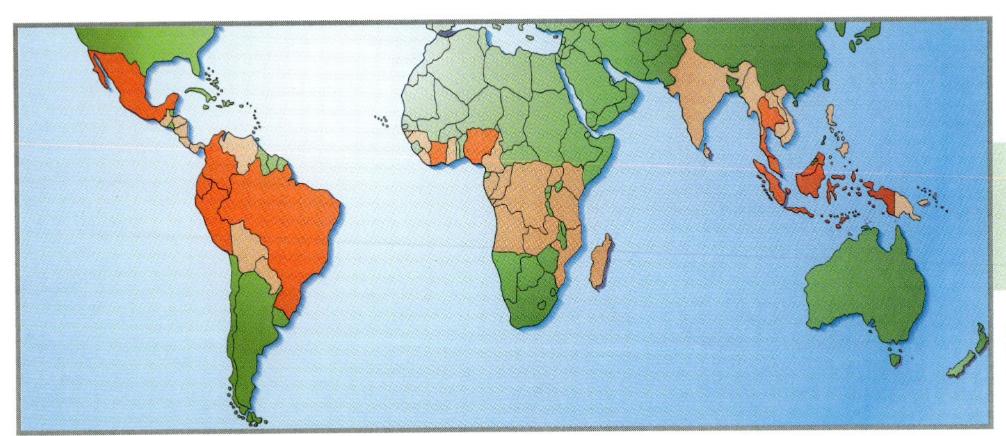

The countries in which the most deforestation is occurring are shown in red and orange.

Rate of deforestation:
Red = 2,000–14,280 km^2 per year
Orange = 100–1900 km^2 per year

Fig. 24.9

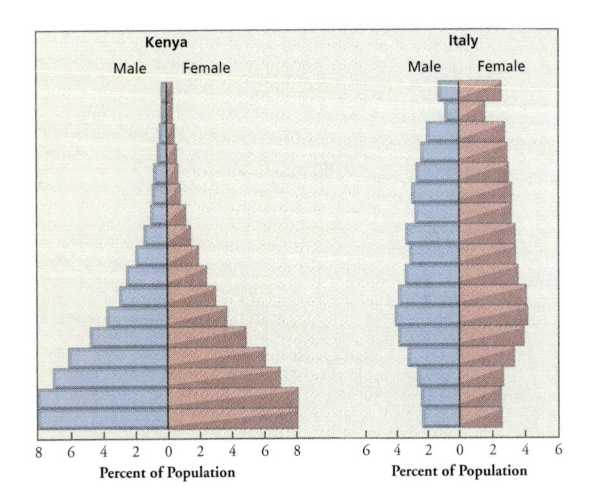